ANALOG CIRCUIT DESIGN

Analog Circuit Design

(X)DSL and other Communication Systems; RF MOST Models; Integrated Filters and Oscillators

Edited by

Willy Sansen
Katholieke Universiteit Leuven,
Belgium

Johan Huijsing
Delft University of Technology,
The Netherlands

and

Rudy van de Plassche
Eindhoven University of Technology,
The Netherlands

SPRINGER-SCIENCE+BUSINESS MEDIA, B.V.

A C.I.P. Catalogue record for this book is available from the Library of Congress.

ISBN 978-1-4419-5101-4 ISBN 978-1-4757-3047-0 (eBook)
DOI 10.1007/978-1-4757-3047-0

Printed on acid-free paper

Table of Contents

Preface

This new book on Analog Circuit Design contains the revised contributions of all the tutorial speakers of the eight workshop AACD (Advances in Analog Circuit Design), which was held at Nice, France on March 23-25, 1999. The workshop was organized by Yves Leduc of TI Nice, France. The program committee consisted of Willy Sansen, K.U.Leuven, Belgium, Han Huijsing, T.U.Delft, The Netherlands and Rudy van de Plassche, T.U.Eindhoven, The Netherlands.

The aim of these AACD workshops is to bring together a restricted group of about 100 people who are personally advancing the frontiers of analog circuit design to brainstorm on new possibilities and future developments in a restricted number of fields. They are concentrated around three topics. In each topic six speakers give a tutorial presentation. Eighteen papers are thus included in this book.

The topics of 1999 are:

(X)DSL and other communication systems
RF MOST models
Integrated filters and oscillators

The other topics, which have been coverd before, are:

1992	Operational amplifiers
	A-D Converters
	Analog CAD
1993	Mixed-mode A+D design
	Sensor interfaces
	Communication circuits
1994	Low-power low-voltage design
	Integrated filters
	Smart power
1995	Low-noise low-power low-voltge design
	Mixed-mode design with CAD tools
	Voltage, current and time references

1996	RF CMOS circuit design
	Bandpass sigma-delta and other data converters
	Translinear circuits
1997	RF A-D Converters
	Sensor and actuator interfaces
	Low-noise oscillators, PLL's and synthesizers
1998	1-Volt electronics
	Design and implementation of mixed-mode systems
	Low-noise amplifiers and RF power amplifiers for telecommunications

I sincerely hope that this book may make an invaluable contribution to the understanding of the art and science of analog design.

Willy M.C. Sansen

(X)DSL and other communication systems

Introduction

Modeling MOST transistors has once more gained importance because of the i(X)DSL and Other Communication Systems

Since its origin the telephone network consisted of a twisted pair of copper wires. The transmission speed over this pair has been limited to voice band. Over the years the needs for higher transmission speeds became evident. Different solutions exist eg. ISDN. Recently improved modem techniques resulted in much higher transmission speeds over the local network. Examples of these modem techniques and components for the implementation of different standards will be presented in this part of the book.

The first paper by Cornil et al gives an introduction to the different standards used and the modulation techniques that can be incorporated in these modem chips. A maximum transmission speed over 52 Mbit/s over a short distance is possible.
The second paper by Hester describes the architectures and the filter implementation problems involved in the design of an ADSL modem. An intelligent choice between a digital implementation and an analog implementation of the required filtering functions is described. The third paper by Geerts et al shows the implementaion and design criteria for high resolution, high speed Delta-Sigma Analog-to-digital converters required for ADSL modem applications. A MASH architecture is used for the A/D converter and the desing space is analyzed to obtain 15 bit resolution with a 1.1 MHz analog bandwidth using a 0.5 micron CMOS technology. The fourth paper describes clipping techniques to be implemented in receiver and transmitter chains for ADSL modems. Clever clipping in the receiver and transmitter chains relax the requirements for the analog circuitry in such systems. Finally Dru Cabler gives a survey of the different techniques used in analog circuit design for ADSL modems. Attentions is paid to the A/D and D/A converters and the requirements needed for modem implementation.

Rudy van de Plassche

"Building an ADSL Modem, the Basics"

Jean-Philippe Cornil

Alcatel
Microelectronic Dpt - VA21
Francis Wellesplein, 1
B-2018 Antwerp, Belgium
Tel: + 32 3 240 78 76

Jean-Philippe.Cornil@ieee.org

Abstract

Asymmetric Digital Subscriber Line (ADSL) is a new broadband communication technology that allows telephone companies to deliver multi-megabit data rate on their existing twisted pairs while simultaneously maintaining a traditional voice service. This new capabilities will boost multimedia internet applications by reducing the download time to a fraction of what it take today using classical voice-band modem.

Slide 2 - Outline

The presentation starts by a few words on the evolution of the local loop, modem technology and network architecture able to provide broadband communication channels to telephone subscribers, and to satisfy future requirements of their multimedia applications.

The next section will describe main twisted pair characteristics and impairments encountered in ADSL systems. It will be followed by a description of the Discrete Multi Tone (DMT) modulation features and will show how DMT combats very efficiently channel non-idealities.

Once these concepts are defined, a step-by-step presentation of all key functions needed to build an ADSL modem will be performed, starting from the digital interface to the copper line.

The conclusion will outline some actual trends concerning ADSL evolution, capacity and deployment.

Introduction

Slide 3 - Local Loop evolution

Twenty years ago, it was believed that fiber optic would replace copper lines for broadband access within a few years. However, the cost associated with the 'last mile' infrastructure has delayed the arrival of the fiber to every home (FFTH) for many years. Another, more economical technology was needed to bridge the gap between growing data rates, pulled now by the Internet, and current modem capabilities.

ADSL (Asymmetric Digital Subscriber Line) provides an answer to that demand. Downstream bitrates of up to 8 Mbits/s and upstream rates of a couple of hundreds kbits/s are feasible. Distances up to a few kilometers can be covered. In this way, ADSL transforms the twisted pair from one limited to voice and low-rate data to a powerful bit pipe. Meanwhile, ADSL leaves the existing (analog) telephone service intact and undisturbed but adds the ability to access a multitude of new multimedia services.

xDSL provides also the interim technology needed to bridge the gap before the advent of FFTH (as commented by Bell Atlantic's CEO , "ADSL is an interim technology for the next 40 years..."). This gives also POTS operators the opportunity to turn their buried lines pairs (almost 1 billion installed lines!) into a new copper mine, with a strategy supporting a more economically viable and progressive introduction of the fiber from the core network to customer premises (ADSL->VDSL->FFTC->FFTH).

Slide 4 - Modem evolution

Since the early 1960's, voice band modems have been developed to transport digital data over the telephone network. These modems modulated the data at the transmitter, then were transported (over one or two pairs) transparently through the telephone network, and finally demodulated at the receiver. Different modulation techniques have been used, starting with FSK (Frequency Shift Keying) in the ITU-T standard V.21, over DPSK (Differential Phase Shift Keying) in V.26, V.27 and V.29 ([1]), to QAM (Quadrature Amplitude Modulation) in V.22bis, V.33 and V.34 ([2]). Also different duplexing techniques have been applied: full duplex transmission

---

1 V.29 makes use of a variant of DPSK: ADPSK (Amplitude Differential Phase Shift Keying).

2 V.33 and V.34 also make use of Trellis coding: 128-QAM with Trellis coding in V.33 and 4-dimensional Trellis coded QAM in V.34.

over a single pair with separated (FDM, Frequency Division Multiplexing) or overlapping (EC, Echo Canceling) frequency bands for both directions of transmission, dual simplex transmission over two pairs and half-duplex transmission over one pair. The bitrates range from 300 bit/s in V.21 to a remarkable 33.6 kbit/s (near the theoretical capacity limit of 10b/Hz/s for voice-band channel) in V.34. Very recently (1997) bitrates of 56 kbit/s have been achieved (V.90). However, these very high rates are only possible in particular configurations where e.g. an Internet service provider has a *digital* connection (= without quantization noise) to the central office, and only in the downstream direction, i.e. towards the customer.

Providing access to the 'information superhighway' for residential customers requires new technologies capable of improving the performance and capacity of these copper lines above the 64kb/s limit. A first step in that direction was the introduction in the 1980s of ISDN-BRA (Integrated Services Digital Network, Basic Rate Access) offering a bi-directional 144 kbit/s, but still based on the PSTN switching.

During the past several years, evolution in Digital Signal Processing techniques coupled with high speed packet switching capabilities enabled the design of several new modem technologies, such as ADSL and very recently VDSL, both able to break the symbolic Mb/s barrier on classical phone lines.

Slide 5 - Network evolution

Slide 5 depicts Internet access by means of a voice band modem (top) and an ADSL modem (bottom). The main difference is that voice-band operates over an end-to-end PSTN connection, whereas ADSL operations are limited to the local loop.

Key for voice band modems is, as the name implies, that the data is modulated in the voice band (from 300 to 3.4 kHz) and carried transparently over the network end-to-end. This requires also higher Quality Of Service (QOS) from the network in order to be able to maintain complex & efficient modulation schemes. However, classical PSTN switches were optimized using voice call duration as parameter for their design, and have 'some' difficulties to avoid blocked calls when faced with "always on" Internet connections. Another disadvantage of this end-to-end connection is that the bandwidth of the resulting communication is not only limited by local loop performances and/or modem type but also by parameters of the network and remote site capabilities.

In order to solve above issues, signals modulated by high speed and/or bursty data have to be limited to the local loop and terminated before entering the local telephone exchange.

This is the case with ADSL, where the transmission on the twisted pair is terminated in the Line Termination (LT) at the local exchange and the Network Termination (NT) at the subscriber side. The 'POTS-splitter' in the LT separates the analog telephone signal from the ADSL data signal. The first one is sent to the PSTN network, the second to the broadband network (e.g. ATM switch or router).

Slide 6 - ADSL Spectrum Allocation

The ADSL transmission system offers an asymmetric capacity to the residential subscriber. In the downstream direction (towards the subscriber), it provides a capacity up to 10 Mb/s (or even more), while in the upstream direction it can go up to 1 Mb/s.

ADSL DMT modulation uses 256 QAM modulated carriers occupying a bandwidth of 1.104MHz. For the separation of the up- and downstream transmissions, two bandwidth allocation schemes are possible. The first one uses overlapping spectra for up- and downstream transmission and applies echo canceling (EC). The second option uses frequency division multiplexing (FDM) in which case no tones are shared by the up- and downstream bands. The latter is illustrated on this slide. The upstream band ranges from about 25 kHz to 138 kHz (carriers 6 to 32) while the downstream band extends up to 1.104 MHz (carrier 256). The lowest carriers are not modulated to avoid interference with POTS.

DMT transmission for ADSL has been standardized in ANSI/T1E1.4 [1] and is supported by ETSI/TM6.

Twisted pair characteristics

In this section we will analyze the main parameters of twisted pair telephone loops. Accurate channel models will enable simulations to be performed in order to understand transmission performances under different system assumptions.

Slide 8 - Loop length distribution

This slide shows the distribution of loop lengths for some representative countries. It can be seen that, in order to reach almost all European, or more than 90% of US potential subscribers (US loop plant distribution tends to be longer than in other countries), ADSL should allow broadband communication within a reach of 6km (18kft). The trends goes also clearly towards a reduction of the average loop length due to increasing Digital Loop Carrier deployment (DLC using CSA rules, see below), mainly in new growth and business areas. Only 15% of telephone lines (loaded coils) in the world will require an upgrade to allow xDSL communication.

Slide 9 – ADSL test loops

Both ANSI and ETSI have specified a set of worst-case test loops (including their RLCG parameters) with their associated noise and crosstalk distributions for the qualification of ADSL modems. But most of modern loops show better characteristics and follow Carrier Serving Area (CSA) guidelines, which specify among others, a maximum distance of 9kft for 26AWG and 12kft for 24AWG.

Slide 10 - Distributed model

Twisted-pair phone lines may be accurately modeled for frequencies up to a few tenth of MHz using classical transmission line techniques. A infinitesimally small portion of the line is characterized, as illustrated on this slide, by 4 frequency dependent primary parameters: R(f), L(f), C(f), G(f) representing resistance, inductance, capacitance and conductance per unit length respectively. The transfer function of the line can then be obtained by cascading these small portions and solving the associated differential equations.

This results in a frequency dependent propagation constant $\gamma(f,d)$ and characteristic impedance $Z_0(f)$. The real part of $\gamma(f,d)$ is very important for xDSL design and will determine the attenuation of the line (far from being lossless !). The frequency dependency of the imaginary part leads to dispersion of the signal energy in time and generates Inter Symbol Interference (ISI=overlap of successive symbols).

Slide 11 - Transfer function

The characteristic impedance $Z_0(f)$ is obtained by taking the ratio of the forward-going and backward-going waves (see figure on the right). This impedance has to be used to terminate the line in order to avoid reflection (echo) and to optimize energy transfer towards the receiver. Typical values for frequency higher than a few 100KHz (downstream band) are real and around 100 ohms (reference value used in ADSL), while for lower frequencies the line impedance increases and becomes complex.

The figure on the left illustrates the attenuation of the signal, as a function of frequency, after propagation over a 1 km, 26 AWG ([3]) twisted pair.

Slide 12 - Loop impairments (1/2)

The transmission channel capacity depends highly on the twisted pair characteristics and suffers from a number of impairments. This section enumerates the major loop impairments that an ADSL modem design has to taken into account. The way they are tackled will be the subject of the next section.

- ***Background noise:***
  The analysis of transmission performance over twisted-pair is based on the assumption that the background noise is gaussian. Analysis performed by Bellcore during the standardization process validated the above assumption and resulted in a worst case Aditive White Gaussian Noise (AWGN) level of –140dBm/Hz. This gives approximatively 30 nV/√Hz on a 100 ohms line.

- ***Crosstalk:***
  Between different wires in the same cable there exists capacitive and inductive coupling. The coupling increases, when the wires are closer to each other's. It causes unwanted crosstalk between the pairs. The crosstalk is typically worse between two pairs in the same binder than for

---

[3] AWG: Average Wire Gauge. 24 AWG and 26 AWG are equivalent to 0.5 mm and 0.4mm cable respectively.

wires in adjacent binders ([4]). It can be reduced by the optimization of the twist of the individual pairs and by the topology of the cable.

Two types of crosstalk can be distinguished. Near-End crosstalk (NEXT) occurs at a receiver that is collocated with the disturbing source, while Far-End crosstalk (FEXT) occurs at a remote receiver. NEXT increases at a typical rate of 4.5 dB/octave ($f^{3/2}$) while FEXT increases by 6 dB/octave (f^2).

FEXT is attenuated by propagation through the loop while NEXT is not. Therefore, NEXT dominates FEXT by far for echo canceled systems. Using a different frequency band for upstream and downstream (FDM) reduces the (self-)NEXT impact.

Slide 13 - Loop impairments (2/2)

- ***Bridged Tap :***
  Some subscriber loops have open-circuited wire pairs tapped onto the main wire pair, called bridged taps. The existence of bridged taps in the loop plant differs from country to country and depends upon the cabling rules used in the past. Their presence causes reflections (echo) and affects the frequency response of the cable, leading to pulse distortion and inter-symbol interference combined with frequency dependent attenuation. A loop can also be built up of wires with different diameters (referred to as gauge transitions), leading to reflections and distortion as well.
- ***Impulsive Noise:***
  Copper transmission suffers from impulsive noise that is characterized by high amplitude bursts of noise, with duration of a few microseconds to hundreds of microseconds. It can be caused by a variety of sources such as central office switching transients, ringing, ring trip, dial pulses and lightning.
- ***RFI:***
  The telephone network consists of copper pairs that are usually twisted. The twist improves the egress and ingress properties of the wire: it reduces the electromagnetic radiation as well as the pick-up of unwanted signals if submerged in an electromagnetic field. As a result of the cable unbalance, RF (Radio Frequency) signals can be picked-up during propagation over the wire and interfere with the transmitted data at the

---

[4] A 10 dB crosstalk reduction is often assumed in capacity simulations to reflect the adjacent binder effect.

receiving side. The balance of the cable decreases as the frequency goes up. The aerial drop-wires, the vertical cables in high-rise buildings and the in-house wires are the most vulnerable for RF ingress. This ingress can come from a variety of sources such as AM broadcast (LW, MW and SW), radio amateur communication and public safety & distress bands.

- ***POTS signaling:***
  High frequency components of POTS signaling voltages (ringing, on/off-hook, metering, battery reversal, ...) generate high level impulsive noise which may corrupt data transmission. Also, seen from the line, off-hook and on-hook states show different impedance level in a wide frequency band and therefore destroy the channel model of the modem (re-training mandatory). The use of a POTS filter between telephone sets and the line ('POTS splitter') as well as between the line and the ADSL modem ('ADSL splitter') removes interactions between POTS and ADSL frequencies and avoids these problems.
- ***Load coils:***
  They have been used in the early days of telephony to boost and flatten the frequency response of the line at the upper edge of the voice band. As these coils give an unacceptable attenuation at higher frequencies they must be removed, if still present, to allow for DSL services.

Because of the wide differences in network topology, installation practices and cable types, the impact of these impairments can vary significantly from operator to operator. Anyhow, a highly adaptive transmission system is needed to cope with the above imperfections.

Signal Description

Slide 15 - Modulation

The maximum capacity (in bps) of an AWGN channel (with no ISI) is given by the Shannon-Hartley relation (1) illustarted on the slide, where SNR is the ratio of the transmitted energy per symbol to the noise power spectral density [2].

For frequency dependent noise or dispersive channels, the same optimal performance level can be obtained by partioning the transmission channel into a large number of narrowband subchannels having little ISI and approximating an AWGN ('divide to conquer' strategy). The capacity of such multicarrier (or multitone) modulation is the sum (integral) of the individual capacities. Therefore, due to its fine granularity, DMT approaches the above theoretical limit very closely.

Slide 16 - Frequency allocation strategy

For the separation of the up- and downstream transmission, two bandwidth allocation policies were accepted. The first one uses overlapping spectra for up- and downstream transmission and applies echo canceling (EC). The second option uses frequency division multiplexing (FDM) in which case no tones are shared by the up- and downstream bands (see also next slide).

Because of its lower bit rate, the upstream channel is placed in the lower part of the spectrum in order to limit attenuation range, NEXT crosstalk and impact of self-NEXT for EC.

FDM offers better upstream performances than EC system because there is no limitation due to self-crosstalk (self-NEXT). However, because of its wider downstream bandwidth and absence of frequency guard band, EC reaches higher downstream performances on short loops. EC system requires also more complex digital signal processing and analog dynamic ranges (A/D converter) than its FDM counterpart, but simplifies the analog front-end.

Slide 17 - Spectum Allocation

The basic principle of DMT is to transmit the information bits in parallel over a large number (256) of carriers (tones), each of which is QAM modulated. The carrier frequencies are multiples of some basic frequency (4.3125 kHz).

The upstream band ranges from about 25 kHz to 138 kHz (carriers 6 to 32) while the downstream band extends up to 1.104 MHz (carrier 256). The start frequency of the downstream spectrum can be anywhere above 25 kHz and is manufacturer discretionary. The lowest carriers are not modulated to avoid interference with POTS.

The transmit power spectrum is almost flat over all used tones. For up- and downstream transmission, the average nominal PSD (Power Spectral Density) is respectively -38 dBm/Hz and -40 dBm/Hz across the whole band. In downstream direction a power cutback is applied on short lines to avoid saturation of the remote receiver. The passband ripple shall not be greater than ±3.5 dB. The number of bits that is assigned to a tone and its precise transmit power is determined during system initialization as a function of the SNR (Signal to Noise Ratio) on that tone and the requested overall bitrate. During operation, adaptation of the bit assignment or corrections to the transmittted power are possible to compensate for alterations in line conditions, due to a variation of the noise or a (slow) drift of the cable transfer function (e.g. because of temperature variations). These on-line adaptations do not interrupt the data flow.

Slide 18 – Constellation

A constellation is just a simple and easy way to represent the amplitude and the phase of a sine wave using a rotating vector in a complex plane. The signal is obtained by taking the real part of that complex number.

Slide 19 - QAM modulation

Quadrature Amplitude Modulation (QAM) encodes a set of bits into positions inside a constellation (2^n-QAM, with n=1..6 are illustrated). One QAM signal can be seen in the time domain as a sine with a phase and amplitude changing at the symbol rate. QAM constellation size are limited by the noise of the channel, but also by the complexity of the equalizer needed to compensate for the channel dispersion (for narrow-band modulation, channels characteristics are 'soft' enough to allow high constellation size with an acceptable equalizer complexity).

Slide 20 – BER

This slide illustrates the Signal to Noise Ratio (SNR) required to meet a given Bit Error Rate (BER) for a QPSK (4-QAM) signal. It can be seen that in order to reach a BER of 10^{-7}, a SNR of 14 dB is needed (without margin). For higher order constellation sizes, because of the quadrature modulation, each additional bit requires 3dB more SNR.

Slide 21 - DMT Characteristics, Dynamic range, SNR

A DMT signal consists of multiple QAM channels equally spaced in the frequency domain (=sum of narrow-band QAM subchannels).

In order to avoid Inter-Carrier Interference (ICI), bandpass filtering and separate processing for each subchannels at the receiver, orthogonality and synchronicity between these QAM tones is required. This is realized by using a symbol length equals to the period of the frequency spacing (4.3215KHz for ADSL DMT) in order to place the zeroes of the power spectrum (sinc shape) on the carrier position. However, as we will see later, synchronous symbol transitions allows the usage of a very effective (de)modulation schemes.

Slide 22 - Crest Factor

The total average power of a DMT signal is the sum of the individual power (independent signals). The maximum amplitude is obtained when all carriers are aligned and is equals to the sum of individual amplitudes. The Power to Average Ratio (PAR) or Crest Factor (CF) is given by the ratio of the maximum and average amplitude.

For 200 QAM carriers (downstream band), the PAR reaches a value as high as 25 dB. This increases the number of bit required by the A/D converter by 4 and reduces the power efficiency of the line driver. Similar PAR issues are not DMT specific, but can also be found in applications like audio (voice), wireless (CDMA), cable (aggregate of TV channels), ...

A possible solution to limit the impact of the PAR on the dynamic range and power consumption is to clip the signal above well-defined amplitude. This will introduce a kind of impulsive noise in the system, which has to be kept low enough to avoid too much impact on the system performances.

Slide 23 – Clipping probability

When the number of carriers increase, the amplitude of a DMT signal approaches a Gaussian distribution [3]. The graph shows the clipping rate probability of a DMT signal in function of the selected PAR. A clip probability of 10^{-8}, resulting in a loss of only a few b/s, is obtained with a PAR of 15 dB, a value typically used in ADSL design.

A number of PAR reduction techniques (clip-mitigation) have also been developed which, for an acceptable complexity and latency, could be good candidates to further reduce of the PAR in future ADSL implementations.

Slide 24 - Building the modem

In this section, a basic ADSL modem architecture will be build step by step. All functions needed between the ATM interface and the twisted pair will be presented and/or discussed, but discussions on specific mapping of these functions into hardware/software or Digital/Analog implementations will be avoided.

Side 25 - ATM interface

This module implements the ATM Transmission Convergence (TC) sublayer, where ATM cells coming(going) from(to) the interface are delineated, checked and adapted (idle cell generation) to the ADSL data rate.

Slides 26 & 27 - Coding

In order to improve the Bit Error Rate (BER) or (equivalently) to increase the system performance, Forward Error Correction (FEC) is applied (resulting in a +/- 3dB of coding gain).

ANSI specifies the use of Reed-Solomon (RS) coding combined with interleaving in order to improve FEC robustness against impulsive noise (distributes erroneous bit on multiple RS codewords). The ANSI standard allows for simultaneous transport of non-interleaved (for delay sensitive data) and interleaved data.

The RS coding allows an improvement of the BER from 10^{-7} in the PMD layer to 10^{-10} in the TC layer.

The additional use of Trellis coding (Wei's 16-states 4D code) is optional but may further reduce the BER or increase transmission capacity by improving the SNR by (+/-) 4 extra dB (particularly useful on long loop where SNR are very low). The viterbi processing is performed on carriers and therefore included in the 'mapper' module described hereafter.

Slides 28 & 29 - Mapping/Demapping

The process of optimizing bit distribution to each of the carriers is called bit loading in DMT terminology. The number of bit (Bi) assigned to one particular carrier is a function of the measured SNR for that channel. The number of bits being integer, the optimal signal level for each carrier may only be reached using carrier specific gains (Gi). This allows an optimal energy distribution between each carrier by allocating excess of energy from one channel to another one.

This is illustrated on the left figure, where theoretical (horizontal lines) and measured SNRi are plotted (top) together with their related Gi's. The figure on the right shows results from a real line, with respectively from top to bottom : SNRi, Margin (the 'peak' with almost 35dB margin is the 'pilot' tone used for timing recovery), Bi and Gi.

Mapper and Demapper modules perform the allocation/extraction of input bit to/from each 256 carriers (up to 15 bit/carrier). This occurs at a rate of 4000 Symbols/s. Each 68 symbols a special synchronization symbol is inserted for training purpose, resulting in an effective throughput of 69/68*4000 Symbols/s.

Slides 30 & 31 - FFT/IFFT

(De)modulation of up to 256 QAM tones can be implemented very efficiently using Fast Fourier Transform (FFT) running at the symbol rate. This is possible without ICI if carrier frequencies are integer multiples of the symbol rate and synchronously modulated (see Slide 21).

FFT is the key function of an ADSL modem. It performs the conversion between the complex frequency domain signal and its real time domain representation. On each transmitted symbol (256 complex carriers), a Fourier transform is performed resulting in 512 real samples (and inversely for the receive direction).

Slides 32 & 33 - Line monitoring & PLL

The output from the 'demapper' is not only used to extract bit from every carriers, but also to monitor individual carrier characteristics (SNR, noise,...) or to extract specific informations (signaling channels, pilot,...).

For timing recovery, DMT transmitter and receiver uses a special tone (called 'pilot' tone) to maintain clock synchronization. This tone is monitored from the FFT output, compared with the reference (pilot tones have a fixed amplitude and fixed or alternating phases) and after filtering used to drive digital or analog VCOs.

Slides 34 & 35 - Cyclic Prefix

In order to reduce the impact of energy leaking from other symbols (ISI) as well as the accuracy of the FFT window on DMT performances, a 'FFT friendly' guard band has been introduced. This guard band is realized by repeating the first 32 samples after the end of symbol.

The cyclic prefix :

- allows a correct FFT computation, because the signal appears periodic and in that case a shifted sampling window generates just a predictable rotation of each carriers, which can easily be compensated in the frequency domain (just complex multiplication).
- reduces the equalizer requirements by allowing a remaining channel impulse response (after equalization) equals to the guard band length instead of ideally zero.

After the cyclic prefix introduction, the number of bit/symbol is now 512+32=544 bit/symbol, resulting in an effective bit rate of (512+32)*69/68*4000 = 2.208Mb/s. This fixes the definitive carrier spacing at 2.208/2/256 = 4.3125KHz.

Slides 36, 37 & 38 - Time-domain Equalizer (TEQ)

The frequency dependent attenuation and dispersion of the channel leads to pulse distortion and inter-symbol interference and also, for multi-carrier systems (like DMT), to inter-carrier interference (ICI). It can be seen like a leakage of energy from one symbol or carrier to an adjacent ones with as consequence a far more difficult symbol detection. This, combined with severe attenuation and crosstalk, are the dominant impairements in DSL transmission and have to be compensated.

The purpose of the TEQ is to cancel ISI by reducing the equalized channel response to the length of the 'cyclic prefix' guard band.

The value of the TEQ filter coefficients are easily derived by performing an FFT on a reference DMT signal (known amplitudes and phases for each carrier). The output of this FFT contains all the information needed to reconstruct the complex channel transfer function, and can therefore be used to estimate an ideal set of TEQ coefficients minimizing ISI.

Slides 39 & 40 - Frequency-domain Equalizer (FEQ)

As illustrated on the slide, residual channel attenuation and delay distortion are not compensated by the TEQ, and individual phase and amplitude corrections have to be performed for each carrier.

The function of the FEQ is to adjust continuously (using an adaptive update algorithm) each received carriers on their respective constellation grids. The FEQ algorithm can be implemented by a complex multiplication in the frequency domain.

After FEQ processing all carriers, being normalized in amplitude and phase, are sent to the 'de-mapping' unit, which will perform the bit extraction.

Slides 41 & 42 - Filtering & Gain Control

- Transmitter digital filtering:
  In order to attenuate the spectrum of image frequencies after digital to analog conversion, the sampling rate at the Digital-to-Analog Converter (DAC) has to be high enough to limit analog filtering complexity (because of its complexity, IFFT is often performed at Nyquist rate). Moreover, DMT modulation generates slowly decreasing side-lobes (sinc spectrum) which have also to be filtered to limit out-of-band energy or self-NEXT. Therefore, digital transmit filters have to interpolate the signal and to provide enough (digitally generated) energy attenuation outside the transmission band.
- Transmitter analog filtering
  Digital-to-analog conversion generates quantization noise and reveals the image spectrum. Only analog filters can realize the suppression of these analog-related sources of noise, and in order to reduce their complexity, it should also be their only function. To reduce the power of the signal on short loop ('politeness' rule), a coarse gain control may also be implemented in the analog transmitter
- Receiver analog filtering
  In the receive direction, the design target is to pass trough the Analog-to-Digital Converter (ADC) without corruption of the signal. This is obtained by providing enough energy suppression from aliased frequency bands and out-of-band signals (like echo). Energy suppression of unwanted signals must be high enough to allow amplification of the desired signal above the ADC noise floor.

- RX digital filtering
  As for the transmitter, in order to limit analog filter requirements and to improve converter sensitivity, oversampling of the signal above the Nyquist frequency is often used. This implies the use of a digital decimator filter before further digital processing (TEQ, FFT, ...) to limit computation burden to the strict minimum required.

Slides 43 & 44 – Analog<->Digital converters

The dynamic range for a DAC used in an ADSL modem is determined by the following factors: the noise margin, the constellation sizes, the BER and the PAR. For example, if we target an ADSL modem able to transport 15 bits on each carrier with a BER lower than 10^{-7}, a noise margin of 6dB and a PAR of 15dB, we need a converter with a noise level (integrated in the signal band) 74 dB lower than the maximum signal amplitude. The linearity of the converter must be large enough such that distortion components of the signal stay under the above noise level. The example above requires a Multi-Tone Power Ratio (MTPR, difference between carrier and distortion levels) greater than 60 dB.

If the transmission channel has only a flat loss and no echo, ADC and DAC requirements would be the same (almost the case for short loops). For long loops, the signal level is dominated by the echo power and the differential attenuation between frequency band edges can be as high as 40dB. The minimum ADC dynamic range is obtained when the echo level is assumed sufficiently attenuated to allow enough amplification of the wanted signal. A reduction of the echo supression will directly increase the ADC dynamic range by the same amount (EC requires better ADC than FDM because the echo signal cannot be attenuated at all).

Slides 45 & 46 - Line Driver & Hybrid

Line Driver:

- *Signal level*
  The ADSL standard specifies a Power Spectral Density of -40dBm/Hz on 100 ohms lines. For the downstream signal (+/- 1MHz bandwidth), this corresponds to a signal level of 3.2Vrms or almost 18Vp (& 180mA) if we consider a PAR of 15dB. But at the line driver output, the power dissipated in the adaptation impedance double these values (6.3Vrms, 36Vp, 180mA). These high signal levels combined with high signal bandwidth are difficult to achieve with current technologies and imply

the use of a transformer. With a 1:2 ratio, we get at the line driver output 3.2Vrms/18Vp/360mA that can be achieved using a +/- 12V supply.

- *Noise & distortion levels*

  This line driver has to amplify the signal coming from a low-voltage device (typically 1Vp), and in order to avoid too much impact on the receiver sensitivity (via echo), it must also limit its out-of-band noise and distortion. If we consider a gain of 25dB (18/1) and worst case Trans-Hybrid Return Loss (THRL) of 12 dB, the input noise and MTPR of the line driver must respectively be lower than 7nV/sqrt(Hz) (or 4nV/sqrt(Hz)/amplifier if a dual configuration is used) and 88dB in order to be negligible when compared with the –140dBm/Hz AWGN level.
- Power efficiency

  The high PAR of the signal affect badly the efficiency of the driver due to the fact that high level supplies are required to generate the peak values. Typically, in order to generate 100mW on line, an ADSL line driver may dissipate as much as 1W additional power !

Hybrid:

The function of the hybrid is to separate the transmitter and receiver circuits from the line (four-wires to two-wires interface). The hybrid is of crucial importance for long loops in order to avoid that strong transmitted signals desensitize (or saturate) the receiver. This may be caused by impedance mismatch between the hybrid nominal impedance and the line complex impedance leading to reflection. This problem can partially be solved either by echo-cancellation or by separation of upstream and downstream transmission by means of Frequency Division Multiplexing (FDM).

Slide 47 – Conclusions

DSL technology has been the subject of a great deal of media attention in the past year, both in technical magazines and in non-technical press. ADSL will indeed play an important role in the coming years, for the customer at home, for the telephone operators worldwide and for the telecommunications and microelectronics industry.

Slide 48 - Trends

Full ADSL, originally intended for video applications with up to 12 Mbps downstream, is now targeted for bringing high-speed Internet access to residential and small business. For this new market, access speed at 10% of full ADSL capabilities are enough for surfing today's Internet which support actually hardly more than a few hundreds of kilobits. Initial focus moved from maximum throughputs towards minimum guaranteed bandwidth (like a T1 or E1 rate) combined with maximal possible reach.

A new version of ADSL, entitled 'ADSL-lite', supporting this concept has also popped up recently and rapidly gained much attention. ADSL-lite propose a low complexity ADSL modem with reduced features and a limited set of parameters (e.g. a reduced number of tones, limited FEC capabilities, etc.), this in order to ease installation (splitterless) and to reduce as much as possible power dissipation, complexity and its associated bill of materials. Performance figures that are recommended by the SIG (Special Interest Group) are limited upstream rates in the range of 32 to 512 kbit/s and downstream rates between 32 kbit/s and 1.5 Mbit/s.

Slide 49 – Performance

Typical performance of a full rate ADSL modem on NEXT-free 26AWG loops is illustrated on this slide. It can be observed than for both CSA (9kft) and ECSA (15kft) loop lengths the modem is able to achieve data transmission rate up to 40 time faster than conventional analog voice-band modem.

Slide 50 - Facts

As illustrated on the slide, ADSL has now reached its maturity phase, and commercial deployment is now a fact. Forecasts project more than 1.000.000 installed line by 2000 and this for a montly fee as low as 25-50$ for a near-megabit basic service.

Reference

[1] ANSI T1.413-1999, Asymmetric Digital Subscriber Line (ADSL) Metallic Interface.

[2] Shannon, C.E., "A Mathematical Theory of Communication", Bell System Technical Journal, Vol. 27, 1948, pp. 379-423 (Part I), pp.623-656 (Part II).

[3] D. Mestdagh, P. Spruyt, B. Biran, "Analysis of Clipping Effect in DMT-Based ADSL Systems", IEEE ICC'94 Conference, May 1-5

[4] D.Macq, "DMT ADSL Circuit & Systems", IEEE ISSCC'98 Conference, San Francisco, CA, Apr. 98, xDSL Broadband interactive communication via POTS (short course).

[5] Chang, Z.Y., et al, "A CMOS Analog Front-End Circuit for and FDM-Based ADSL System", IEEE JSSC, vol. 30, pp. 1449-1456, Dec. 1995.

[6] Lanford, D.S., et al, "A BiCMOS Analog Front-End Circuit for an FDM-Based ADSL System", IEEE JSSC, vol. 33, pp.1383-1393, Sept. 98.

[7] L. Kiss, et al, "Sachem, a versatile DMT-based modem transceiver for ADSL", IEEE ASIC'98 Conference, Sept. 98, Rochester, NY.

[8] Cornil, J.P., et al, "A 0.5um CMOS ADSL Analog Front-End IC", ISSCC Digest of Technical Papers, San-Francisco, CA, pp. 238-239, Feb. 1999.

[9] Conroy, C., et al, "A CMOS Analog Front-End IC for DMT ADSL ", ISSCC Digest of Technical Papers, San-Francisco, CA, pp. 240-241, Feb. 1999

[10] Hester, R., et al, "CODEC for Echo-Canceling Full-Rate ADSL modems", ISSCC Digest of Technical Papers, San-Francisco, pp. 242-243, Feb. 1999

Further readings

T. Starr, J.M. Cioffi, P. Silverman, "Understanding Digital Subscriber Line technology", Prentice-Hall, Communications Engineering and Emerging Technologies Series, ISBN 0-13-780545-4

W.Y. Chen, "DSL Simulation Techniques and Standards Development for Digital Subscriber Line Systems", Macmillan Technical Publishing, Technologies Series, ISBN 1-57870-017-5

ALCATEL

"Building an ADSL Modem"
The Basics

AACD'99
23-25 March 1999

Jean-Philippe Cornil

Alcatel, Microelectronic Dpt - VA21
Francis Wellesplein, 1
B-2018 Antwerp, Belgium
Tel: + 32 3 240 78 **76**

Jean-Philippe.Cornil@ieee.org

ALCATEL

Outline

- Introduction
- Twisted pair characteristics
- Signal Description
- Building the modem
- Conclusions

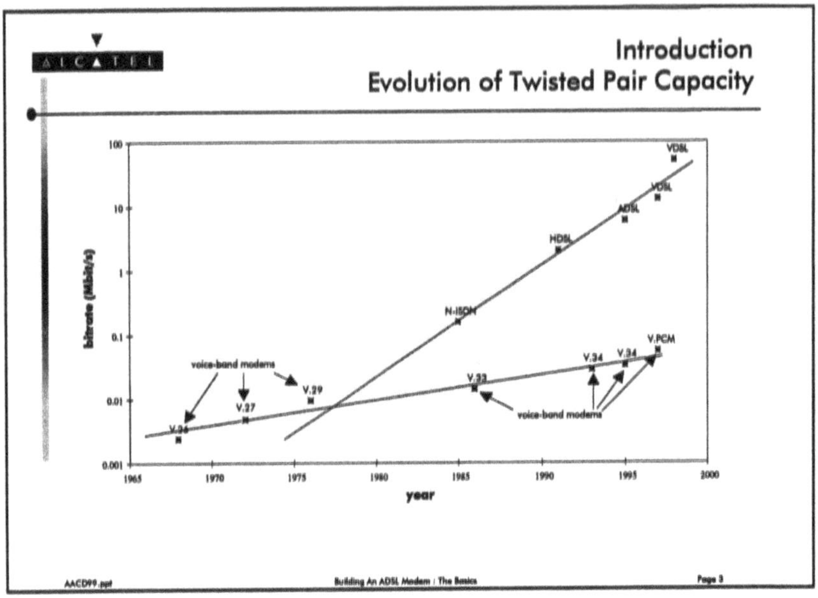
ALCATEL
Introduction
Evolution of Twisted Pair Capacity
bitrate (Mbit/s)
year
voice-band modems
V.27
V.29
V.32
V.34
V.PCM
N-ISDN
HDSL
ADSL
VDSL
AACD99.ppt
Building An ADSL Modem : The Basics
Page 3

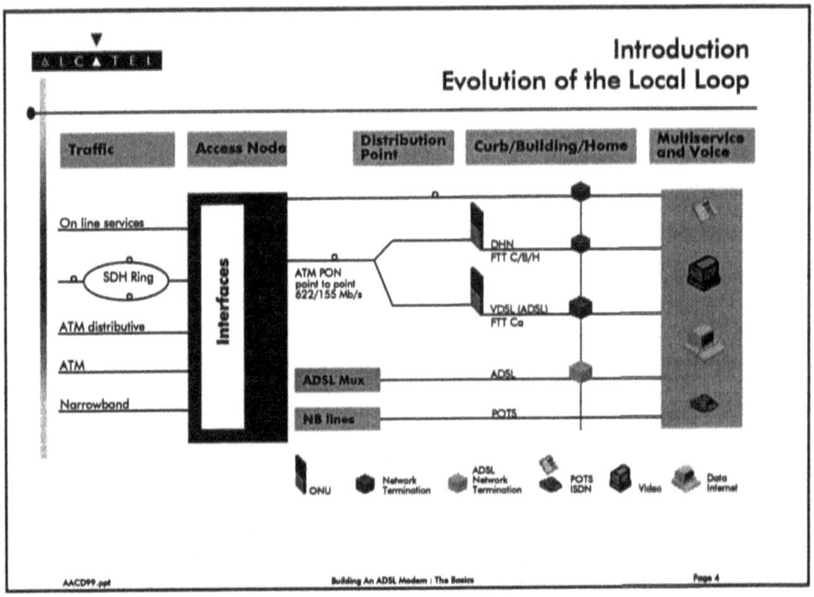
ALCATEL
Introduction
Evolution of the Local Loop
Traffic
Access Node
Distribution Point
Curb/Building/Home
Multiservice and Voice
On line services
SDH Ring
ATM distributive
ATM
Narrowband
Interfaces
ATM PON
point to point
622/155 Mb/s
DHN
FTT C/B/H
VDSL (ADSL)
FTT Ca
ADSL Mux
ADSL
NB lines
POTS
ONU
Network Termination
ADSL Network Termination
POTS ISDN
Video
Data Internet
AACD99.ppt
Building An ADSL Modem : The Basics
Page 4

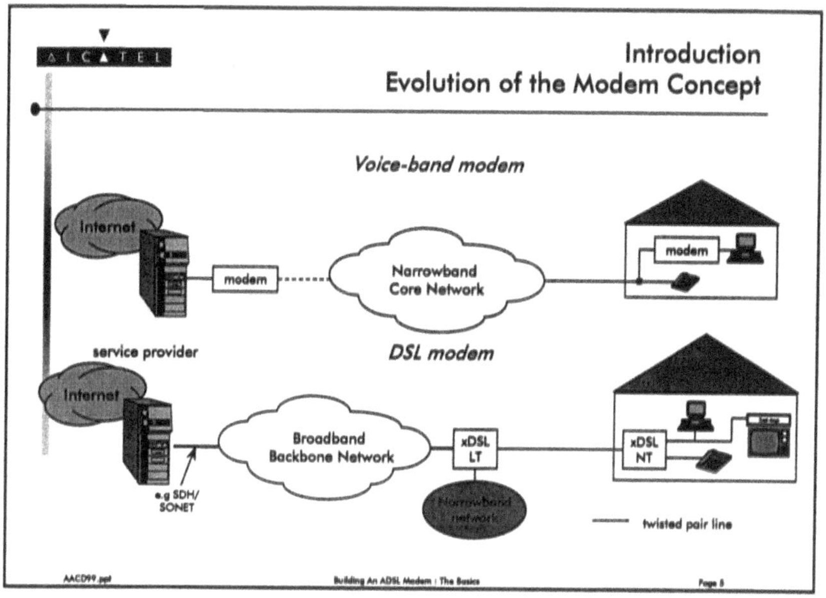
Introduction
Evolution of the Modem Concept
Voice-band modem
Internet
modem
Narrowband
Core Network
modem
service provider
DSL modem
Internet
Broadband
Backbone Network
xDSL
LT
xDSL
NT
e.g SDH/
SONET
twisted pair line
AACD99.ppt
Building An ADSL Modem : The Basics
Page 5

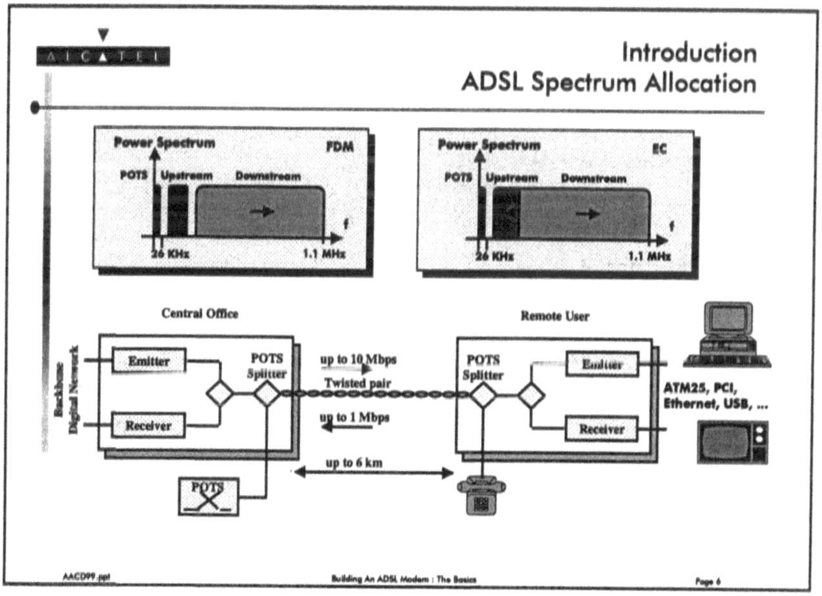
Introduction
ADSL Spectrum Allocation
Power Spectrum
FDM
POTS
Upstream
Downstream
f
26 KHz
1.1 MHz
Power Spectrum
EC
POTS
Upstream
Downstream
f
26 KHz
1.1 MHz
Central Office
Remote User
Backbone
Digital Network
Emitter
Receiver
POTS
Splitter
up to 10 Mbps
Twisted pair
up to 1 Mbps
up to 6 km
POTS
POTS
Splitter
Emitter
Receiver
ATM25, PCI,
Ethernet, USB, ...
AACD99.ppt
Building An ADSL Modem : The Basics
Page 6

ALCATEL

Outline

- ▼ Introduction
- ▼ Twisted pair characteristics
- ▼ Signal Description
- ▼ Building the modem
- ▼ Conclusions

ALCATEL

Twisted Pair Characteristics
Loop Length Length Distribution

- ▼ Carrier Serving Area (CSA)
  - ⇨ 12kft on 24AWG, 9kft on 26AWG
  - ⇨ 75% of all US customer
- ▼ Extended CSA (ECSA)
  - ⇨ 18kft on 24AWG, 13.5kft on 26AWG
  - ⇨ 90% of all US customer

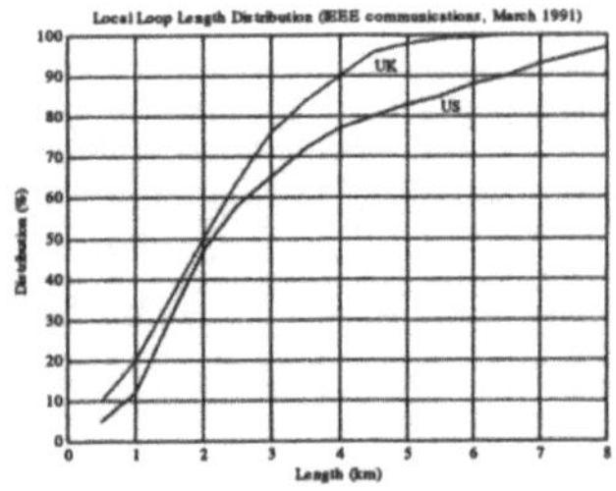

Twisted Pair Characteristics
Loop types

▼ ANSI/ETSI loops

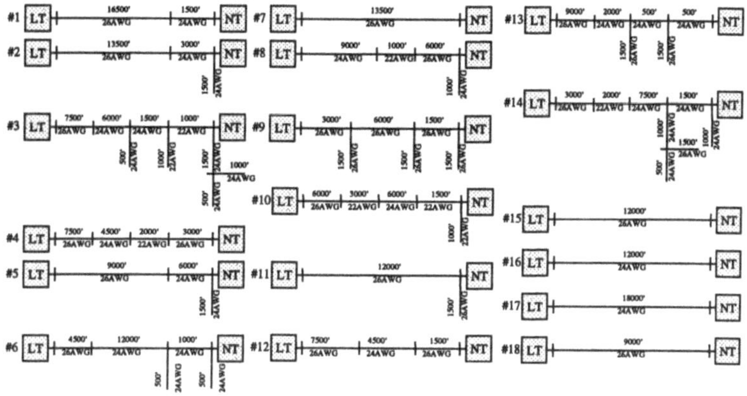

Twisted Pair Characteristics
Model

▼ Distributed passive filter

⇨ Primary parameters

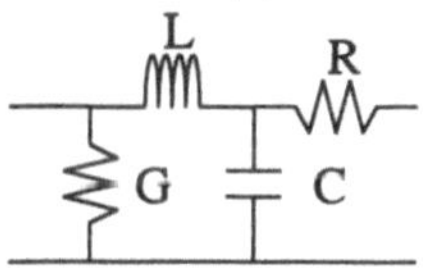

$$R(f)=\sqrt[4]{r_{OC}^4+a_c\cdot f^2},\; L(f)=\frac{l_0+l_\infty\left(\frac{f}{f_m}\right)^b}{1+\left(\frac{f}{f_m}\right)}$$

$$C(f)=c_\infty+c_0\cdot f^{-c_e},\; G(f)=g_0\cdot f^{g_e}$$

⇨ Secondary parameters

$$Z_0(f)=\sqrt{\frac{R(f)+s\cdot L(f)}{G(f)+s\cdot C(f)}},\; \gamma(s)=\sqrt{(G(f)+s\cdot C(f))\cdot(R(f)+s\cdot L(f))}$$

Twisted pairs Characteristics
Transfer function

- Transfer Function

$$H(d,f) = e^{-d \cdot \gamma(f)}$$

- ⇨ Attenuation (d in km, f in Hz)

$$A_{dB} = -d \cdot \mathrm{Real}(\gamma(f)) \approx -k \cdot d \cdot \sqrt{f} \cong -0.025 \cdot d \cdot \sqrt{f} \quad (26AWG)$$

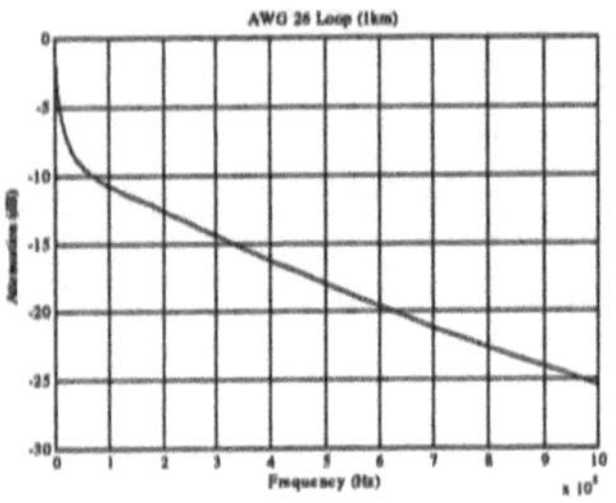

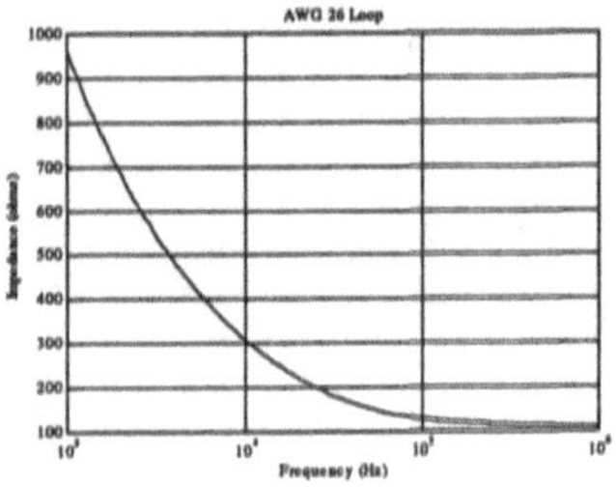

Twisted pairs Characteristics
Loop impairments (1/2)

- Thermal Noise : -140 dBm/Hz
- Crosstalk (NEXT & FEXT)
  - ⇨ Coupling between wires in same and adjacent binders
    - Twisting mandatory
  - ⇨ Self-NEXT

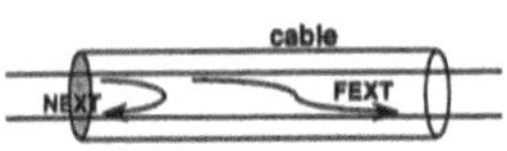

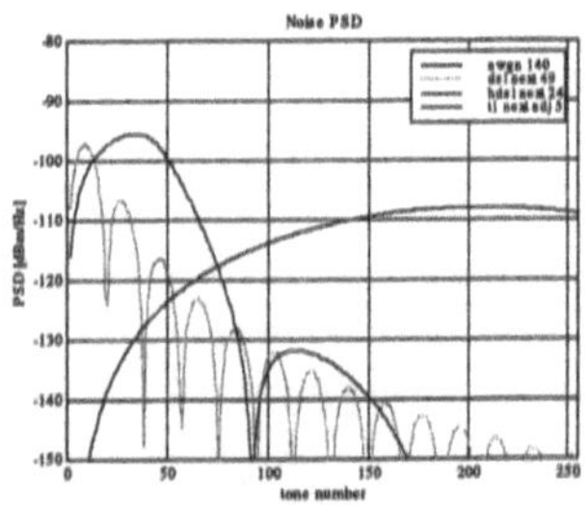

ALCATEL

Twisted pairs Characteristics
Loop impairments (2/2)

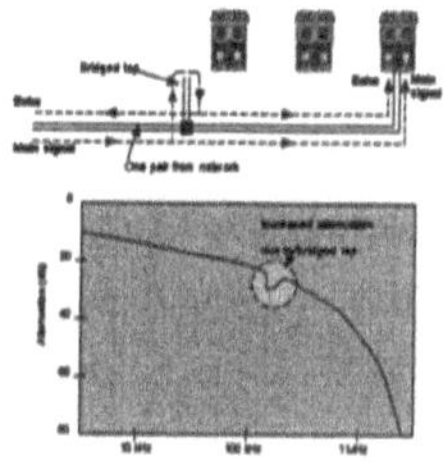

- Bridged taps and gauge transitions
  - ⇨ Reflection : Impedance ripple
  - ⇨ Echo
- Impulsive noise
  - ⇨ High bursts of short duration (10 µs - 1 ms)
- RFI interference
  - ⇨ AM & amateur radio
  - ⇨ Ingress & egress
- POTS signaling (Ringing, on/off hook)
- Loading coils
  - ⇨ Not allowed !

ALCATEL

Outline

- Introduction
- Twisted pair characteristics
- Signal Description
- Building the modem
- Conclusions

ALCATEL

Signal Description
Modulation : "Divide to conquer"

▼ The challenge : Survive the loop impairements !

⇨ Shannon-Hartley

$$Capacity(bps) = W \cdot \log_2(1+SNR) \approx W/3 \cdot SNR_{dB} \quad (1)$$

⇨ For frequency-dependent channel, (1) becomes

$$Capacity(bps) = 1/3 \cdot \int_{f_{min}}^{f_{max}} SNR_{dB}(f) \cdot df$$

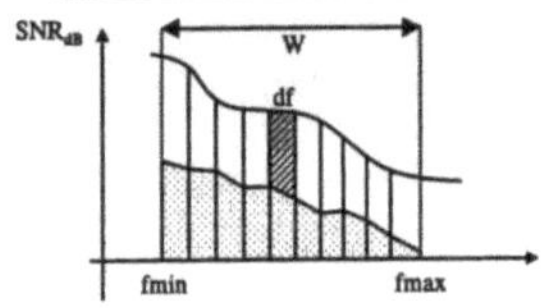

☞ Due to its fine frequency granularity, DMT approaches theoretical limit very closely.

ALCATEL

Signal Description
Frequency allocation strategy

▼ Frequency Position

⇨ Upstream position

- Smaller bit rate than Downstream (Adsl)
- Positioned to reduce self-NEXT (increasing with $f^{3/2}$)
- For EC (@ CO side), limits the RX loss & frequency

⇨ Downstream position

- Wideband ...

⇨ Upstream placed at lower frequency with or without overlap (FDM)

▼ EC vs FDM

⇨ Lower Upstream, Higher Downstream capacity (Self-NEXT/Bandwidth)

⇨ More DSP intensive

⇨ Simple discrete AFE, more analog dynamic range

ALCATEL

Signal Description
ADSL DMT Spectrum

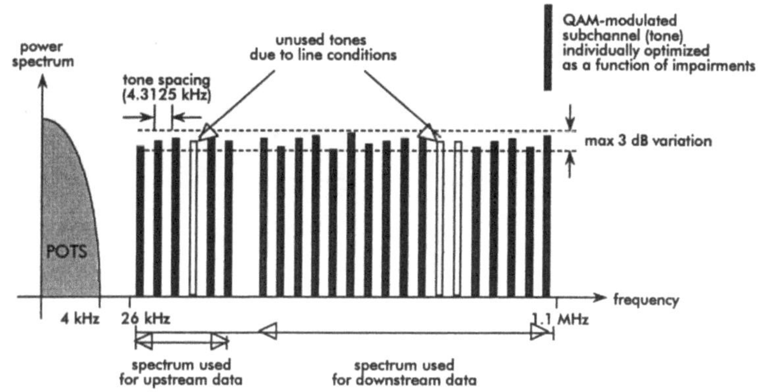

AACD99.ppt Building An ADSL Modem : The Basics Page 17

ALCATEL

Signal Description
Constellation

▼ What's a constellation ?

⇨ A simple way to represent a sine wave using a rotating vector.

⇨ $A \cdot \cos(\omega.t) = \mathrm{Re}(A \cdot e^{i.\omega.t})$

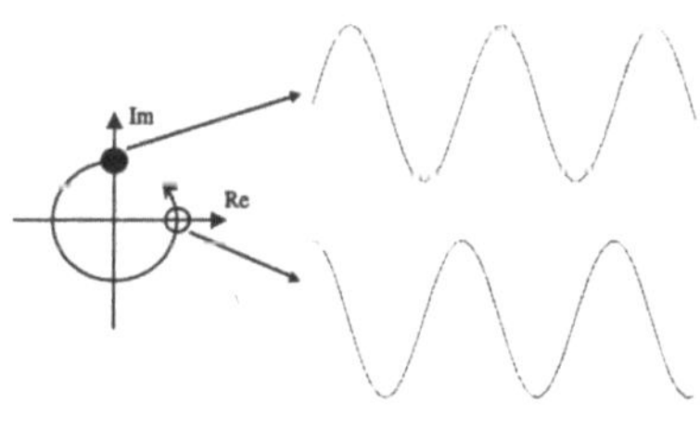

AACD99.ppt Building An ADSL Modem : The Basics Page 18

ALCATEL

Signal Description QAM

- What's QAM (Quadrature Amplitude Modulation)
  - ⇨ A set of bits are used to code one position in a constellation

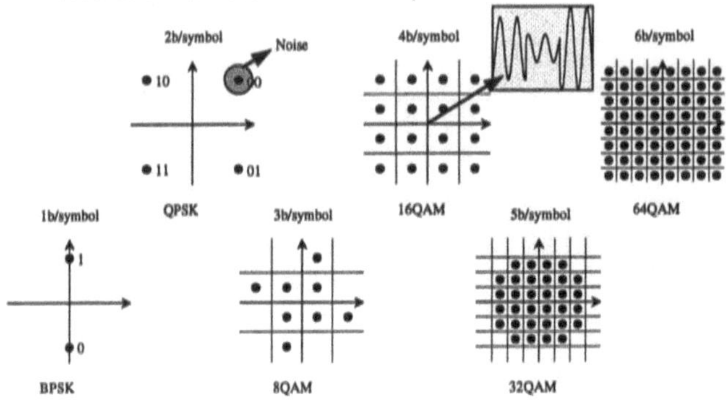

ALCATEL

Signal Description BER

- Bit Error Rate of a 4QAM signal
  - ⇨ 1 additionnal bit = 3dB more
  - ⇨ Example :
    - BER = 10^{-7}
    - 10b QAM
      - ➠SNR=38 dB

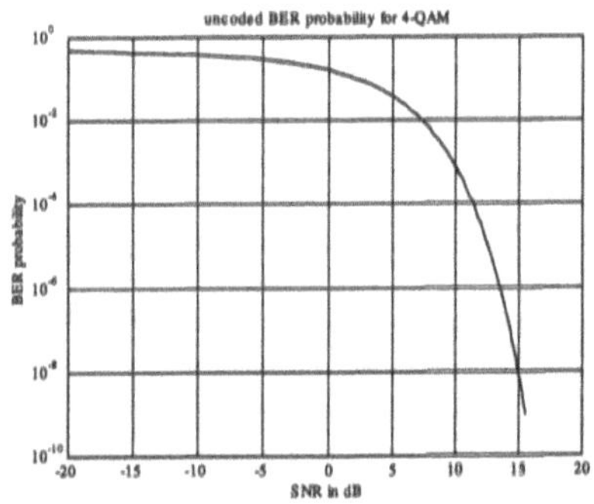

Signal description
DMT

- **Signal description**
  - ⇨ Discrete Multitone Modulation
  - ⇨ QAM carriers evenly spaced in frequency (n x 4.3215kHz)
    - Synchronous in time
    - Orthogonal

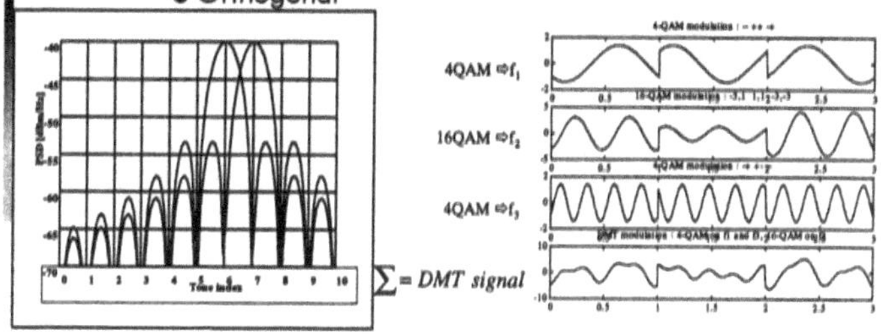

Signal Description
Dynamic range

- **Sum of N independent QAM signals**
  - ⇨ RMS value = sqrt(2/3 N) Amax
  - ⇨ Peak value = N . Amax
  - ⇨ Peak-to-Average (PAR) = sqrt(3N/2)
    - N=200 -> CF = 25dB
  - ⇨ CLIPPING
    - Optimize ADC DR
    - Reduce Power
    - Introduce noise !

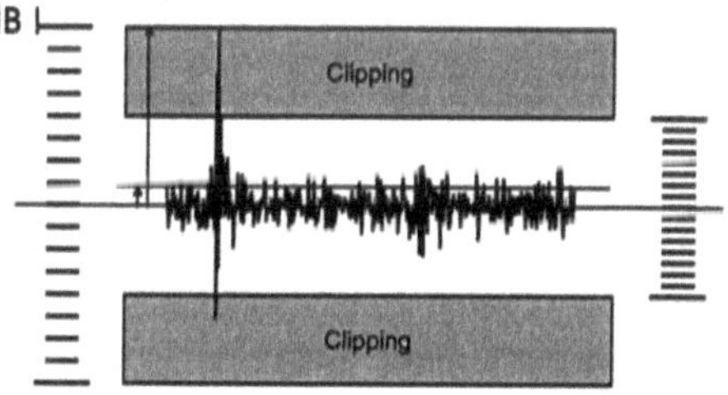

ALCATEL

Signal Description
Clipping probability

- ▼ How much ?
  - ⇨ Gaussian distribution
  - ⇨ PAR=15dB (5.6), CR=10^{-8}
- ▼ Clipping Noise
  - ⇨ Clipping Rate (CR)
  - ⇨ # samples involved
- ▼ PAR reduction techniques

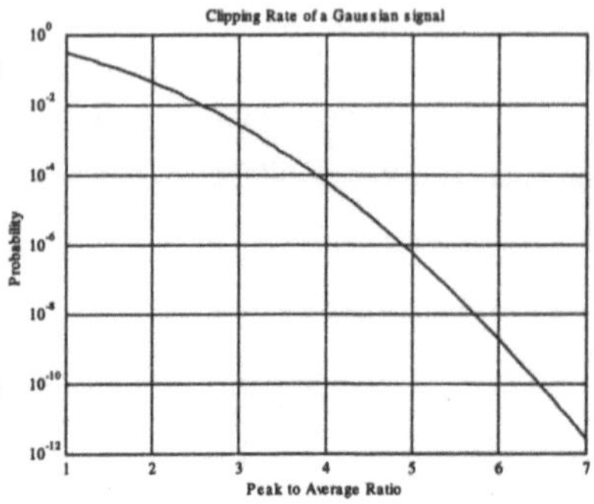

ALCATEL

Outline

- ▼ Introduction
- ▼ Twisted pairs characteristics
- ▼ Signal Description
- ▼ Building the modem
- ▼ Conclusions

Building the modem
Architecture

ADSL Reference Model

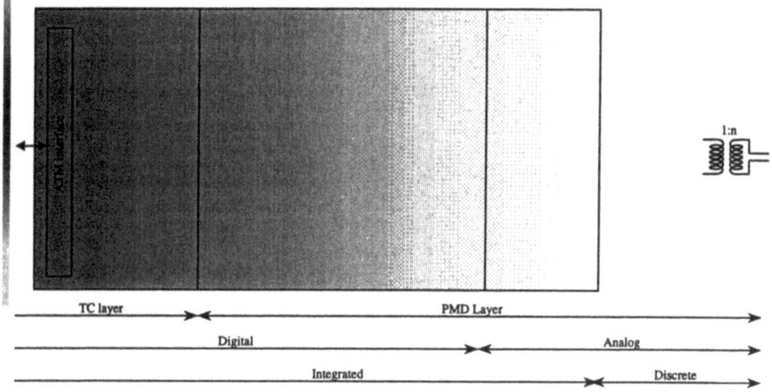

Building the modem
Coding

- ▼ Trellis-coding
  - ⇨ Wei's 16-state 4D code
  - ⇨ ≈4dB of coding gain (BER of 10^{-7} with less SNR)
  - ⇨ Improve performance for long loop
  - ⇨ Increase transmission capacity
- ▼ FEC
  - ⇨ Reed-Solomon (max. codeword length 256)
  - ↳ ≈3dB of coding gain
  - ⇨ Reduce BER from 10^{-7} to 10^{-10}
- ▼ Interleaving
  - ⇨ Distribute Impulse Noise on multiple RS codewords

Building the modem
Architecture : Coding

ADSL Reference Model

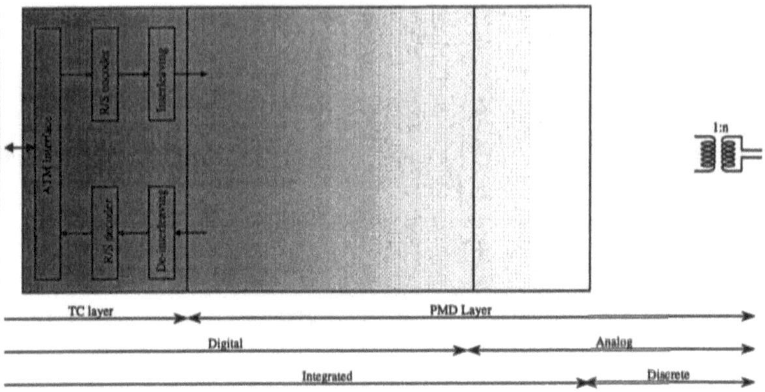

Building the modem
Bit (de)mapping

- ▼ Bits (Bi) & Gain (Gi) allocation
  - ⇨ Bi based on SNRi
  - ⇨ Improved channel efficiency using optimal power/subchannel (Gi)

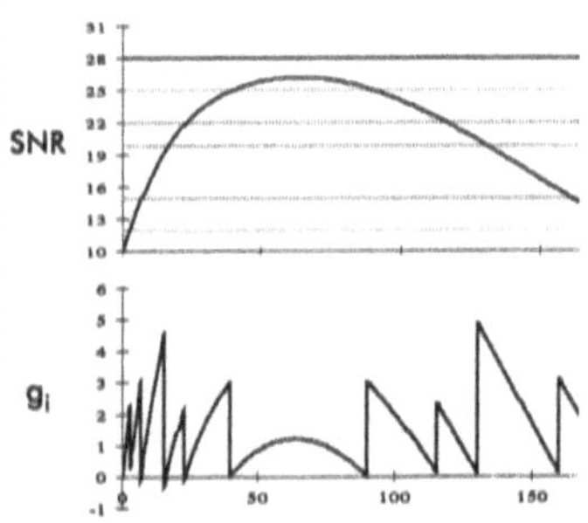

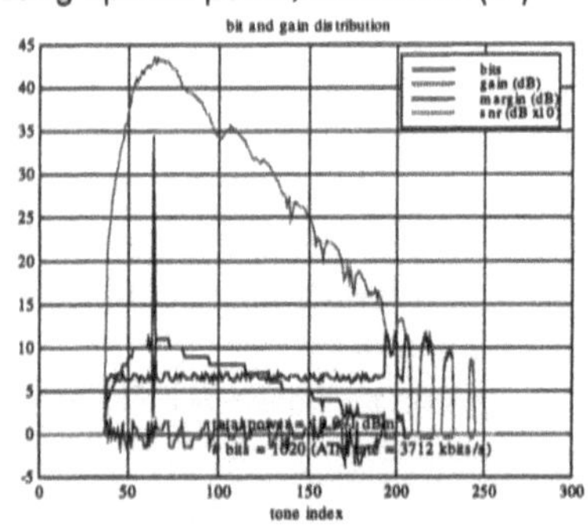

Building the modem
Architecture : Bits (de)mapping

ADSL Reference Model

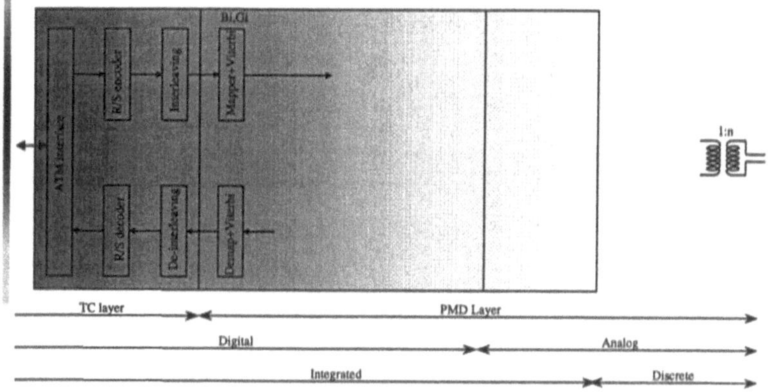

Building the modem
FFT/IFFT

- (De)Modulator : Fast Fourier Transform (FFT)
  - The FFT is an easy way to compute the time domain signal from the constellation points on all tones simultaneously.
  - Size : 512 (CO) or 64 (CPE) points / symbol
  - Rate : 4.3125 KSymbol/s

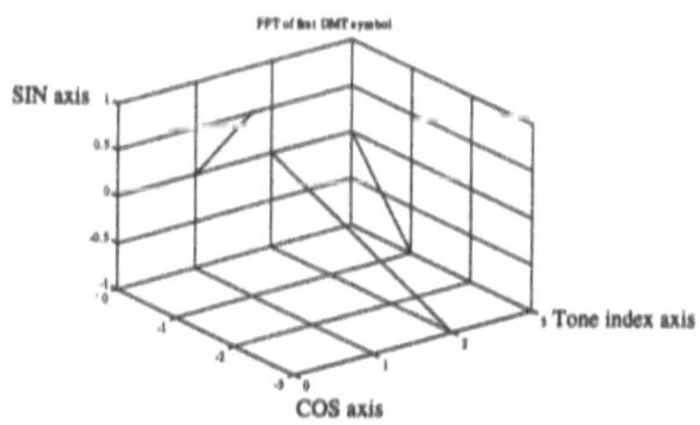

ALCATEL

Building the modem
Architecture : FFT/IFFT

ADSL Reference Model

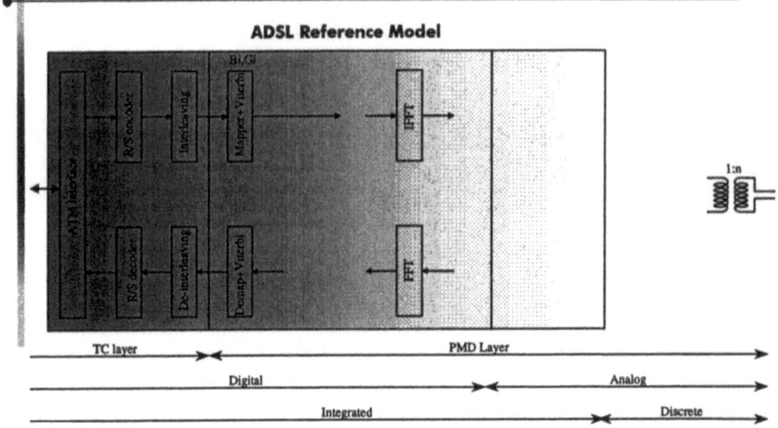

ALCATEL

Building the modem
Line Monitoring & PLL

- ▼ 'Quality' control
  - ⇨ SNR (size of the clouds)
    - ● Bit swaps
  - ⇨ Noise measurements
- ▼ Timing recovery
  - ⇨ 'Pilot' tracking
  - ⇨ DPLL controls:
    - ● Digital Rotor
    - ● Interpolator
    - ● XTAL
- ▼ EOC/AOC extraction
- ▼ Signal detection

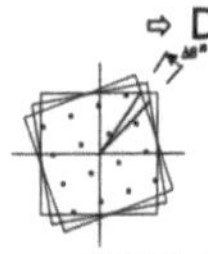

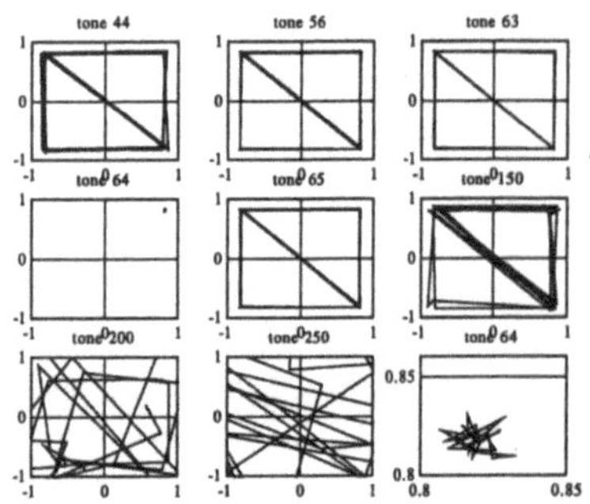

ALCATEL

Building the modem
Architecture : Line Monitoring & PLL

ADSL Reference Model

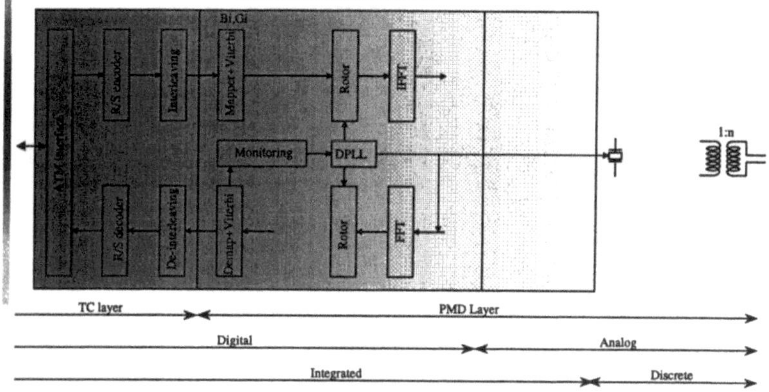

ALCATEL

Building the modem
Cyclic Prefix

- ▼ FFT of the signal, but:
  - ⇨ Channel distortion causes energy leaks between symbols
  - ⇨ Sampling window has to be perfect (ICI & ISI)

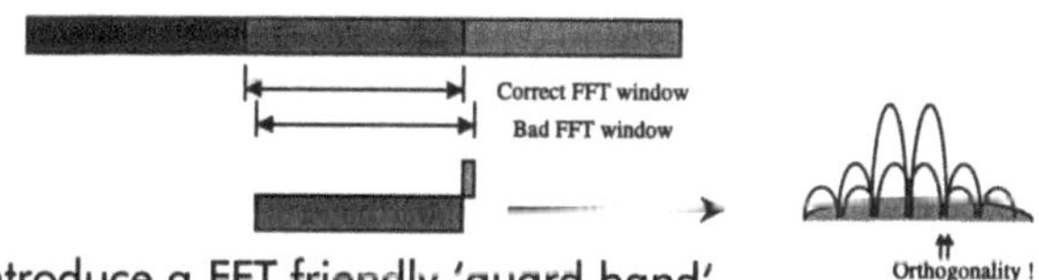

- ▼ Introduce a FFT friendly 'guard band'
  - ⇨ Improve robustness against ISI & ICI

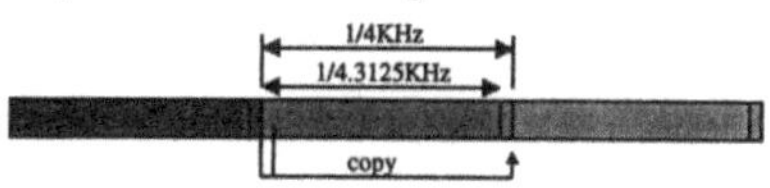

ALCATEL

Building the modem
Architecture : Cyclic Prefix

ADSL Reference Model

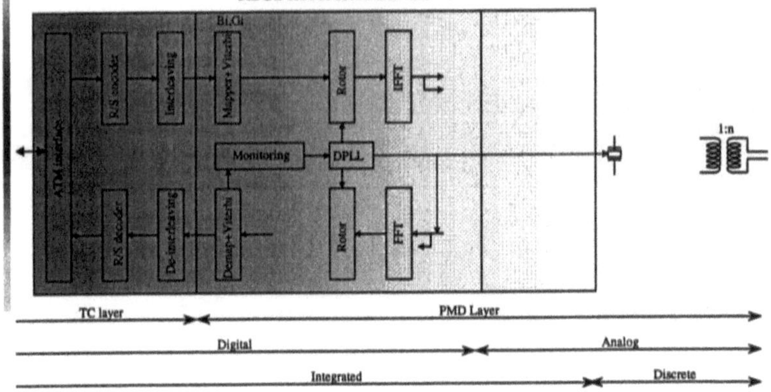

ALCATEL

Building the modem
Time Domain Equalizer (TEQ)

- Frequency dependent attenuation and dispersion
  - ⇨ Inter Symbol & Inter Carrier Interference
- Bridged taps and gauge transitions
  - ⇨ Echo
- ... is killing for FFT processing (ISI, ICI)
  - ⇨ Equalization (reduce ISI)
  - ⇨ Cyclic prefix for the remaining ISI (reduce TEQ complexity)

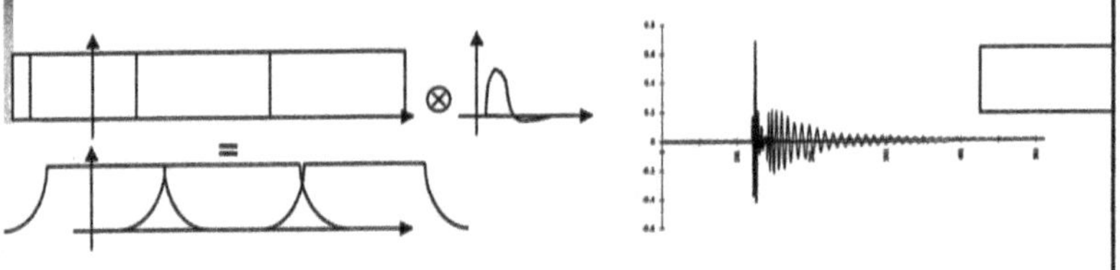

Building the modem
TEQ

- **Filter characteristics**
  - ⇨ By performing an IFFT of the channel frequency response, we get the line impulse response
  - ⇨ TEQ = Minimize Energy outside the cyclic prefix window

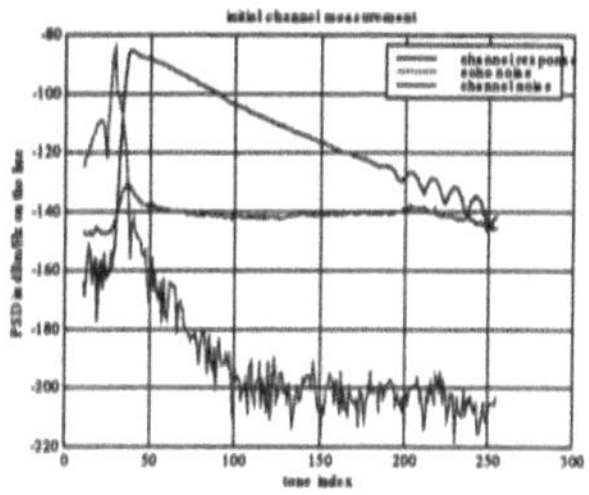

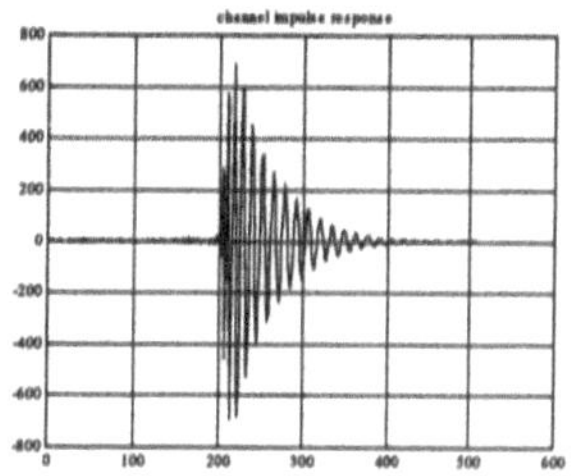

Building the modem
Architecture : TEQ

ADSL Reference Model

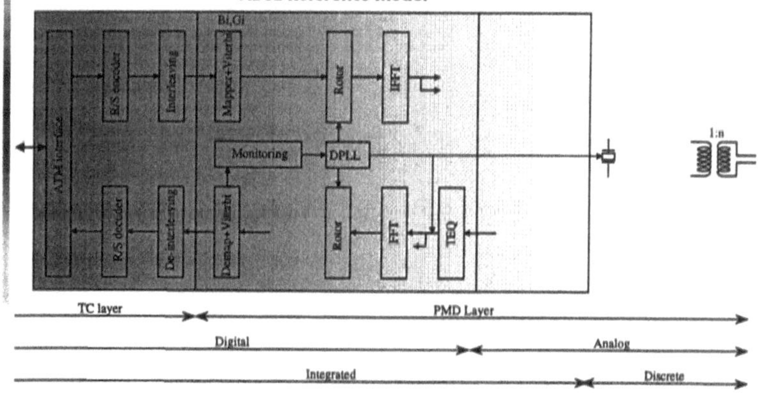

ALCATEL

Building the modem
Frequency domain Equalizer (FEQ)

- Frequency domain equalization
  - ⇨ TEQ & cyclic prefix reduce the channel dispersion but are far to produce a flat channel output !
  - ⇨ Adaptive FEQ needed to flatten & track the channel variations before demapping

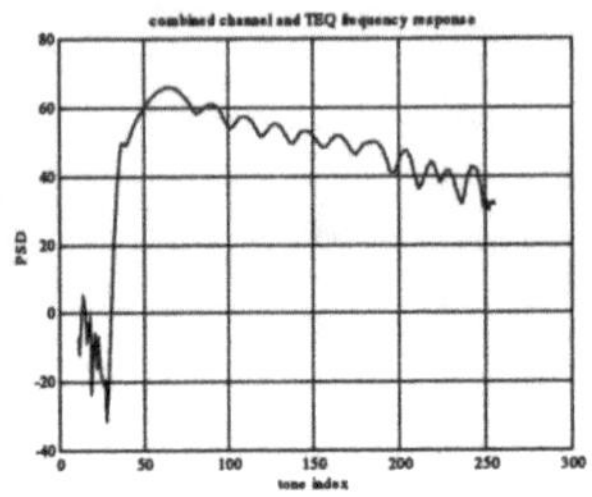

ALCATEL

Building the modem
Architecture : FEQ

ADSL Reference Model

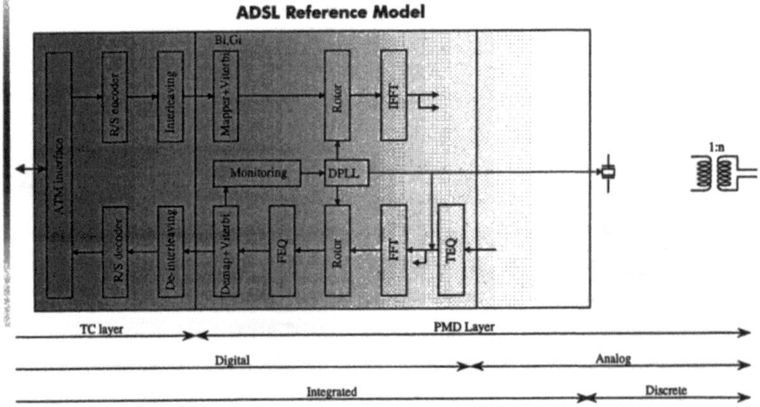

ALCATEL

Building the modem
Filtering & Gain control

- **TX Digital filtering**
  - ⇨ Remove DMT sidelobes (sinc)
  - ⇨ Interpolation to ease filtering of analog image frequencies
- **TX Analog filtering**
  - ⇨ Remove remaining image frequencies (see ANSI PSD mask)
  - ⇨ Remove out-of-band noise (=RX echo)
  - ⇨ PGA implements 'politeness'' rule
- **RX Analog filtering & Gain control**
  - ⇨ Echo reduction
  - ⇨ Anti-Alias Filtering before ADC
  - ⇨ LNA needed to match ADC & Signal dynamics
- **RX digital filtering**
  - ⇨ Decimation to FFT input rate

ALCATEL

Building the modem
Architecture : Filtering

ADSL Reference Model

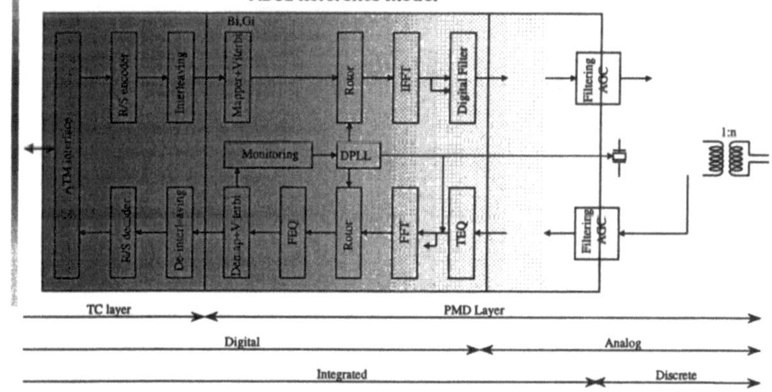

ALCATEL

Building the modem
A2D & D2A converters

- ▼ Minimum requirements
  - ⇨ Digital to Analog Converter (DAC)
    - 15b (I) constellation, 15dB PAR, 6dB margin
      - ➨ 14+(15-2)*3+15+6=74dB
      - ➨ 12b (DNL) Nyquist converter
    - Linearity : MTPR > 60dB
    - But margin (or filtering) needed to garanty RX noise floor
  - ⇨ Analog to Digital converter (ADC)
    - Short Loop : same as above (Energy comes from RX)
    - Long Loop : determined by the ECHO level (filtering strategy)
      - ➨ [11b - 15b] noise level
      - ➨ Linearity : Echo distortion dominates

ALCATEL

Building the modem
Architecture : Filtering

ADSL Reference Model

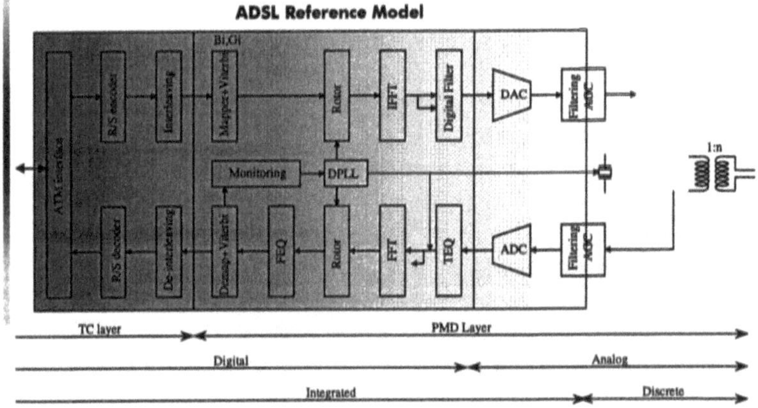

Building the modem
Line Driver & Hybrid

- CO line driver example:
  - ⇨ -40 dBm/Hz, 220 carriers -> 20dBm (100 ohms)
  - ⇨ Vrms = 3.16 V, PAR=15dB ->Vpeak = 18V
  - ⇨ -140dBm/Hz -> Vnoise = 30nV/sqrt(Hz)
  - ⇨ Supply > +/- 12V
  - ⇨ Ipeak > 400mA
  - ⇨ Gain : 25dB
  - ⇨ with a THRL=12dB
    - Inoise : < 7nV/sqrt(Hz)
    - MTPR > 90dB !
  - ⇨ Dynamic Power (+PAR)

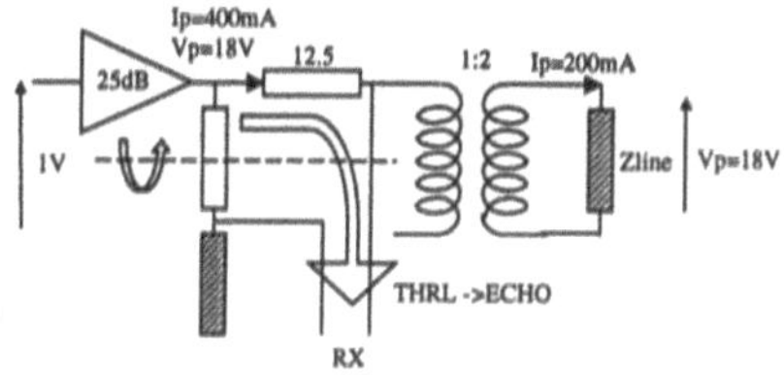

Building the modem
Architecture : FFT/IFFT

ADSL Reference Model

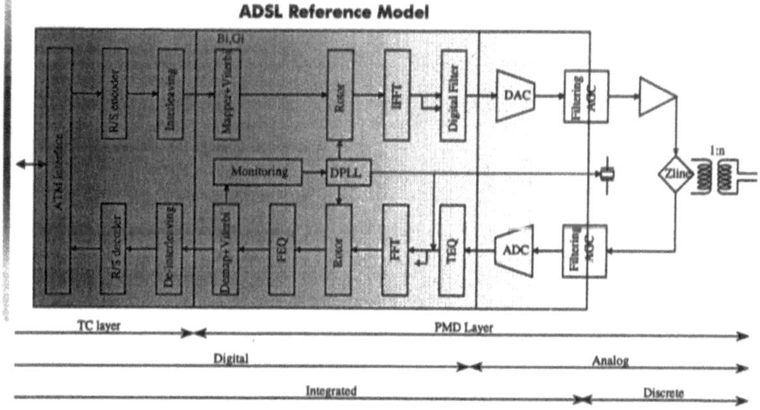

Outline

- Introduction
- Twisted pairs characteristics
- Signal Description
- Building the modem
- Conclusions

AACD99.ppt Building An ADSL Modem : The Basics Page 47

Conclusions
Trends

- Mature modem technology
- Easy to deploy (low marginal cost)
- Optimization
  - ⇨ Cost
  - ⇨ Reach
  - ⇨ Board area
  - ⇨ Power
  - ⇨ Data rate

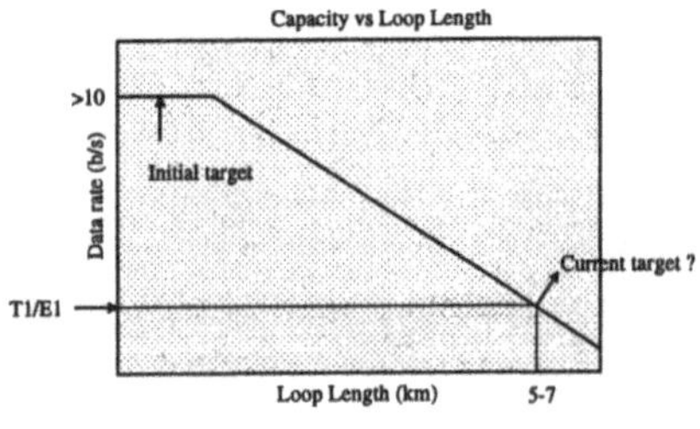

AACD99.ppt Building An ADSL Modem : The Basics Page 48

Conclusion
Typical performances

26AWG; 6dB noise margin; Fast mode; Trellis; no disturbers

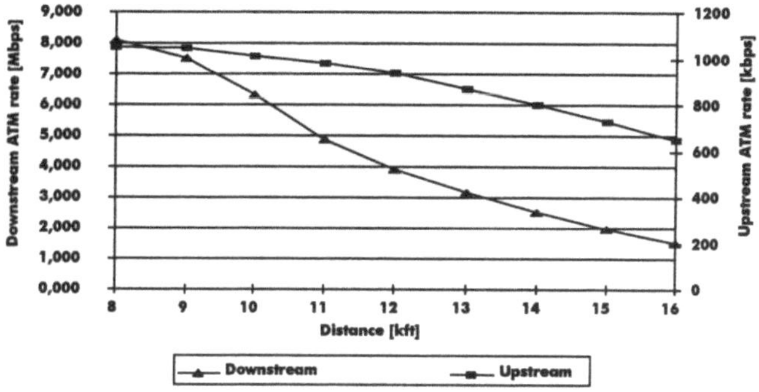

Conclusion
ADSL : A reality !

<u>MAGIX ADSL service from SINGTEL</u>

- One time registration fee: S$30
- Monthly subscription: S$25
- Access charge: S$0.05 per minute

ADSL CODEC ARCHITECTURE THAT MINIMIZES DSP COMPUTATIONAL BURDEN

Richard K. Hester

Texas Instruments Incorporated
Dallas, Texas, USA

ABSTRACT

The optimum analog front-end architecture for ADSL modems results from careful consideration of international ADSL standards, the attenuation characteristics of subscriber loops, the echo return loss of typical hybrid networks and the duplexing method employed. The architecture proposed here is based upon oversampled data converters. It relies on digital filters for standards compliance and minimizes analog filters.

1. INTRODUCTION

Asymmetric Digital Subscriber Loop modems provide high-speed data over the telephone twisted wire pair up to 5.5km. The system, illustrated in figure 1, employs discrete multi-tone (dmt) modulation with carrier tone frequencies located at multiples of the baud frequency, n* 4.3125kHz. In full rate ADSL systems coexisting with plain old telephone service

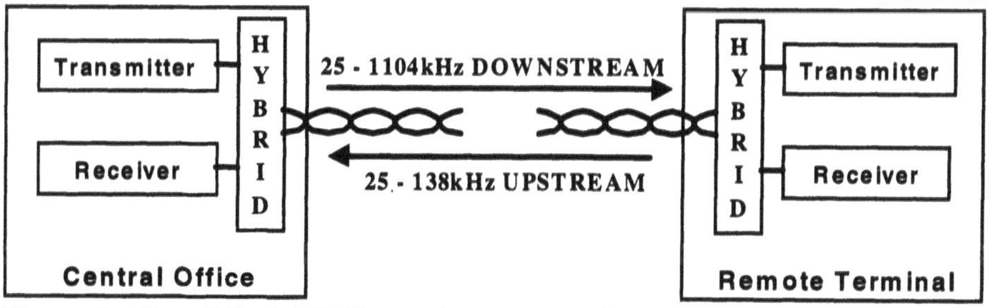

Figure 1. Simplified ADSL modem connection.

(POTS), downstream (from central office to remote terminal) data, ≤8Mb/sec, may use tones, n=6,..., 256. In ADSL "Lite" systems, downstream data, ≤1.5Mb/sec, may use tones, n=6,..., 128. Upstream data, ≤800kb/sec, may use tones, n=6,..., 32.

The overlap of the upstream and downstream bands is left to the discretion of the modem vendor. When there is overlap, echo-cancellation must be used to salvage the SNR of the signal in the presence of the echo. When there is no overlap (frequency division duplexing, or FDD), filters may be used to reject the echo, and echo-cancellation is unnecessary. A third approach that employs both echo-canceling and non-overlapping bands relaxes the required performance of both the echo-canceling algorithm and the filter rejection. This is sometimes called *pseudo-FDD*.

The system's asymmetric nature means that the analog front end at each end of the loop is unique. Transmitters are designed to shape their power spectral density (psd) with two criteria in mind, (1) compliance with limits dictated by the ITU standard[1], and (2) minimization of the corrupting effect of the transmitter echo on the receiver SNR. Receivers are designed to accommodate the wide range of signal and echo power encountered when connected to the universe of subscriber loops.

In the remainder of this text, we will describe the analog front-end design issues in some detail. We will propose specifications that make the front-end operation transparent to the modem performance. Signal characteristics and psd limits are described in section 2. Sections 3 and 4 present proposed transmitter and receiver designs.

2. SHAPING THE TRANSMITTER PSD

A. THE SOURCE

The transmitter signal is produced by an inverse FFT computation every baud period, 231.9μsec. The minimum sized full rate IFFT results in upstream and downstream sample rates of 276kS/sec and 2208kS/sec, respectively. Figure 2 shows the psd corresponding to this minimum-sized IFFT using all available tones and converting outputs to the analog domain using zero-order-hold digital-to-analog converters. Two aspects are noteworthy. One is that each DAC produces significant (-3.9dB) droop at the high frequency end of its respective base-band. The other is that the signal images produced by the DAC have relatively high psd far from the base-band. Neither is acceptable.

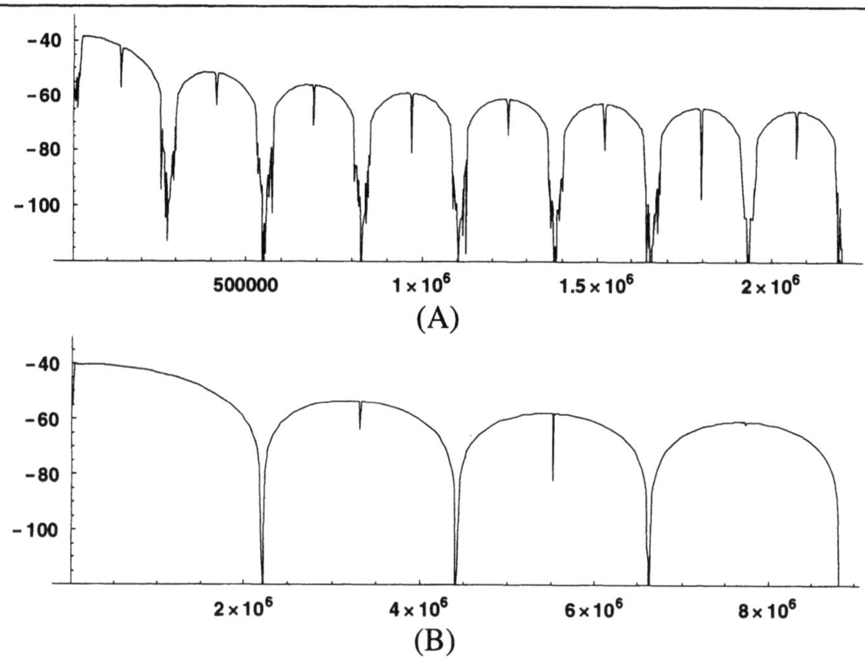

Figure 2. PSD of (A) upstream and (B) downstream psd produced when each IFFT output is converted to the analog domain using a zero-order-hold DAC. Units are dBm/Hz.

B. THE ECHO

Ideal hybrid operation means that all of the transmitter energy reaches the subscriber loop and none leaks into the receiver. Unfortunately, hybrids are far from ideal. The hybrid return loss, defined as the ratio of power transmitted on the loop to power "echoed" into the receiver, is a measure of the hybrid quality.

It is not unusual to find both a low hybrid return loss (big echo) and large loop attenuation (small signal), so avoiding the SNR reduction caused by the echo is critical to modem performance. FDD-based modems reduce the echo power by employing band-splitting filters in both the transmitter and receiver at each end of the loop. Echo-canceling-based modems reduce it algorithmically in the digital domain and do not need extensive filtering, but as we shall see, they require higher resolution transmitter DACs to avoid SNR degradation by the echo of transmitter quantization noise.

C. THE STANDARD

The signal psd limits are provided in the ITU standard. Plots of the upstream and downstream limits are shown in figure 3. The downstream pass-band peak psd must be less than –36.5dBm/Hz and average less than –40dBm/Hz over the 25-1104kHz band. The upstream signal may be 2dB greater in its 25-138kHz band. Outside the pass-bands, the psd limits decrease at rates determined to minimize ADSL impairments to other telephone company services.

We should note that figure 3 does not quite represent the standard. The minimum peak psd in the standard is –90dBm/Hz, but the figure indicates otherwise above 3093kHz downstream and 1221kHz upstream. The figure represents an overly conservative interpretation of the fact that the average psd (1MHz bandwidth) may not exceed –110dBm/Hz at frequencies above 4545kHz downstream and 1630kHz upstream.

In the central office there is concern that the near end cross talk (NEXT) from the transmitter of one subscriber loop into the central office receiver of a second loop will degrade the upstream SNR and data rate over the second loop. This central office concern is the motivation behind the "reduced NEXT" psd mask shown in figure 3A. Compliance, while not mandated by the standard, is important to central office operators in North America and requires that the psd decrease by approximately -36dB/octave

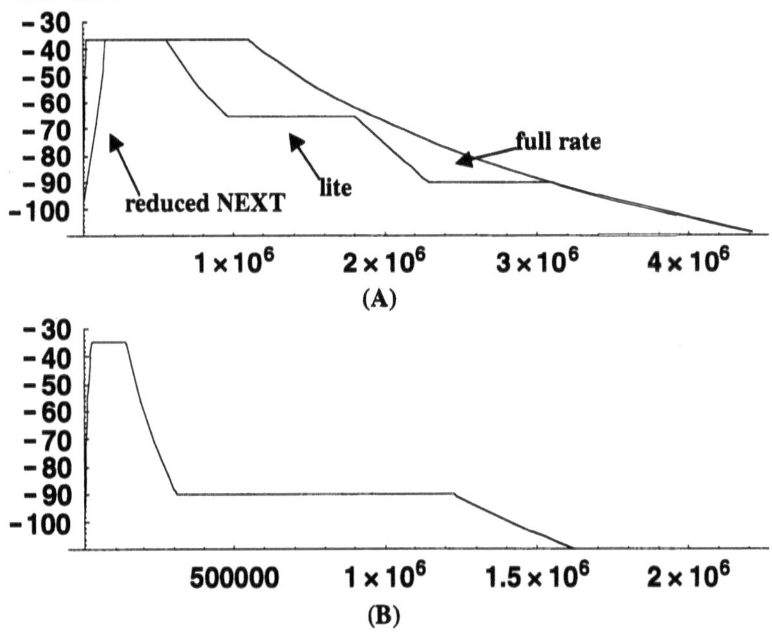

Figure 3. ITU downstream (A) and upstream (B) psd masks. Note that downstream masks have reduced NEXT option for both G.lite and G.dmt operation.

at frequencies below 138kHz. In FDD-based modems, the same filters used to separate upstream and downstream signals will also reject the NEXT. Echo-canceling-based modems with their overlapping bands, by definition, cannot comply. The reduced NEXT psd mask is important primarily to pseudo-FDD-based modems where the band splitting filters alone may not be adequate to prevent NEXT corruption of adjacent loop upstream performance.

D. ADDITIONAL PSD SHAPING FOR FDD

While the psd shaping in echo-canceling modems need only comply with the requirements of the standard, the FDD modem operation requires additional psd shaping constraints. At each end of the connection, the transmitter psd must be shaped such that its echo psd, when referred to the loop, is no greater than the loop noise psd within the band of the receiver. The loop noise is usually taken to be -140dBm/Hz or 31.6nV/√Hz.

Consider an FDD modem that employs the narrow 25-138kHz band upstream (carriers 6-31) and the broad 172.5-1104kHz band downstream (carriers 40-255). At first sight, one might argue that this assignment wastes the 138-172.5kHz band. In fact, however, the subcarriers are so significantly broadened by the data they carry that high order filters are required to suppress the transmitter echoes. The dead band is a compromise in spectral efficiency that reduces the filter complexity to a semi-reasonable level. If we assume a worst-case hybrid echo return loss of 12dB, then the downstream transmitter psd must decrease from –40dBm/Hz at 172.5kHz to -128dBm/Hz at 138kHz. This is an 88dB roll-off in about 0.32 octaves, or 273dB/octave, and it exceeds by far the low frequency psd roll-off imposed by the ITU standard. Likewise, given the same hybrid echo return loss, the remote terminal transmitter psd must decrease from –38dBm/Hz at 138kHz to –128dBm/Hz at 172.5kHz. This roll-off rate is 280dB/octave, far greater than that dictated by the standard. These FDD psd limits are plotted in figure 4.

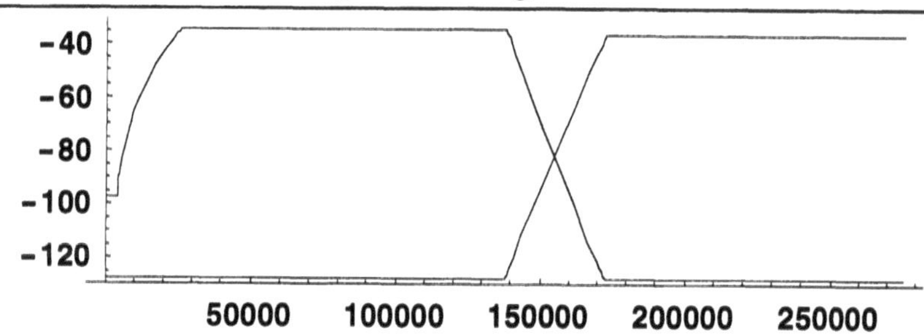

Figure 4. FDD psd limits.

3. TRANSMITTERS

A. ARCHITECTURE

The CODEC transmitter that minimizes the computational burden of the DSP converts the minimum-sized IFFT output to the analog domain without significant pass-band droop and shapes the out-of-band psd to comply with either the ITU standard or the FDD-based modem receiver constraint when it is applicable.

One approach to eliminating the pass-band droop employs a filter with a compensating peak at the high frequency end of its pass-band. This filter may be in either the digital domain ahead of the DAC or in the analog domain after it. Of course, such a filter only approximates the desired x/Sin[x] shape, so some undesirable pass-band ripple inevitably results. Additionally, if an analog filter is used, then processing variations exacerbate the ripple.

Another approach to reduce the pass-band droop is to avoid creating it in the first place. This results when a DAC with a higher conversion rate is employed. This strategy does, however, introduce a new problem, that of increasing the sample rate of the DAC input. The higher sample rate may be obtained either by using an oversampled IFFT or by interpolating the sample rate of the minimum sized IFFT output. The latter method, being the basis for the title of this paper when implemented on the CODEC, is advocated because it is the least demanding of the DSP and it minimizes the I/O rate of data transfer between the DSP and the CODEC.

There are other benefits to interpolating the sample rate. One is the opportunity it provides to perform the psd shaping in the digital interpolation filter and thereby reduce the order of the analog filter. In effect, we are trading analog filter complexity for digital filter complexity, and when symmetric FIR interpolating filters are employed, there is a net reduction in group delay distortion.

For all these reasons, the optimal transmitter signal path consists of a digital interpolation filter, an oversampled DAC and a low-order continuous-time analog filter. The digital filter provides both the psd shaping and the interpolation. The analog filter rejects the signal images produced by the DAC. This is illustrated in detail in Figure 5. Some of the specifications for those elements will be highlighted below.

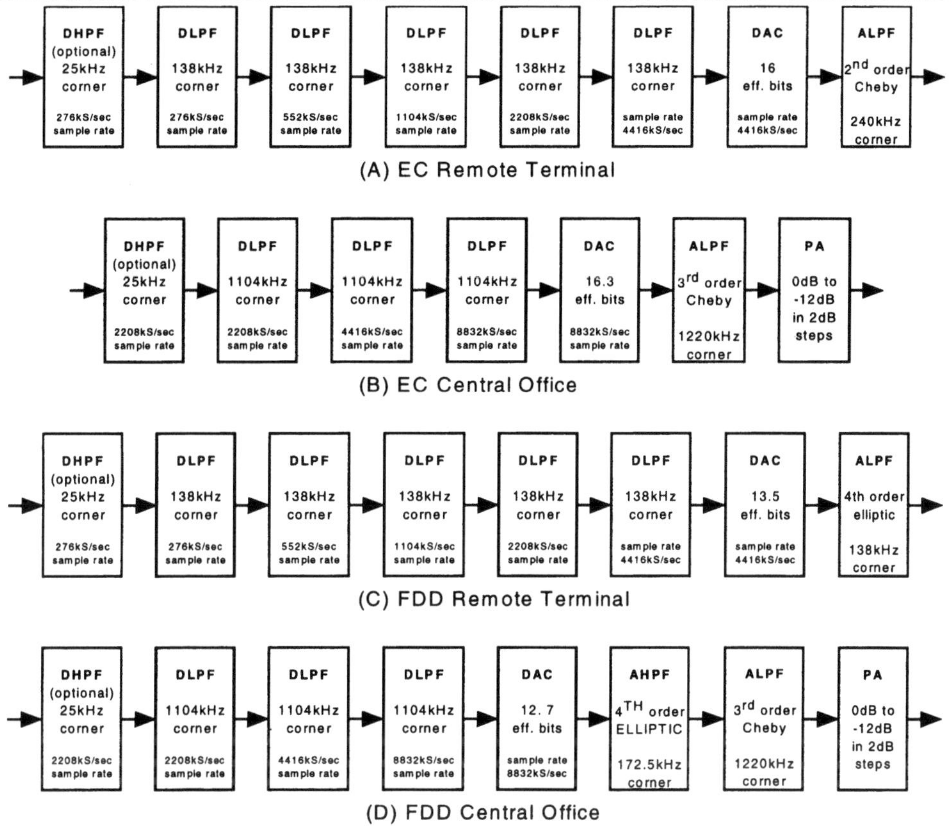

Figure 5. Proposed transmitters.

B. DAC

Consider first the DAC conversion rates. It is highly desirable to minimize the complexity of analog filtering required in order to reduce the number of sources of noise, distortion and power dissipation. From this point of view, very high oversampling factors are desirable. However, considerations of the DAC settling and digital filter complexity suggest a practical limit of about 16X interpolation up to a DAC conversion rate of 10MS/sec. The oversampling factors in the transmitters proposed here are 4X downstream and 16X upstream. The corresponding DAC-induced pass-band droop is less than 0.25dB downstream and 0.014dB upstream.

Finite DAC resolution adds quantization noise to the transmitter signal. There are two ways that this noise can limit the modem performance. Either it limits the SNR of the transmitted signal, or its echo limits the SNR of the receiver.

Consider first the transmitter SNR. The ITU standard indicates that the SNR of a given subcarrier must be (20+3b)dB in order for the subcarrier to support the transmission of a b-bit constellation with 10^{-7} bit error rate. This SNR requirement is considered by many to be excessively conservative, but in the spirit of transparent CODEC operation we shall use it in the following derivation of converter resolution. We assume here that the DAC input signal level is adjusted such that the ratio of full-scale peak signal to rms signal is 5.61 (15dB). This assumption insures that the noise induced by signal clipping is insignificant [2]. The 15dB peak-to-rms ratio means that the average rms subcarrier voltage is then $V_{REF}/(11.22\sqrt{M})$, where V_{REF} is the DAC peak-to-peak output voltage range and M represents the number of subcarriers in the transmitter signal. Ideal N-bit DAC rms quantization noise, $V_{REF}/(2^N\sqrt{12})$, is spread uniformly across the DC – $f_S/2$ frequency band, where f_S is the DAC conversion rate, and the quantization noise within the bandwidth of a subcarrier, f_{BW}, is

$$\frac{V_{REF}}{2^N\sqrt{12}}\sqrt{\frac{2f_{BW}}{f_S}}.$$

Thus, the subcarrier SNR, expressed in dB, is given by

$$SNR_{DN} = 6.02N_{DN} + 13.87 - 10\mathrm{Log}_{10}\left[R_{DN}/M_{DN}\right] \text{ and}$$

$$SNR_{UP} = 6.02N_{UP} + 4.84 - 10\mathrm{Log}_{10}\left[R_{UP}/M_{UP}\right],$$

where R_{UP} and R_{DN} are the upstream and downstream oversampling factors.

When these expressions are equated to 20+3b and the numbers of upstream and downstream subcarriers are inserted, we may solve for the DAC resolution in terms of constellation size (2^b). The results are

$$N_{DN} \cong 5 + \frac{b_{DN}}{2} - 1.66\mathrm{Log}_{10}\left[R_{DN}\right] \text{ and}$$

$$N_{UP} \cong 4.9 + \frac{b_{UP}}{2} - 1.66\mathrm{Log}_{10}\left[R_{UP}\right].$$

(Note that the weak dependence upon the number of subcarriers results in no significant difference between the minimum DAC resolutions for echo canceling, M_{DN} =250, and FDD, M_{DN}=215, modems.) When R_{DN}=R_{UP}=1, 15-bit constellations are supported by DACs with about 12.5 effective bits.

The expression above indicates a half bit DAC resolution reduction is permitted for every oversampling octave. This a little misleading since it is only possible if DAC-induced harmonic distortion can be neglected. Only DACs with ≥12.5-bit integral linearity can be used to take advantage of the oversampling factor. Since DAC linearity is normally no better than its resolution and since (12.5-k)-bit DACs with 12.5-bit linearity are no easier to build than 12.5-bit DACs, we conclude that transmitters require the full 12.5-bit resolution in order to support 15-bit constellations.

Now consider the DAC quantization noise echo. At the loop connection to the transmitter, the signal powers are 20.3dBm (echo-canceling modem), 19.7dBm (FDD modem) downstream and 12.5dBm upstream. We may use these numbers to calculate the DAC quantization noise power measured on the subscriber loop. Given the 15dB signal peak-to-rms ratio, the signal levels are 36.7Vpp (E-C) and 34.3Vpp (FDD) downstream and 15Vpp upstream. If these levels are the amplified full-scale peak-to-peak outputs of N-bit digital-to-analog converters, their corresponding quantization noise powers are (30.5-6N_{DN})dBm, (29.9-6N_{DN})dBm and (22.7-6N_{UP})dBm, respectively. Without oversampling, the psd in each is (–29.9-6N_{DN})dBm (EC), (–30.5-6N_{DN})dBm (FDD) downstream and (–31.7-6N_{UP})dBm upstream. Given 12dB worst case hybrid return loss, the resolutions required to lower the effective DAC quantization noise at their corresponding receivers to –140dBm/Hz are 16.3b (EC), 16.2b (FDD) downstream and 16.0b upstream. As discussed above, oversampling may be used to reduce the resolution by a half bit for every octave, but the >16-bit linearity requirement and the corresponding design difficulty is unaffected.

For transparent operation, echo-canceling modems must employ DACs with >16 effective-bit resolution. In FDD-based modems, however, the DAC quantization noise in the receiver band can be rendered negligible by the stop-band attenuation of the analog filter following the DAC. A design compromise exists between DAC resolution and analog filter complexity. When the quantization noise echo is eliminated the required DAC resolution is the >12.5 effective bits required to achieve the 65dB transmitter subcarrier SNR.

C. FILTERS

Since the digital filter preceding the DAC provides virtually all the filtering necessary for standards compliance, the primary purpose of the analog filters in the transmitter is the removal of the signal image energy produced by the DAC. In FDD modems, they also reduce the DAC quantization noise in the receiver frequency band.

In the upstream transmitter of the echo-canceling modem, roughly 67dB total image attenuation is required for compliance with the ITU standard. The Sin[X]/X attenuation produced by the 16X oversampled DAC is ≥29.84dB in the 4278kHz to 4554kHz frequency band, so the subsequent low-pass filter must deliver about 37.2dB attenuation above 4278kHz. This is provided by a second order Chebyshev with 0.5dB ripple with a nominal corner frequency of 240kHz ±75%. This tolerance to large corner frequency variation means that no special process control or form of filter trimming is required.

The corresponding low-pass of the FDD modem is quite different since it is also used to reduce the required resolution of the DAC. This filter must provide the 37.2dB image attenuation above 4278kHz and at the same time produce at least -6dB per bit of desired DAC resolution reduction at the lower edge of downstream band, nominally 172.5kHz. The elliptic approximation is a good choice for its steep roll-off. A third order elliptic filter with 0.5dB ripple provides roughly 6dB attenuation at 172.5kHz and 37.2dB attenuation above 340kHz, and it reduces the DAC resolution requirement to 15 effective bits. A fourth order elliptic filter, producing 15dB attenuation at 172.5kHz and 37.2dB attenuation above 211kHz reduces the DAC resolution to 13.5 effective bits.

In the downstream transmitter of the echo-canceling modem, roughly 65dB total image attenuation is required for compliance with the ITU standard. The Sin[X]/X attenuation produced by the DAC is ≥17.1dB in the 7728kHz to 9936kHz frequency band, so the subsequent low-pass filter must possess about 48dB attenuation above 7728kHz. This is provided by a third order Chebyshev with 0.5dB ripple with a nominal corner frequency of 1220kHz and variations within ±10%. This intolerance to corner frequency variation means that particularly good process control of resistor and capacitor absolute values or some form of trimming is required.

In addition to a similar third order Chebyshev low-pass filter, an analog high-pass filter is required in the downstream transmitter of the FDD modem to lower the DAC resolution. The high-pass corner frequency is nominally 172.5kHz, and it must produce at least 6dB attenuation below 138kHz per bit that the DAC resolution is reduced from 16.2 effective bits. A third order elliptic high-pass with 0.25dB ripple above its 172.5kHz corner frequency delivers 12.5dB attenuation below 138kHz and reduces the DAC resolution requirement to about 14.1 effective bits. Increasing the filter order to four will result in about 21dB attenuation below 138kHz and enable the use of a 12.7 effective-bit DAC. Some form of trimming is required to maintain the high-pass filter corner frequency within a few percent of its nominal position.

Finally, we will make a few comments about the digital filters. Firstly, in contrast to what we have assumed in the preceding discussion, they need not provide all the psd shaping for standards compliance, but only that portion not provided by the analog filters with their worst case variation of corner frequency. Secondly, they are best implemented in 2X interpolation filter stages. The first stage, clocking at the lowest frequency, provides most of the precision psd roll-off starting at the edge of the pass-band. Each subsequent stage increases that roll-off and eliminates pass-band images of previous stages. Thirdly, if these are implemented as FIR filters, the signal path group delay distortion will be minimized. Fourthly, all these filters are relatively easily implemented without dedicated multipliers, using only shifts and adds to perform multiplication with the fixed filter coefficients that can average less than two non-zero bits each when represented by canonical signed digits. Finally, high-pass filters for standards compliance in the POTS band are not necessarily required if some care is exercised in AC-coupling the line driver to the subscriber loop and to the codec.

4. RECEIVERS

A. THE SIGNAL AND THE ECHO

Figure 6 shows the power levels of the signals and echoes as measured at the end of the loop and plotted versus the loop length (in feet) of a 26awg (0.4mm) wire. Implicit in this plot is an assumption that the echo return loss is 12dB and independent of frequency. The figure shows that the maximum receiver input power is about 12dBm, or 14Vpp, and it varies with loop length. Actually, echo return loss varies significantly with different loops, as well, and there are situations where the total receiver input power may be 10-20dB lower than indicated.

The ideal hybrid gain should be programmable so that as much gain as possible can be applied as early as possible in the receiver path. However, for the purposes of discussing the receiver architecture and avoiding the tedium of power level bookkeeping, we will assume a fixed, frequency-independent net signal gain from the loop to the CODEC of –12dB at both the central office and the remote terminal. This means, according to figure 6, that the maximum CODEC input power is about 0dBm, or 3.5Vpp, a level compatible with modern mixed-signal integrated circuit technology.

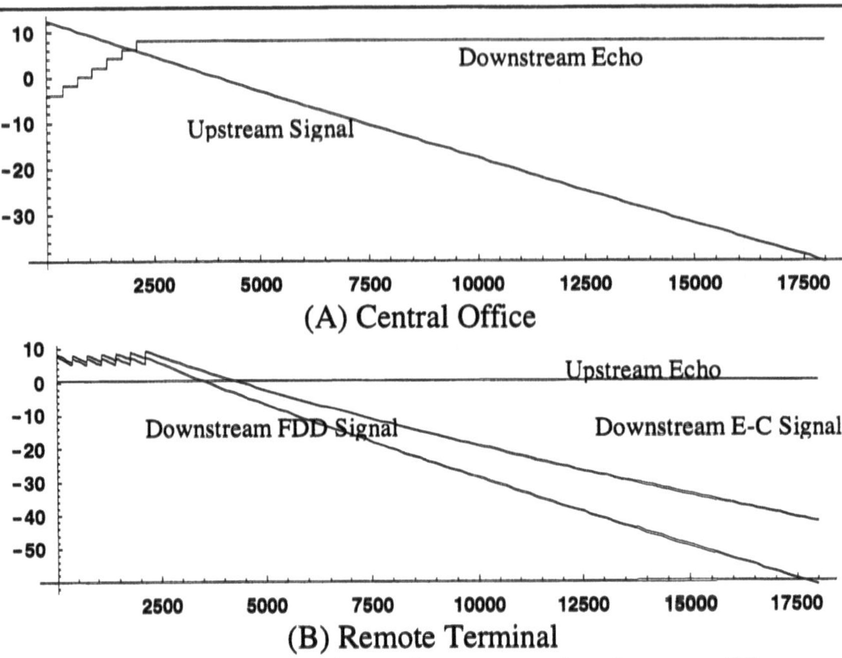

Figure 6. (A) Downstream echo and upstream signal power (B) upstream echo and downstream signal power, echo canceling and FDD, plotted versus loop length (feet of 26awg). Units are dBm. The steps apparent on the downstream signal and its echo are due to the power cutback of the central office transmitter when the loop is short.

B. ADC RESOLUTION

It is instructive to determine what ADC resolution would be required if it were connected directly to the hybrid output. To match loop noise, -140dBm/Hz, the ADC quantization noise would need to be -152dBm/Hz, or 7.9nV/√Hz, measured at the hybrid output. This corresponds to the quantization noise density of an ADC with a 40.9μV least significant bit converting at a 4416kS/sec rate. Thus, an ADC with a 3.5V peak-to-peak input range and 16.4 effective bits of resolution converting at 4416kS/sec is required. While building such an ADC is within the realm of possibility, it is highly desirable to configure the receiver to relax this specification.

Reducing the ADC resolution requires programmable gain amplifiers and strategic filtering, but there is a limit. The same considerations for DAC resolution and sub-carrier SNR apply here, so 12.5 effective bits is the minimum ADC resolution that supports 15b/symbol. In reality, the ADC dynamic range must be greater than this so that there is some headroom for the echo. Otherwise, it would be necessary to eliminate completely all echo power in order to amplify the incoming signal to 0dBm. Thus, the ADC resolution must be in the 12.5 to 16.4 effective bit range. The exact requirement depends upon the degree to which the echo attenuation may be guaranteed.

C. KILLING THE ECHO

A portion of a generic receiver is shown in figure 7. It consists of a block to attenuate echo power, a variable gain amplifier and an N-bit ADC. If N=16.4, then the ADC quantization noise matches the loop noise. If N=16.4-M, then the ADC quantization noise is (6M)dB higher than the loop noise, and subcarrier SNR can be increased by applying VGA gain up to (6M)dB. Gain greater than 6M will not result in significant SNR improvement because the dominant noise source would then be the loop noise that is amplified along with the signal. The "echo killer" block reduces the echo power so that the signal (not signal plus echo) may be amplified to the desired level without saturating the ADC. In most cases the echo killer is a filter. The one exception is the remote terminal receiver in an echo-canceling modem.

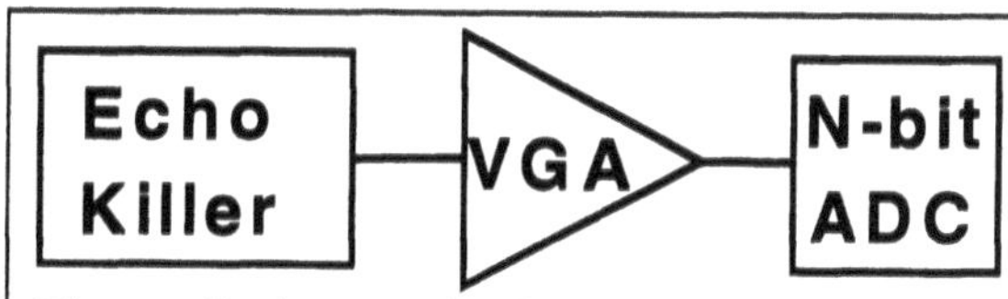

Figure 7. A generic CODEC receiver.

Consider the remote terminal receiver when connecting to the longest loop. The signal power is more than 40dB below the ADC maximum input power, so we should apply at least (6M)dB gain to insure that the SNR is limited by the loop noise and not the quantizer. However, given the 0.5dBm echo and the -12dB hybrid gain, the echo power at the hybrid output is about -11.5dBm, so without killing the echo, we cannot employ more than 11.5dB VGA gain without saturating the ADC.

In FDD modems, the echo killing is done with a high-pass filter designed to reduce the echo power by at least (6M-11.5)dB. The choice of $0 \leq M \leq 4$ is a trade-off between filter and VGA complexity versus ADC resolution. If $M \approx 4$ (a 12.5 effective-bit ADC), the filter must reduce the echo power by at least 12.5dB. A third order elliptic high-pass with 0.25dB ripple and corner frequency of 172.5kHz will suffice.

In echo-canceling modems the echo killing cannot be done with a filter without also killing the low frequency portion of the downstream signal. There is basically only one way to kill the echo, and that is improving the effective hybrid echo return loss. Figure 8 shows a scheme to do just this.

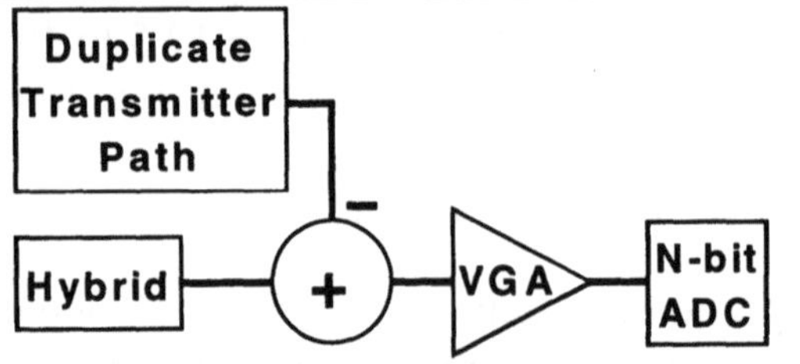

Figure 8. Scheme to increase hybrid echo return loss.

A duplicate of the transmitter path sums into the receive path. During the modem initialization, the hybrid echo return impulse response is measured and the tap weights of an FIR filter are defined. During normal operation, the transmitter signal is passed through this FIR filter and output to the duplicate transmitter path where it is subtracted from the receiver input. A typical echo return loss improvement of about 20dB is obtained, so remote terminal VGA gain up to about 30dB is possible without saturating the ADC input when connecting to long loops. The undesirable costs of this approach are the increases in DSP MIPS, die area, power dissipation and noise.

Now consider the central office receiver when connecting to the longest loop. Since the signal power is small, we should apply the maximum sensible gain, 6M. To do so, since the worst case echo power is –3.7dBm, we want to reduce the echo power by at least (6M-3.7)dB.

In an echo-canceling modem with full spectral overlap, the maximum echo attenuation that an echo killer low-pass filter can provide without affecting the upstream signal is about 10dB. This is because the filter can only be used to attenuate the non-overlapping portion of the echo. Thus, the maximum gain that may be applied is 13.7dB, corresponding to a maximum ADC resolution reduction of M=2.3, meaning that the ADC resolution should be at least 14.3 effective bits. A third order Chebyshev with 0.5dB ripple and 138kHz corner frequency provides the echo power reduction.

In central office FDD modems, the worst case echo power, taking the -12dB hybrid gain into account, is about -4.4dBm. (This is because the downstream power of an FDD modem is about 0.7dB lower than that of an echo-canceling modem.) The echo can be rejected without affecting the signal with a low-pass filter. Once again, the choice of $0 \le M \le 4$ is a tradeoff between low-pass filter and VGA complexity versus ADC resolution. If M=4, the filter must provide about 20dB echo power attenuation. This can be done with the same low-pass filter used in the echo-canceling central office modem.

D. ANTIALIASING

The degree of antialiasing filtering required ahead of the 4416kS/sec ADC depends, of course, on the psd of the receiver input above 2208kHz. In central offices, this is predominately the echo energy that is adequately removed by the echo-killing low-pass filters. In remote terminals, the energy above 2208kHz is primarily the side lobes of the downstream signal

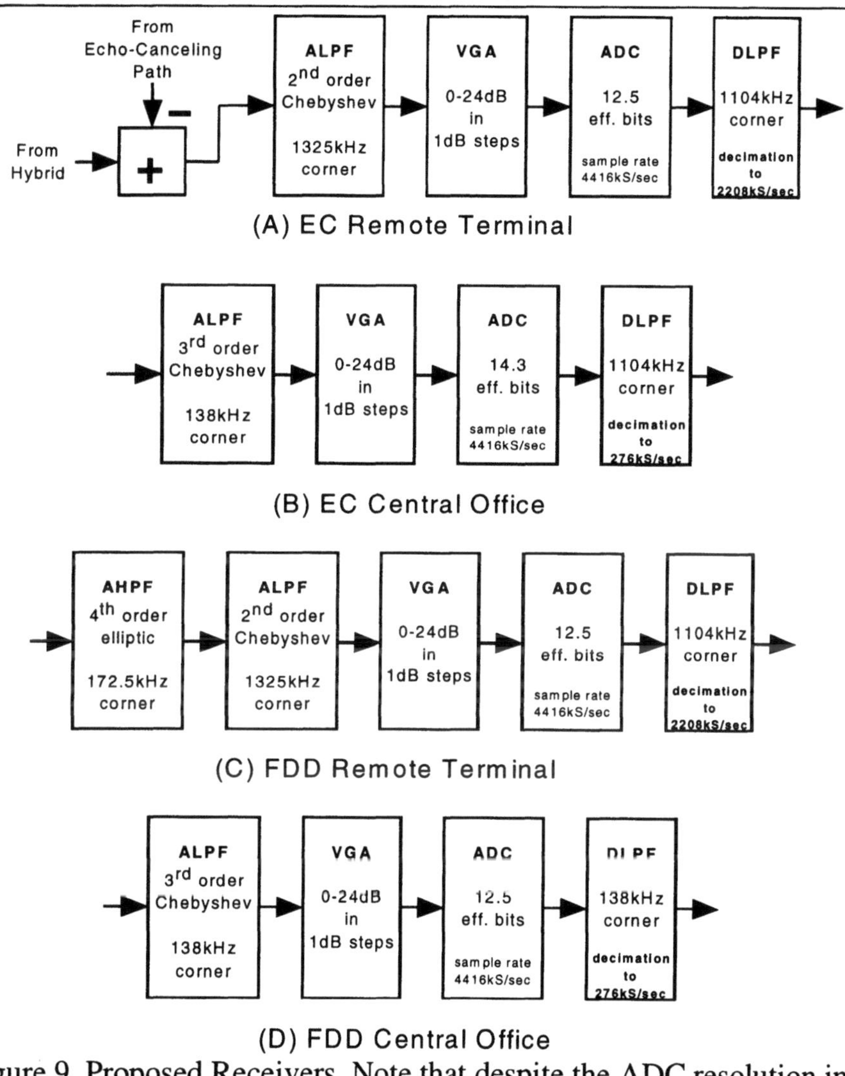

Figure 9. Proposed Receivers. Note that despite the ADC resolution in the echo-canceling central office ADC, the 0-24dB gain range of the VGA is retained to take advantage of situations where both the echo return loss and the loop noise are lower than we have assumed.

itself, and since the ADC is oversampling by 2X, only frequencies above 3312kHz will alias into the downstream band. The ITU standard dictates that the side lobe energy above 3312kHz is no greater than about -93.5dBm/Hz at the transmitter, and on 26awg loops longer than about 3kft, the loop attenuation suppresses this below the loop noise floor. Aliasing is only a problem on very short loops where the signal energy is high and there is little loop attenuation difference between 1104Khz and 3312kHz. In this case, the filter needs only to deliver about 12dB rejection above 3312kHz. This is easily provided by a second order Chebyshev with (1325±20%)kHz corner frequency.

The digital filter following the ADC provides the antialiasing required for decimation of the sample rate to its minimum value without distortion. In the remote terminal this is a simple 2X decimation FIR filter. In the central office it is a 16X decimation that is most effectively implemented in 2X decimation stages.

E. RECEIVER ARCHITECTURE

Figure 9 summarizes this discussion of the proposed receivers and indicates reasonable choices of analog filter VGA gain and ADC resolution for all four receivers.

5. SUMMARY

We have deduced the optimum architecture and sub-circuit specifications for transparent operation of the ADSL modem analog front end (AFE) based upon consideration the international ADSL standards, subscriber loop attenuation and worst case hybrid echo return loss as a function of the duplexing method. We submit that the CODECs proposed here will provide the best possible ADSL modem performance. However, since there is significant margin between the data rate defined by the ITU and the channel capacity, we must note that transparent CODEC operation is not strictly required. Thus there is significant room to make intelligent CODEC design compromises to reduce power or cost or both.

6. REFERENCES

1. Proposed Summaries for new Recommendations G.992.1, G.992.2, G.994.1, G.995.1, G.996.1 and G.997.1, ITU-Telecommunications Standardization Sector, Study Group 15, Geneva, 12 October, 1998.
2. D. Langford, B. Tesch, B. Williams and G. Nelson, "A BiCMOS Analog Front-End Circuit for an FDM-Based ADSL System," IEEE J. Solid-State Circuits, vol. 33, pp. 1383-1393, Sept. 1998.

DESIGNING CODECS TO MINIMIZE DSP COMPUTATIONAL BURDEN

R. K. Hester
Texas Instruments Incorporated
Dallas, Texas, USA

Outline

- ADSL Signal Characteristics and PSD Compliance
- Echo-Canceling and FDD Transmitters
- Subscriber loop Attenuation and Hybrid Echo
- Echo-Canceling and FDD Receivers
- Echo-Canceling CODEC Design Example

Peak-to-RMS Ratio

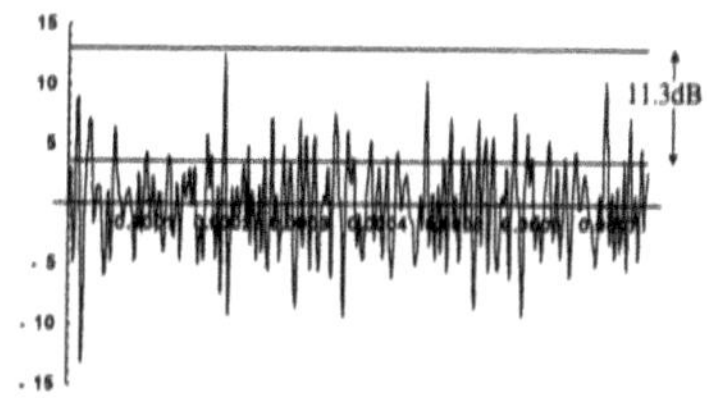

Accommodating 15dB insures error rates $<10^{-7}$

Sub-Carrier Broadening

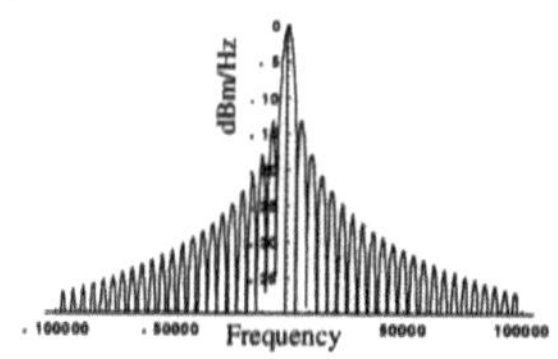

$$Power = \int_{-\infty}^{\infty}\left(\frac{\sin\left[\frac{\pi f}{f_0}\right]}{\frac{\pi f}{f_0}}\right)^2 df = f_0 = 4312.5 \longrightarrow 36.35\text{dB}$$

Unfiltered Downstream PSD (Minimum IFFT Output)

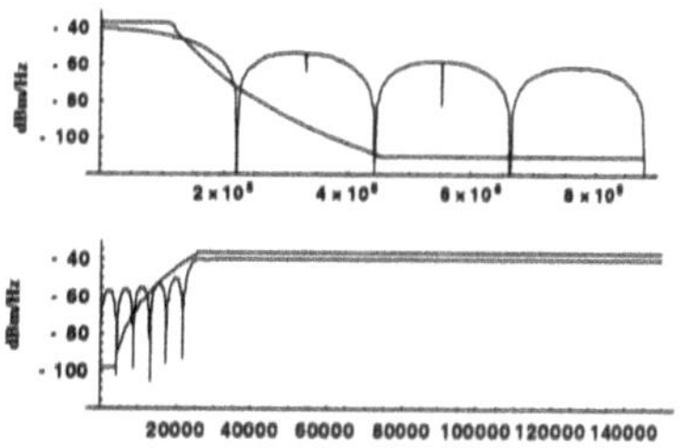

Unfiltered Upstream PSD (Minimum IFFT Output)

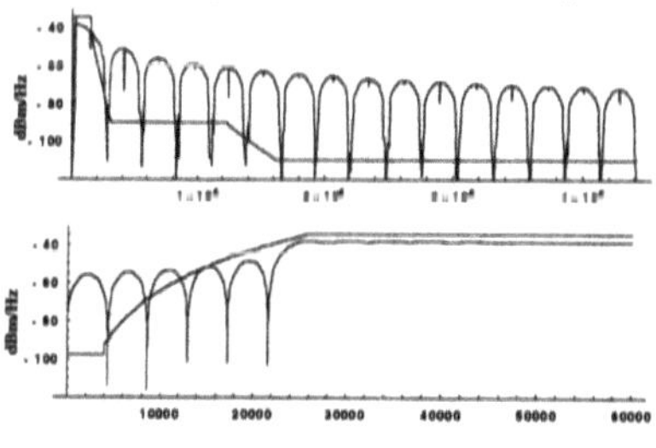

Frequency Division Duplexing

- Transmitter PSD must be attenuated in receiver signal band such that it is below the system noise floor, -140dBm/Hz-ERL.
- Upstream low-pass and downstream high-pass filters are much higher order than required by ITU standard.
- Guard band is compromise between spectrum efficiency and psd-shaping filter complexity.

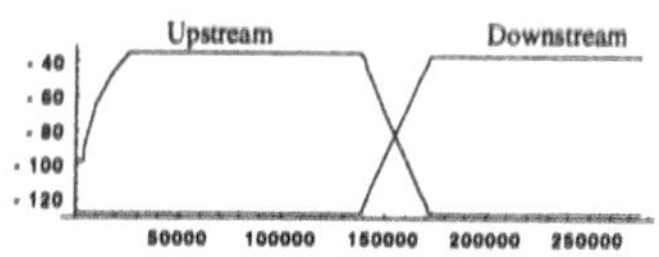

Sub-Carrier SNR and Constellation Size

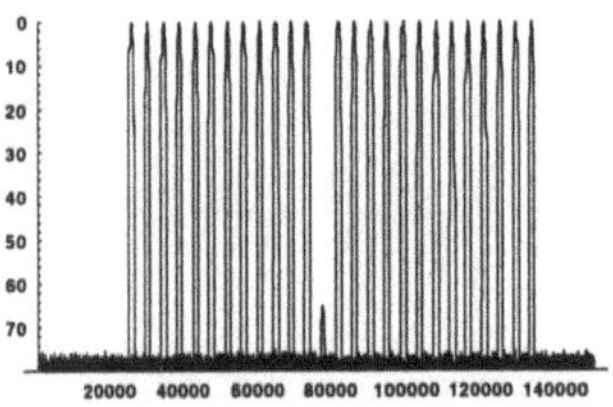

b-bit constellation requires (20+3b)dB sub-carrier SNR

SNR, Constellation Size and Minimum Resolution

- Single sub-carrier with -40dBm/Hz peak psd has -3.65dBm total power.
- Power in signal with M such sub-carriers is (10Log$_{10}$[M] -3.65)dBm.
- Given 15dB peak-to-rms ratio, the peak-to-peak voltage of a downstream signal with M sub-carriers is 2.331M$^{0.5}$ V (upstream is 2dB higher).
- To support a b-bit constellation, the channel noise floor must be (20+3b)dB below sub-carrier psd, or -(60+3b)dBm/Hz.

SNR, Constellation and Resolution Cont'd

- Data converter noise floor psd is $\frac{V_{pp}}{2^N\sqrt{\ldots f_S}}$ V/$\sqrt{\text{Hz}}$
- If $f_S = R_{DN}$*2208kS/sec and $Vpp = 2.331 M_{DN}^{0.5}$, the result of equating the data converter noise floor to -(60+3b)dBm/Hz is

$$N_{DN} = \log_2\left[2.025 * 10^{0.15b}\sqrt{\frac{M_{DN}}{R_{DN}}}\right] \xrightarrow{M_{DN}=250} N_{DN}(R_{DN}) = 5 + \frac{1}{2}b - Log_2(R_{DN})$$

- If $f_S = R_{UP}$*276kS/sec and $Vpp = 2.935 M_{UP}^{0.5}$ and M_{UP}=26, the result of equating the data converter noise floor to -(58+3b)dBm/Hz is

$$N_{UP} = \log_2\left[5.728 * 10^{0.15b}\sqrt{\frac{M_{UP}}{R_{UP}}}\right] \xrightarrow{M_{UP}=26} N_{UP}(R_{UP}) = 4.868 + \frac{1}{2}b - Log_2(R_{UP})$$

Converter Resolution Versus Over-Sampling Ratio, b=15

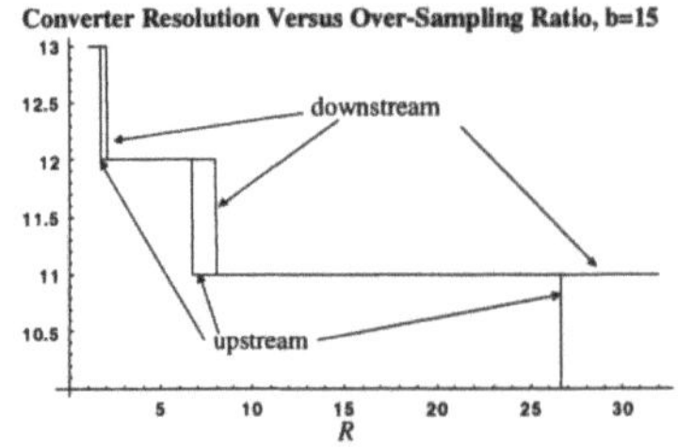

In any case, converter must provide 13b signal-to-distortion ratio.

Transmitter Quantization Noise Echo Can Limit Receiver

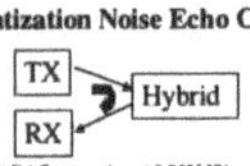

- The quantization noise of an N-bit DAC converting at 8.832MS/sec is 0.0050454*2^{-N} V/Hz$^{0.5}$, measured on the subscriber line.
- If the worst case hybrid return loss is 12dB, then the minimum DAC resolution that produces -140dBm/Hz (31.6nV/ Hz$^{0.5}$) receiver input referred noise is
  downstream N_{MIN}=Log$_2$[(1/4)(4.54/3.16)*10^5]= 15.1 bits
- The corresponding upstream minimum DAC resolution is
  upstream N_{MIN}=Log$_2$[(1/4)(2.91/3.16)*10^5]= 14.5 bits
- The transmitter DAC resolution can be reduced by L bits in FDD systems if the downstream DAC (upstream DAC) is followed by a high-pass filter (low-pass filter) with at least (6L)dB rejection in the upstream (downstream) band.
- In an echo-canceling modem, while no analog filtering reduces the downstream quantization noise or the upstream in-band quantization noise, the upstream low-pass filter will reduce quantization noise over most of the downstream receiver band.

Transmitter Architectures

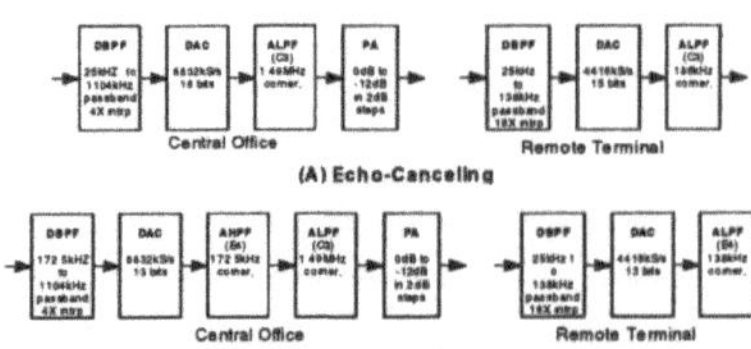

(A) Echo-Canceling

(B) Frequency Division Duplexing

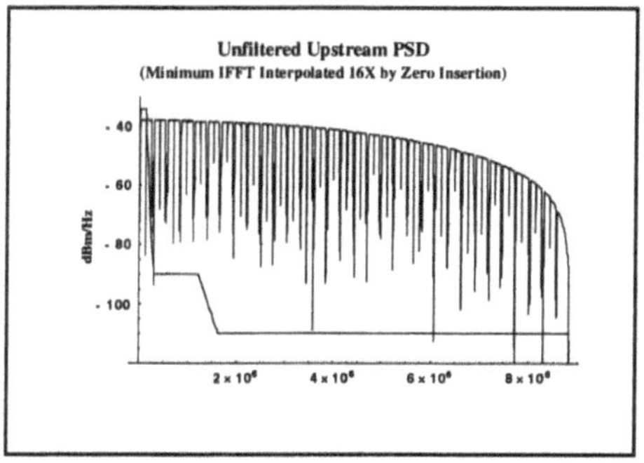
Unfiltered Upstream PSD
(Minimum IFFT Interpolated 16X by Zero Insertion)
dBm/Hz
- 40
- 60
- 80
- 100
2 × 10⁶
4 × 10⁶
6 × 10⁶
8 × 10⁶

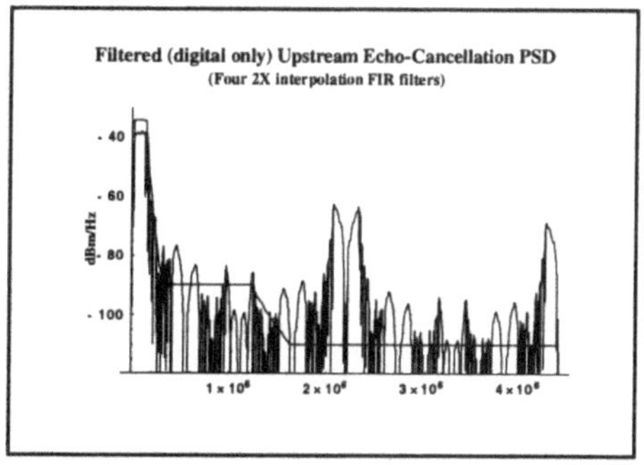
Filtered (digital only) Upstream Echo-Cancellation PSD
(Four 2X interpolation FIR filters)
dBm/Hz
- 40
- 60
- 80
- 100
1 × 10⁶
2 × 10⁶
3 × 10⁶
4 × 10⁶

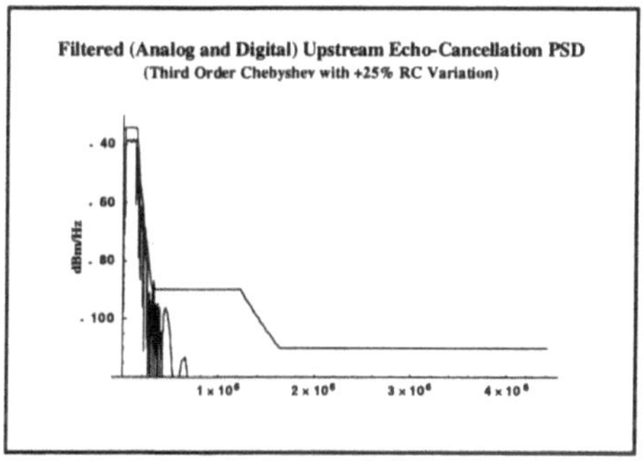
Filtered (Analog and Digital) Upstream Echo-Cancellation PSD
(Third Order Chebyshev with +25% RC Variation)
dBm/Hz
- 40
- 60
- 80
- 100
1 × 10⁶
2 × 10⁶
3 × 10⁶
4 × 10⁶

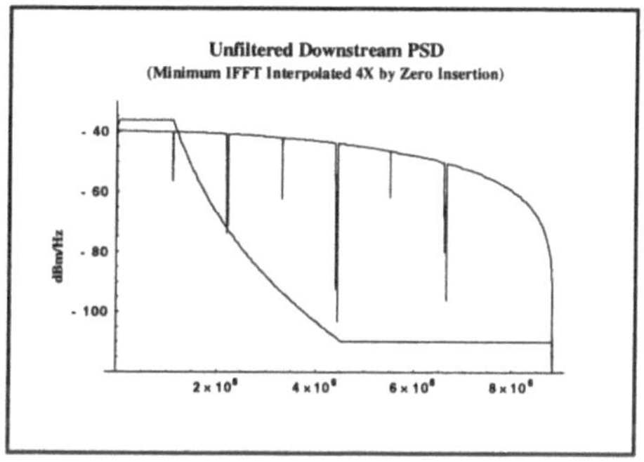
Unfiltered Downstream PSD
(Minimum IFFT Interpolated 4X by Zero Insertion)
dBm/Hz
- 40
- 60
- 80
- 100
2 × 10⁶
4 × 10⁶
6 × 10⁶
8 × 10⁶

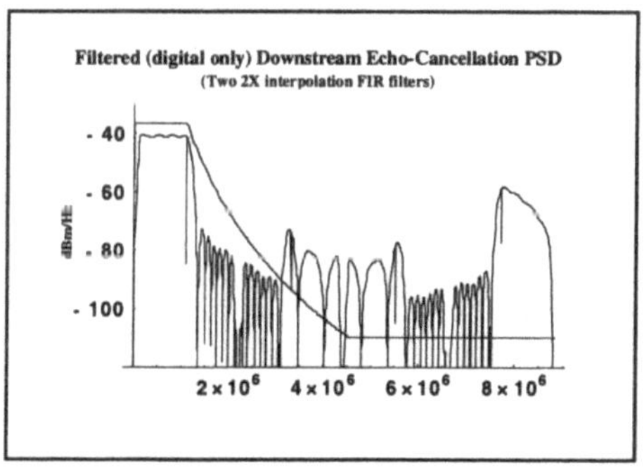
Filtered (digital only) Downstream Echo-Cancellation PSD
(Two 2X interpolation FIR filters)
dBm/Hz
- 40
- 60
- 80
- 100
2 × 10⁶
4 × 10⁶
6 × 10⁶
8 × 10⁶

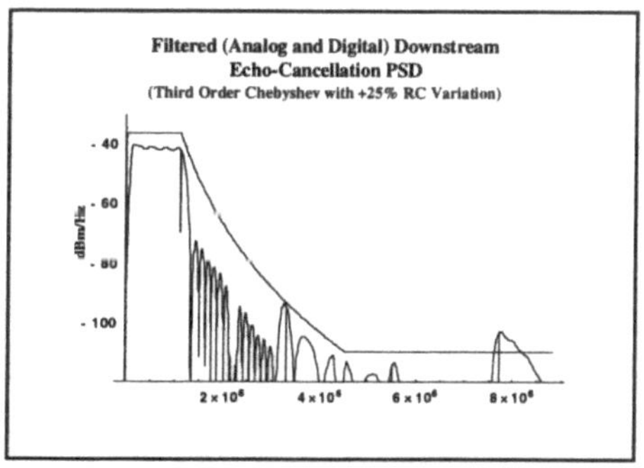
Filtered (Analog and Digital) Downstream
Echo-Cancellation PSD
(Third Order Chebyshev with +25% RC Variation)
dBm/Hz
- 40
- 60
- 80
- 100
2 × 10⁶
4 × 10⁶
6 × 10⁶
8 × 10⁶

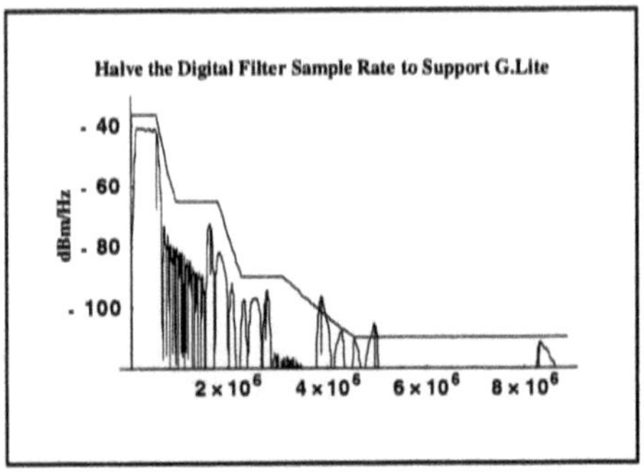
Halve the Digital Filter Sample Rate to Support G.Lite
dBm/Hz
- 40
- 60
- 80
- 100
2×10^6
4×10^6
6×10^6
8×10^6

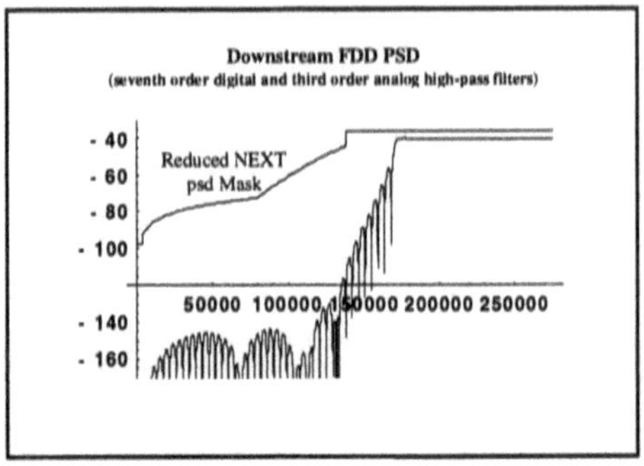
Downstream FDD PSD
(seventh order digital and third order analog high-pass filters)
Reduced NEXT
psd Mask
- 40
- 60
- 80
- 100
- 140
- 160
50000 100000 150000 200000 250000

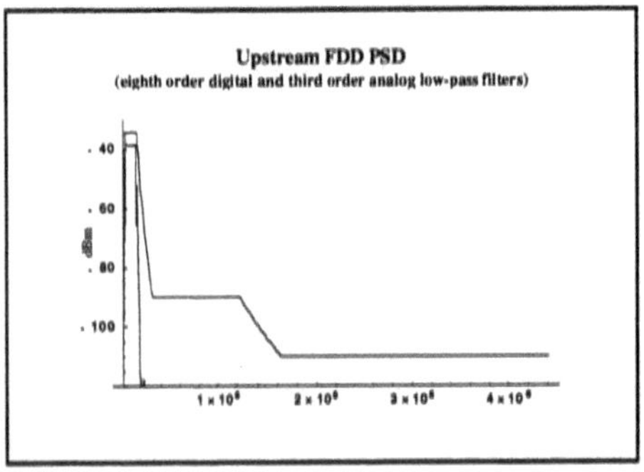
Upstream FDD PSD
(eighth order digital and third order analog low-pass filters)
dBm
- 40
- 60
- 80
- 100
1×10^6
2×10^6
3×10^6
4×10^6

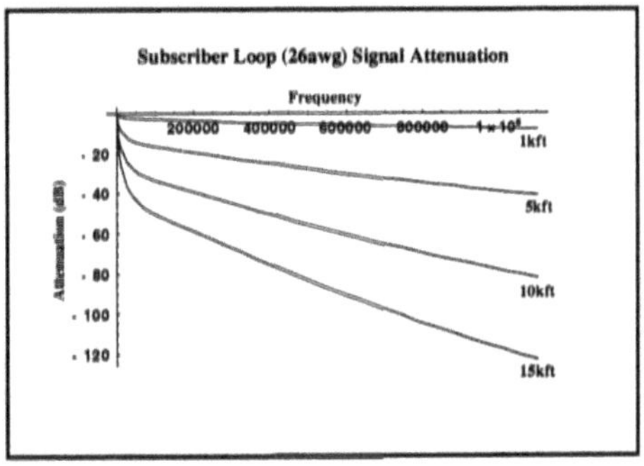
Subscriber Loop (26awg) Signal Attenuation
Frequency
200000 400000 600000 800000 1×10^6
Attenuation (dB)
- 20
- 40
- 60
- 80
- 100
- 120
1kft
5kft
10kft
15kft

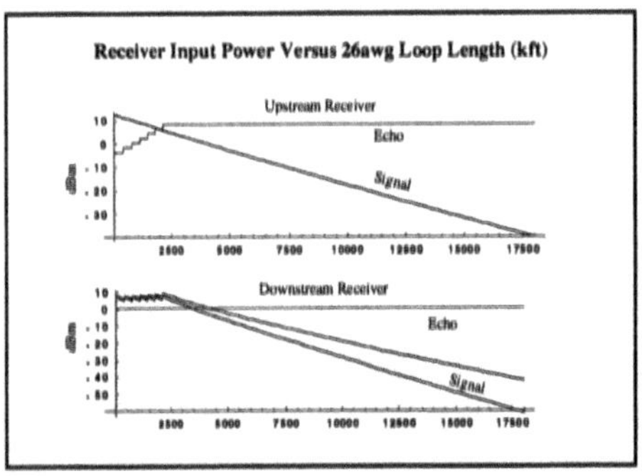
Receiver Input Power Versus 26awg Loop Length (kft)
Upstream Receiver
Echo
Signal
dBm
2500 5000 7500 10000 12500 15000 17500
Downstream Receiver
Echo
Signal
dBm
2500 5000 7500 10000 12500 15000 17500

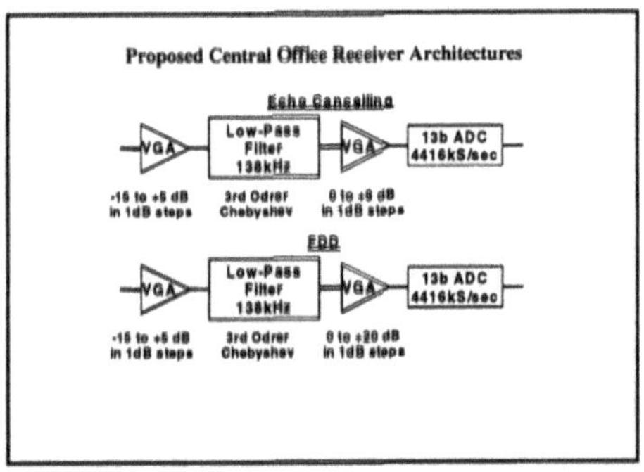
Proposed Central Office Receiver Architectures
Echo Cancelling
VGA
Low-Pass
Filter
138kHz
VGA
13b ADC
4416kS/sec
-15 to +5 dB
in 1dB steps
3rd Odrer
Chebyshev
0 to +9 dB
in 1dB steps
FDD
VGA
Low-Pass
Filter
138kHz
VGA
13b ADC
4416kS/sec
-15 to +5 dB
in 1dB steps
3rd Odrer
Chebyshev
0 to +20 dB
in 1dB steps

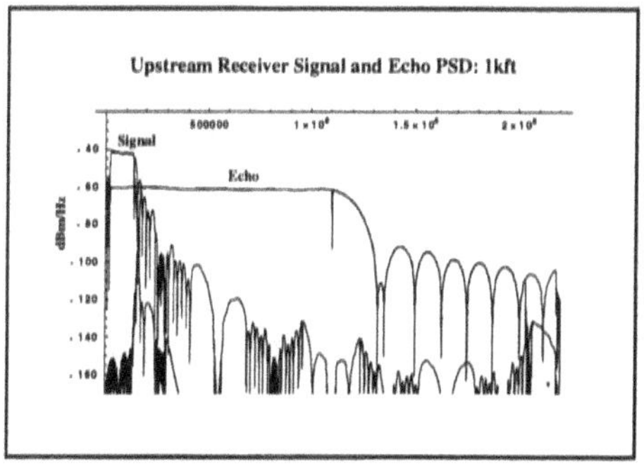
Upstream Receiver Signal and Echo PSD: 1kft
Signal
Echo
dBm/Hz

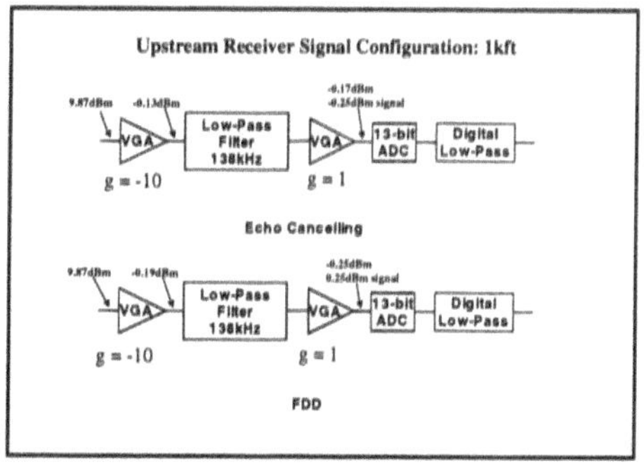
Upstream Receiver Signal Configuration: 1kft
9.87dBm
-0.13dBm
VGA
Low-Pass Filter 138kHz
VGA
-0.17dBm
-0.25dBm signal
13-bit ADC
Digital Low-Pass
g = -10
g = 1
Echo Cancelling
9.87dBm
-0.19dBm
VGA
Low-Pass Filter 138kHz
VGA
-0.25dBm
0.25dBm signal
13-bit ADC
Digital Low-Pass
g = -10
g = 1
FDD

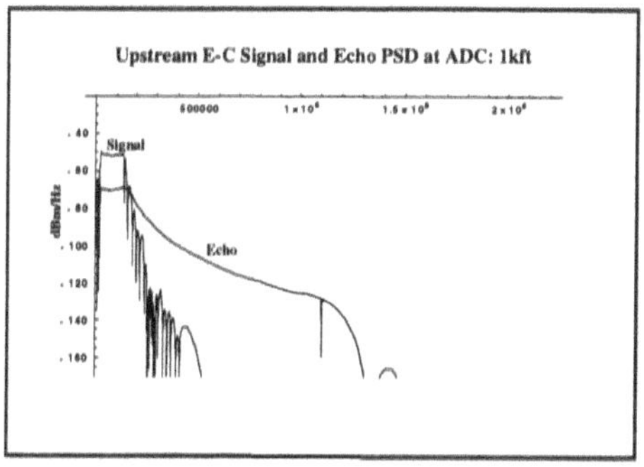
Upstream E-C Signal and Echo PSD at ADC: 1kft
Signal
Echo
dBm/Hz

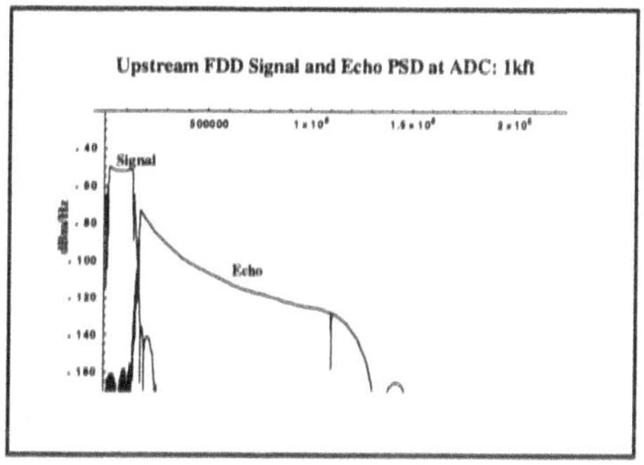
Upstream FDD Signal and Echo PSD at ADC: 1kft
Signal
Echo
dBm/Hz

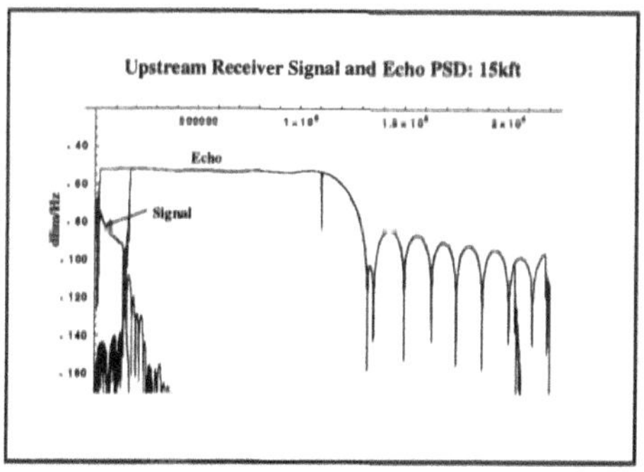
Upstream Receiver Signal and Echo PSD: 15kft
Echo
Signal
dBm/Hz

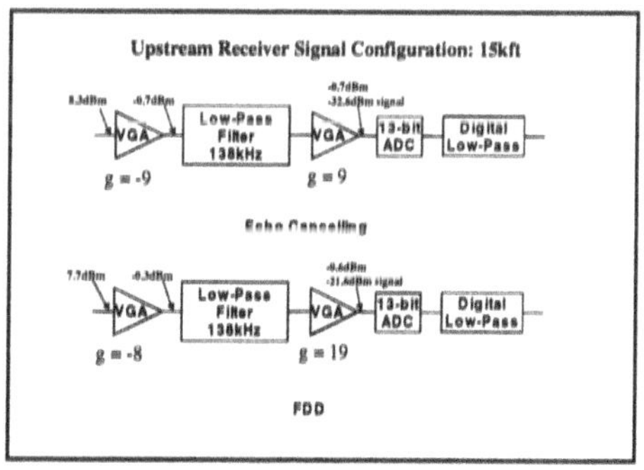
Upstream Receiver Signal Configuration: 15kft
-0.7dBm
VGA
Low-Pass Filter 138kHz
VGA
-0.7dBm
-32.6dBm signal
13-bit ADC
Digital Low-Pass
g = -9
g = 9
Echo Cancelling
7.7dBm
-0.3dBm
VGA
Low-Pass Filter 138kHz
VGA
-0.6dBm
-21.6dBm signal
13-bit ADC
Digital Low-Pass
g = -8
g = 19
FDD

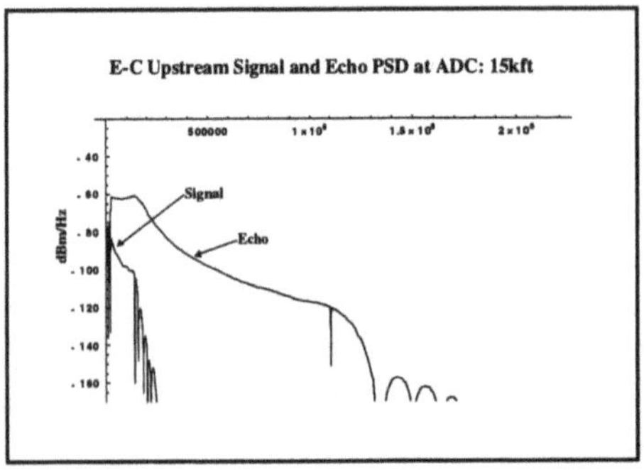
E-C Upstream Signal and Echo PSD at ADC: 15kft
Signal
Echo
dBm/Hz

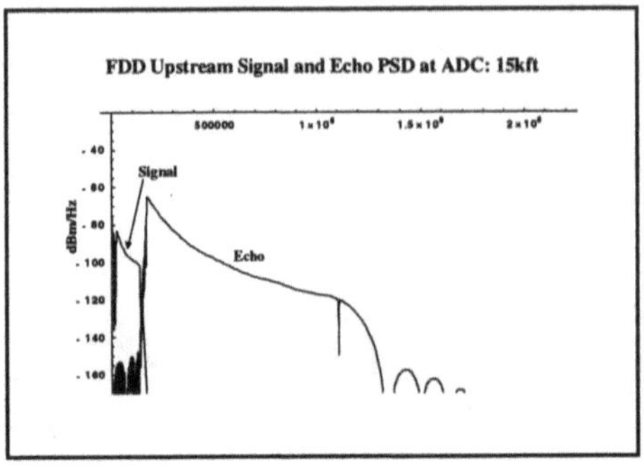
FDD Upstream Signal and Echo PSD at ADC: 15kft
Signal
Echo
dBm/Hz

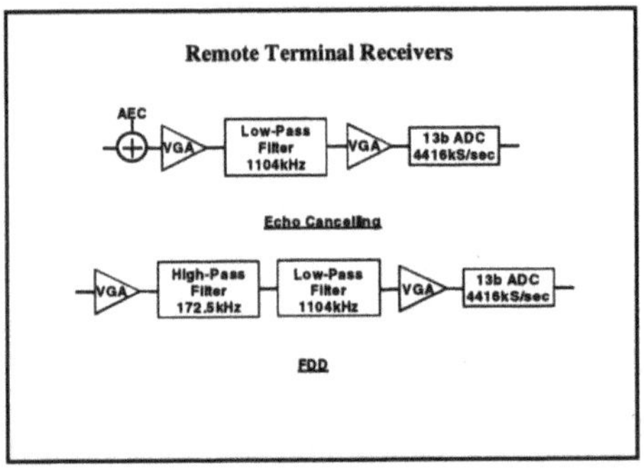
Remote Terminal Receivers
AEC
VGA
Low-Pass Filter 1104kHz
VGA
13b ADC 4416kS/sec
Echo Cancelling
VGA
High-Pass Filter 172.5kHz
Low-Pass Filter 1104kHz
VGA
13b ADC 4416kS/sec
FDD

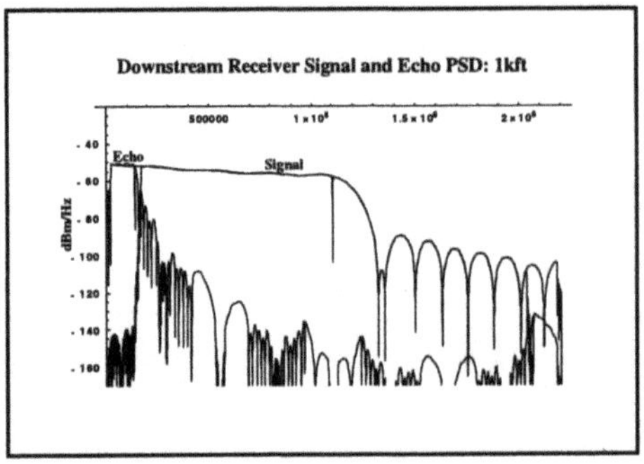
Downstream Receiver Signal and Echo PSD: 1kft
Echo
Signal
dBm/Hz

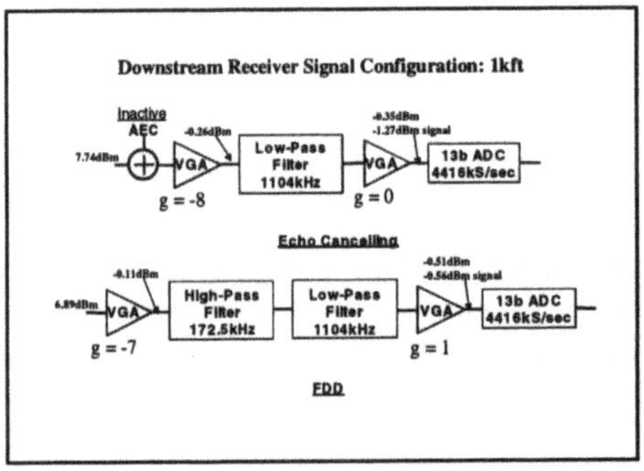
Downstream Receiver Signal Configuration: 1kft
Inactive
AEC
7.74dBm
-0.26dBm
VGA
Low-Pass Filter 1104kHz
g = -8
-0.35dBm
-1.27dBm signal
VGA
13b ADC 4416kS/sec
g = 0
Echo Cancelling
6.89dBm
-0.11dBm
VGA
g = -7
High-Pass Filter 172.5kHz
Low-Pass Filter 1104kHz
-0.51dBm
-0.56dBm signal
VGA
g = 1
13b ADC 4416kS/sec
FDD

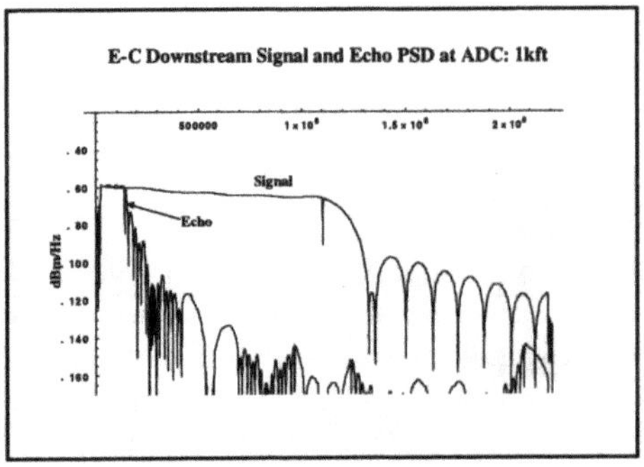
E-C Downstream Signal and Echo PSD at ADC: 1kft
Signal
Echo
dBm/Hz

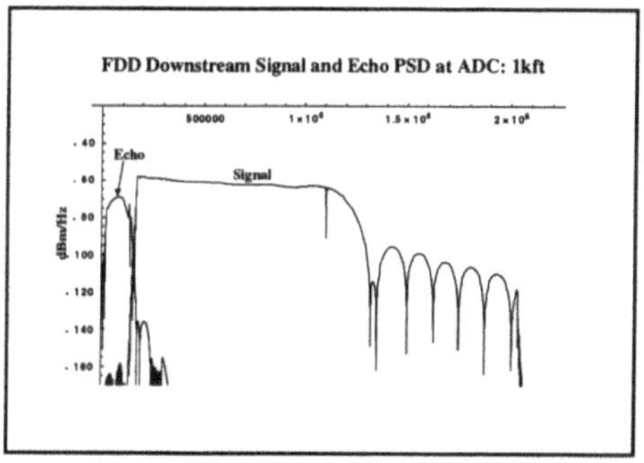
FDD Downstream Signal and Echo PSD at ADC: 1kft
Echo
Signal
dBm/Hz

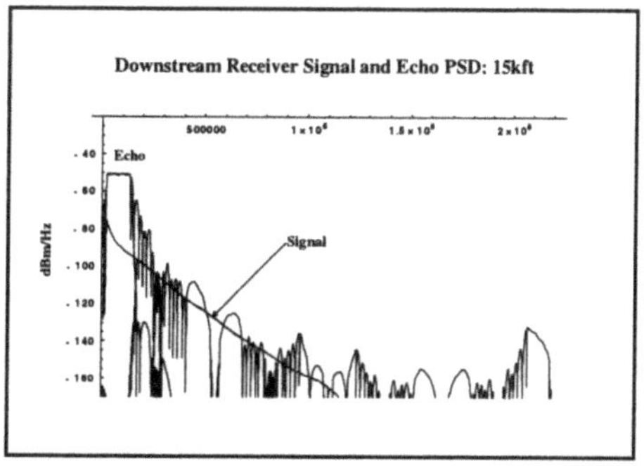
Downstream Receiver Signal and Echo PSD: 15kft
Echo
Signal
dBm/Hz

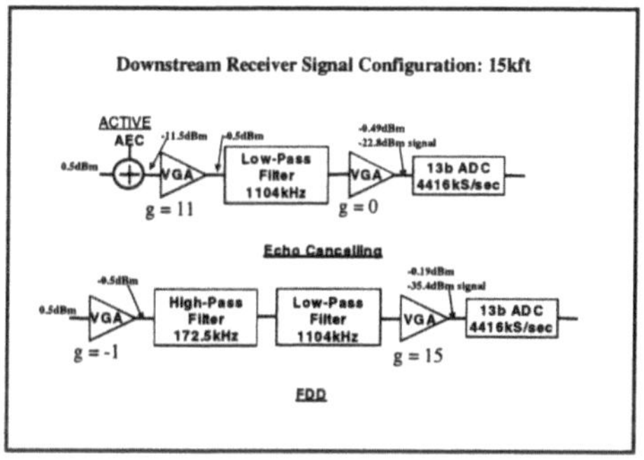
Downstream Receiver Signal Configuration: 15kft
ACTIVE
AEC
0.5dBm
-11.5dBm
-0.5dBm
VGA
Low-Pass Filter 1104kHz
-0.49dBm
-22.8dBm signal
13b ADC 4416kS/sec
g = 11
g = 0
Echo Cancelling
0.5dBm
-0.5dBm
High-Pass Filter 172.5kHz
Low-Pass Filter 1104kHz
-0.19dBm
-35.4dBm signal
13b ADC 4416kS/sec
g = -1
g = 15
FDD

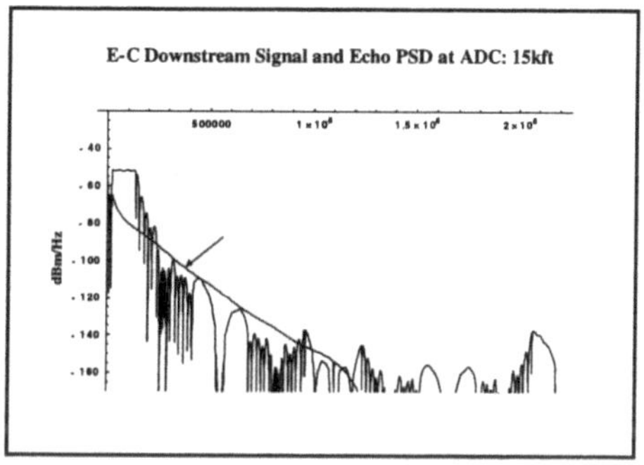
E-C Downstream Signal and Echo PSD at ADC: 15kft
dBm/Hz

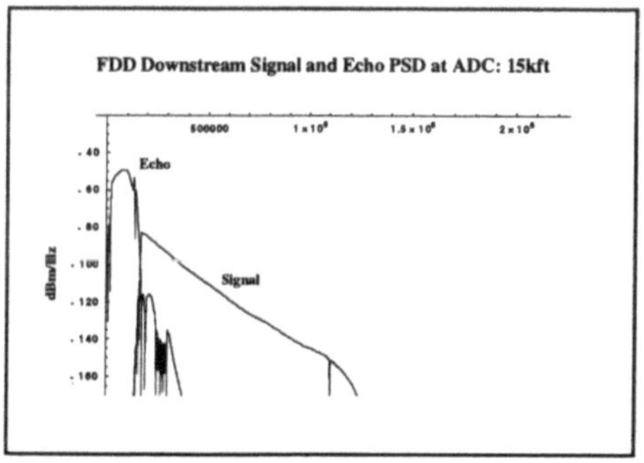
FDD Downstream Signal and Echo PSD at ADC: 15kft
Echo
Signal
dBm/Hz

A 3.3 V 15-bit Delta-Sigma ADC with a Signal Bandwidth of 1.1 MHz for ADSL-Applications

Yves Geerts, Michiel Steyaert and Willy Sansen
KU Leuven, ESAT-MICAS, Heverlee, Belgium

ABSTRACT

The design of a high-resolution high-speed delta-sigma analog-to-digital converter is presented. An overview of the performance of different topologies is shown. For the selected 2-1-1 cascade topology, a system level optimisation is performed. Several circuit non-idealities are analysed to arrive at building block specifications. The design of the building blocks is treated. Measurement results of the 3.3 V 15-bit $\Delta\Sigma$ with a signal bandwidth of 1.1 MHz are discussed.

1 Introduction

The growing market for ADSL-applications is a driving force for the development of high-speed high-resolution delta-sigma analog-to-digital converters (ADC's).

In Discrete multi-tone (DMT) ADSL, two techniques can be used to separate the receive and transmit channel: frequency division multiplexing (FMD) and echo cancellation (EC). The frequency of the channel is divided into different regions. Both implementations block off the lower 25 kHz of the signal band for POTS. The next part is taken up by the upstream channel, which has a bandwidth of approximately 135 kHz. In FMD-DMT, the downstream channel starts above the upstream channel, at 240 kHz. In EC-DMT, the downstream channel overlaps with the upstream channel. Frequencies up to 1.1 MHz are used for the downstream channel in both implementations [1].

These up and downstream channels are divided into small bins with a bandwidth of 4 kHz. Each of these bins carries a QAM signal at the centre of the bin. Since these bins have a small bandwidth, the need for equalisation is minimal. The data is spread over all the channels depending on the SNR in each of them. Therefore, DMT makes optimum use of all the channels.

A fast Fourier transform (FFT) to used to generate the data for these multiple channels. The DMT transmitter encodes the data, performs an inverse FFT and an D/A conversion. At the receiver side, some bandpass filtering and an AGC is needed before the AD converter. In the digital domain an FFT is performed to reassemble the serial bitstream which was transmitted.

The main requirements imposed by ADSL on the AD-converter are a signal bandwidth of 1.1 MHz and a resolution of 13 bits [2]. Since going to deeper sub-micron technologies offers a speed advantage, this seems a logical evolution in the design of these high-speed data converters. However, to reduce the high electrical fields in these sub-micron technologies, it is necessary to operate the circuits from a reduced supply voltage. Since even a 5 V voltage supply imposes severe constraints on the achievable performance of analog circuits, it is a great challenge to maintain the same level of performance with a reduced voltage supply of 3.3 V.

2 System Level Considerations

2.1 Topology selection

The combined goal of a high-speed (1.1 MHz bandwidth) and high-resolution (80 dB) converter puts several demands on the chosen topology for the converter.

First of all, in order to maximize the signal bandwidth of an oversampled converter, it is necessary to increase the sampling frequency of the converter, as well as to reduce the oversampling ratio. Since the maximum sampling frequency of CMOS switched-capacitor circuits is ultimately limited by the speed of the technology, it is necessary to choose a low oversampling ratio for the converter. Therefore, the oversampling ratio should not exceed 32.

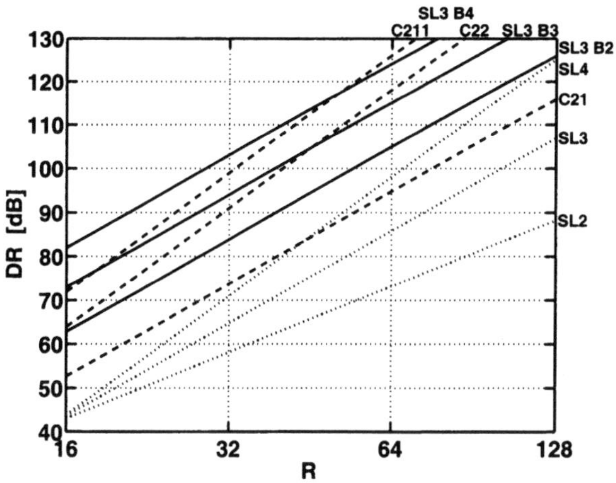

Fig. 1 : Performance of several ΔΣ architectures versus the oversampling ratio. SLi = i^{th} order single-loop, Cijk = Cascaded topologie i-j-k, SL3 Bi = 3rd order multi-bit single-loop topologie with i-bit quantiser.

The second issue results from the high-resolution requirement. Using the methods discribed in [3], optimal parameters for different architectures can be found, resulting in Fig. 1, which shows the maximum dynamic range of several ΔΣ architectures. This graph reveals that the previous derived condition of a low oversampling ratio significantly reduces the SNR_{peak} of the converter. The different architecture types will now be discussed briefly:

- Single-loop topologies: the performance is not adequate at low oversampling ratios to achieve the specifications for ADSL. This is mainly due to stability problems of higher order single loop topologies which limits the effectiveness of increasing the order.
- Multi-bit topologies: these topologies offer a good improvement of the resolution, also at a low oversampling ratio. The multi-bit quantiser also improves the stability of the ΔΣ converter, thereby allowing more aggressive noise transfer functions. The main drawback is the linearity requirement of the digital-to-analog converter in the feedback loop [4]. Possible solutions for this problem fall beyond the scope of this paper, therefore, multi-bit topologies are no further considered.
- Cascaded-topologies: these topologies combine the stability of second-order single-loop converters with the high performance of higher-order topologies [3,5]. The disadvantages of these topology are the high building-block specifications and the capacitance

matching requirements in order to avoid quantisation noise leakage from the first stage to the output.

These considerations lead to the selection of a 2-1-1 cascaded topology. It can achieve a resolution of 15-bits for an oversampling-ratio of 24. This results is a sampling frequency of 52.8 MHz for a signal bandwidth of 1.1 MHz.

2.2 Architecture choices

An essential issue in the design of high-speed ΔΣ ADC's is to reduce the capacitive load of the OTA's as much as possible, since this leads to a larger achievable dominant closed loop pole of the OTA ($f_{d,cl}$) and a reduced power consumption. This increased $f_{d,cl}$ results in a higher sampling frequency and signal bandwidth since the required $f_{d,cl}$ is proportional to the sampling frequency in order not to deteriorate the resolution of the converter (considering a certain topology and oversampling ratio). This is illustrated in section 3.2.

This reduction of the capacitive load of the OTA's can be achieved by reducing the sampling capacitance. In high-resolution ΔΣ converters, this sampling capacitance is determined by noise requirements due to the kT/C noise of the switches. Therefore, it is of the utmost importance to reduce the kT/C noise as much as possible, since this immediately results in a reduction of the capacitive load of the OTA's and thereby improves the power-speed-resolution trade-off of the converter.

A first way to achieve this is by employing fully differential circuits. This improves the signal to white noise ratio ($SNR_{kT/C}$) by 3 dB compared to a single ended implementation, which means that the sampling capacitances can be reduced by a factor of two for the same $SNR_{kT/C}$. Differential circuits also offer a better cancellation of clock-feedthrough, reduced even harmonic distortion and an improved PSRR.

A second important architecture choice concerns the feedback of the delta-sigma converter. This can be implemented with double or single polarity reference voltages. In the case of double polarity reference voltages, the same capacitance can be used to perform the sampling operation of the input signal and to subtract the feedback signal from the comparator. This reduces the kT/C noise again by 3 dB for the same input sampling capacitance, implying that this capacitance can be reduced by a factor of two. Furthermore, the feedback factor also increases and the capacitive load of the OTA decreases even further.

These choices lead to a fully differential ΔΣ converter with double polarity reference voltages. For this converter, the SNR due to kT/C noise can be written as

$$SNR_{kT/C} = \frac{\left(2 \cdot OL \cdot V_{ref}\right)^2}{2} \cdot \frac{C_s \cdot R}{4 \cdot k \cdot T} \qquad (1)$$

where OL represents the overload level of the converter.

This formula reveals a third way of reducing the size of the sampling capacitance for the same $SNR_{kT/C}$: increasing the signal power at the input by increasing the reference voltage. However, increasing the reference voltage also causes an increase of the required signal swing at the outputs of the integrators. Since this circuit operates from a 3.3 V supply, it is very important to tackle this problem at the architecture level. This is done by choosing the coefficients of the converter with the proper considerations.

First of all, these coefficients should provide a stable converter with a high-resolution and a high overload-level. Furthermore, by an appropriate choice of the coefficients, the required signal swing at the outputs of all the integrators can be reduced to the reference voltage. Finally, at the input of each integrator, the switched-capacitors are shared as much as possible, thereby decreasing the capacitive load and the kT/C contributions to the noise floor even further. By using these guidelines, the block diagram of Fig. 2 results.

These coefficients assure that the signal swings at the output of all the integrators are smaller than the reference voltage. Therefore, the reference voltage is determined by the maximum available output swing of the OTA's. This swing is to be maximized (see section 4.1) for the previously mentioned reasons (eq. 1) and this results in a reference voltage of 900 mV.

The white noise floor of the converter is mainly determined by the first integrator sampling capacitance. From (1) it can be calculated that a 3 pF sampling capacitance results in $SNR_{kT/C} = 97$ dB. This is sufficient for a resolution of 15 bits and still leaves some margin for the OTA and quantization noise. The sampling capacitances of the last stage are determined by matching considerations and the capacitances in the other stages are scaled down progressively. This implies that the capacitive load of the OTA's decreases gradually, making it possible to scale the OTA's down, thereby significantly reducing the power consumption and the die area. The scaling factors of the OTA's are 1:0.5:0.35:0.35.

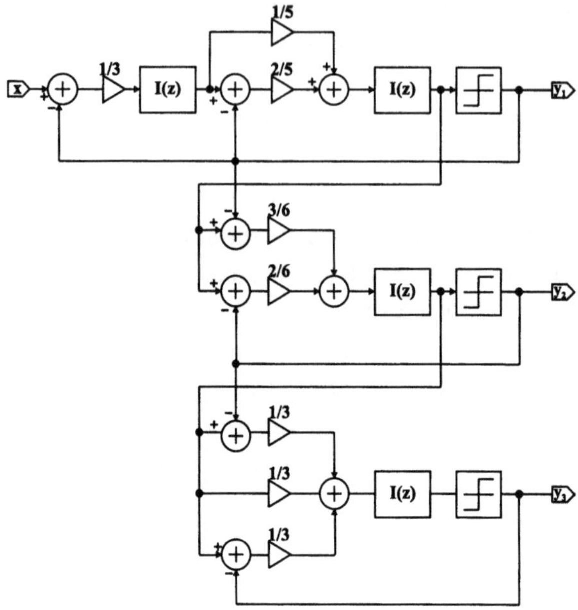

Fig. 2 Block diagram of the converter.

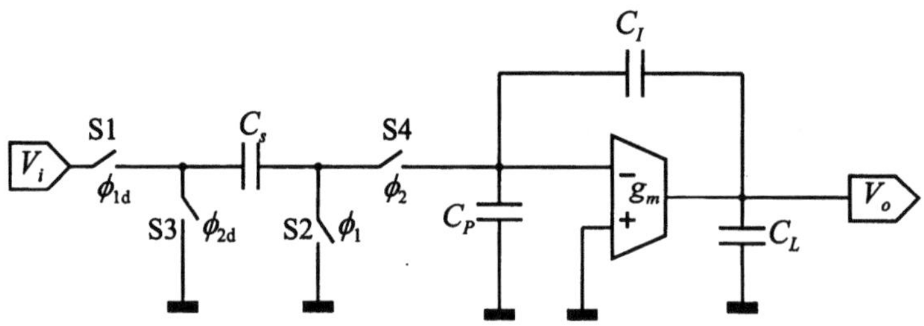

Fig. 3 Switched capacitor integrator.

3 Study of the non-idealities

The performance of a cascaded $\Delta\Sigma$ converter is known to be sensitive to circuit non-idealities [5,6], since quantization noise leakage of the first stage immediately results in a performance degradation. Therefore, several non-idealities of the switched-capacitor integrator shown in Fig. 3 are studied in this section. The transfer functions of the integrators are calculated for a finite gain and finite closed loop pole of the OTA. Also the influence of the non-zero resistance of the switches is considered. These derived transfer functions are then used to determine the minimum required specifications for the OTA and the switches through behavioral simulations. Furthermore, the distortion introduced by the sampling operation at the input of the converter is also studied.

3.1 Finite Operational Amplifier Gain

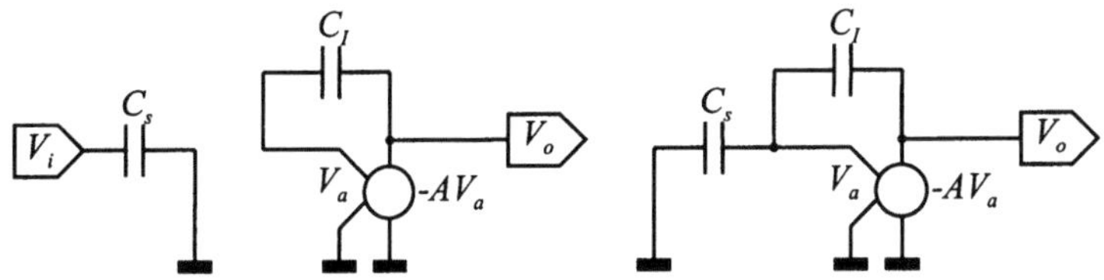

Fig. 4 Switched capacitor integrator considering an amplifier with finite gain.

Fig. 4 presents a model for a switched-capacitor integrator during both the sampling and the integration phase. In this model, the amplifier is simply represented by a voltage-controlled voltage-source. The transfer function of this integrator structure can be calculated [7] and results in

$$I(z) = \frac{C_S}{C_I} \cdot \frac{r_2 z^{-1}}{1 - \frac{r_2}{r_1} z^{-1}} \tag{2}$$

where r_1 and r_2 are the closed loop static errors, and f_{dc1} and f_{dc2} are the DC feedback factors during respectively the sampling and the integration phase. They are given by

$$r_1 = \frac{f_{dc1} A}{1 + f_{dc1} A} \qquad r_2 = \frac{f_{dc2} A}{1 + f_{dc2} A} \tag{3}$$

$$f_{dc1} = \frac{C_I}{C_I} = 1 \qquad f_{dc2} = \frac{C_I}{C_s + C_I} \tag{4}$$

This model can further be extended by adding a parasitic capacitance at the input (C_P) and output (C_L) terminals of the OTA. Note that C_L also includes the bottom plate parasitic of the integration capacitance. For this extended model equations (2) and (3) are still valid, only the feedback factors are different:

$$f_{dc1} = \frac{C_I}{C_P + C_I} \qquad f_{dc2} = \frac{C_I}{C_s + C_P + C_I} \tag{5}$$

As can be seen, the gain and pole errors of the integrator transfer function depend on the gain of the amplifier and on the DC feedback factors. This transfer function is used in behavioral simulations of the 2-1-1 cascade structure to determine the minimum required OTA gain to avoid a performance degradation. A gain of 60 dB is sufficient, however, in order to provide some margin for the degradation due to other non-idealities, a minimum gain of 80 dB is selected for the OTA.

3.2 Finite OTA Closed Loop Pole

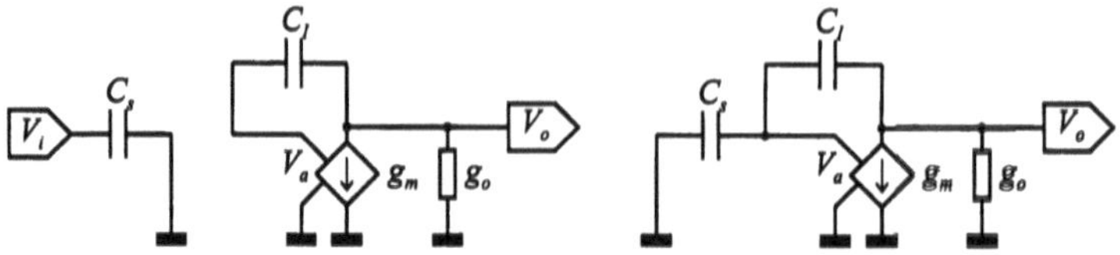

Fig. 5 Switched capacitor integrator using an OTA structure.

Another non-ideality that introduces gain and pole errors in the transfer function is the finite closed loop pole of the OTA. The OTA is modelled with a finite output conductance g_o and a input-output transconductance g_m as shown in Fig. 5. Following an approach similar to [7] the transfer function of the integrator can be calculated. It is given by

$$I(z) = \frac{C_S}{C_I} \cdot \frac{r_2(1-k_2)z^{-1}}{1 - \frac{r_2}{r_1}\left(1 - k_2\left(1 - \frac{r_1}{r_2}\right)\right)z^{-1}} \tag{6}$$

where r_1, r_2 and the DC-feedback factors are still given by (3) and (4). The amplifier gain used in these equations is evidently

$$A = \frac{g_m}{g_o} \tag{7}$$

The parameter k_2 represents the settling error in the integration phase. It is expressed by

$$k_2 = \exp\left(-\frac{g_m}{C_S} \cdot \frac{\tau_2}{r_2}\right) \tag{8}$$

where τ_2 represents the time available for settling during the integration phase.

This equation shows that the transconductance of the amplifier combined with the sampling capacitance creates a closed loop pole that limits the speed of the settling process. The effect of this closed loop pole on the integrator transfer function is to increase even further the gain and pole errors already caused by the finite amplifier gain.

One should note that this model does not include the parasitic capacitance at the input of the OTA and the load capacitance at the output. However, if one takes these two capacitances into account, the obtained expressions are very complicated and consequently a significant loss in understanding occurs.

One can however reason in a different way. One can preserve the previous simple model and change only the most inaccurately modeled factor. On one hand, the introduction of these capacitances changes all the linear factors in the transfer function. But these do not strongly influence the settling process. On the other hand, the exponential factor is strongly affected: the parasitic capacitances increase the equivalent closed loop capacitance, reducing the closed loop pole, and strongly influencing the exponential factor.

Consequently, the equation that expresses the settling error should be modified. In general, (8) can be expressed as

$$k_2 = \exp\left(-p_{cl} \cdot \frac{\tau_2}{r_2}\right) \tag{9}$$

To include the effect of C_P and C_L, r_2 is now expressed by equations (3) and (5) and the closed loop pole is simply expressed by

$$p_{cl} = \frac{g_m}{C_{eq,cl}} \tag{10}$$

The equivalent closed loop capacitance depends on the equivalent open loop capacitance and the DC feedback factor during the integration phase. Considering the parasitic capacitances C_P and C_L, it is given by

$$C_{eq,cl} = \frac{C_{eq,ol}}{f_{dc2}} = C_S + C_P + \frac{(C_S + C_P + C_I)C_L}{C_I} \tag{11}$$

Substituting (10) and (11) into (9) results in a more accurate expression for the settling error. It is given by:

$$k_2 = \exp\left(-\frac{g_m}{C_S + C_P + \frac{(C_S + C_P + C_I)C_L}{C_I}} \cdot \frac{\tau_2}{r_2}\right) \tag{12}$$

Assuming an OTA gain of 80 dB, behavioural simulations show that the frequency of the closed loop pole should be larger than three times the sampling frequency of the converter. This value should be increased by 15% since some time of the clock-period is used to ensure non-overlapping and delayed clock phases. For the clock frequency of 52.8 MHz used in this converter, this results in a closed loop pole of 180 MHz. Circuit simulations (section 4.1) show that it is located at 300 MHz, which corresponds to a model with a closed loop pole at 5 times the sampling frequency, thereby providing some margin for other non-idealities.

3.3 Non Zero Switch-Resistance

In the previous analysis, it is always assumed that the switches used in the integrators have a zero on resistance. Since these switches are generally implemented with nMOS and/or pMOS transistors, the on resistance is non-zero in reality. The on resistance of an n and pMOS transistor used as a switch are respectively given by [8]

$$R_{SW,N} = \frac{1}{KP_n \cdot \frac{W}{L} \cdot \left(v_{GS} - V_{Tn} - \frac{v_{DS}}{2} \right)} = \frac{1}{KP_n \cdot \frac{W}{L} \cdot \left(v_G - V_{Tn} - \frac{v_S + v_D}{2} \right)} \quad (13)$$

$$R_{SW,P} = \frac{1}{KP_p \cdot \frac{W}{L} \cdot \left(\frac{v_{DS}}{2} - (v_{GS} - V_{Tp}) \right)} = \frac{1}{KP_p \cdot \frac{W}{L} \cdot \left(\frac{v_S + v_D}{2} - (v_G - V_{Tp}) \right)} \quad (14)$$

These equations show two important effects. First, a reduction of the supply voltage immediately increases the switch resistance since the overdrive voltage of the transistor decreases. Second, the switch resistance is dependent on the source and drain voltages, or in the case of switch S1 (Fig. 3) on the input signal. This generates harmonic distortion. Both these effects can be reduced by employing transmission gates with n and pMOS transistors in parallel, which also reduces the clock-feedthrough. The resistance of a transmission gate can be expressed as

$$\frac{1}{R_{sw,NP}} = \frac{1}{R_{sw,N}} + \frac{1}{R_{sw,P}} \quad (15)$$

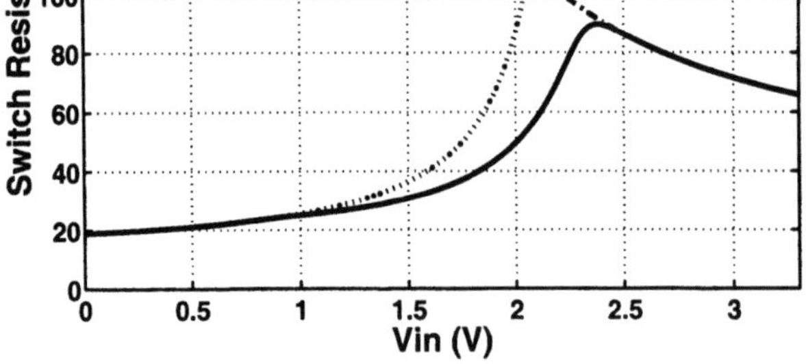

Fig. 6 Simulation of the switch-resistance versus input signal.

The simulated switch-resistance of switch S1 as a function of the input-signal is shown in Fig. 6. In some designs, even with a 5 V supply, special boosting circuits are used [9] in order to increase the overdrive voltage of the switches and/or to make the overdrive voltage independent of the input signal. However, these techniques become more difficult to implement as the clock-frequency increases. Therefore, a thorough analysis of the problems is carried out by modeling the effects associated with the switch-resistance and using behavioral simulations for the sizing of the switches. This will show that it is possible to dimension the switches properly for optimal settling performance of the switched-capacitor integrators and low distortion, even when the circuit operates from a 3.3 V voltage supply.

3.3.1 Influence on the settling

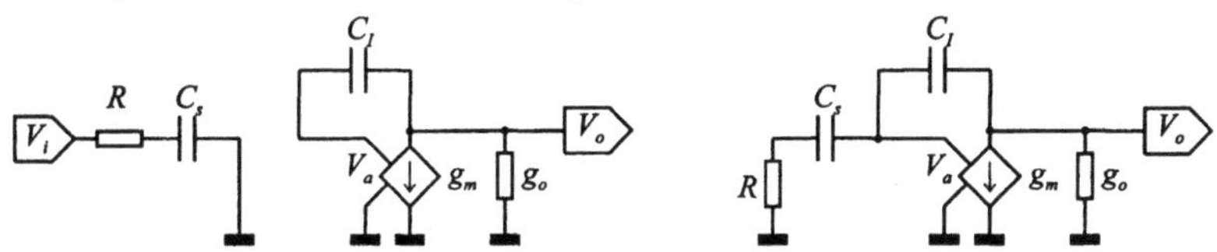

Fig. 7 Switched capacitor integrator including the resistance of the switches and an OTA structure.

To study the influence of the non-zero switch resistance, one can use different models for the OTA: the model with a finite gain of section 3.1, or the model with g_m and g_o of section 3.2. Since both these models assume an immediate charge redistribution between the sample and integration capacitances, adding the non-zero switch resistance influences the gain and pole errors because the charge transfer is now also determined by the RC time constant of the sampling network. However, the second model is more appropriate, since it also reveals the influence of the RC network on the settling characteristics of an OTA with a finite closed loop pole. Fig. 7 shows the model including the non-zero resistance for the switches. Note that the switches S1 and S2 of Fig. 3 are lumped into one element with a resistance R (the same applies for switches S3 and S4). The transfer function of this system can be calculated as

$$I(z) = \frac{C_S}{C_I} \cdot \frac{r_2(1-k_1)(1-k_2)\,z^{-1}}{1-b_1 z^{-1}+b_2 z^{-2}} \tag{16}$$

with following parameter definitions

$$b_1 = \frac{r_2}{r_1} + k_1\lambda_2 + k_2\left(1 - \frac{f_{dc2}\lambda_1}{1+\lambda_1} - \frac{r_2}{r_1}\,\frac{1+r_1\lambda_1\left(1-f_{dc2}\right)}{1+\lambda_1}\right) + k_1 k_2 \frac{r_2}{r_1} \tag{17}$$

$$b_2 = \frac{k_1 k_2}{1+\lambda_1}\left[k_2 \frac{r_2}{r_1}\left(1-\left(\frac{r_2}{r_1}+\lambda_2\right)\right)+\lambda_2+\left(\frac{r_2}{r_1}+\lambda_2\right)\left(\frac{r_2}{r_1}+\lambda_1\left(\frac{r_2}{r_1}+\lambda_2\right)\right)\right] \qquad (18)$$

where the settling errors during the sampling and integration phase are respectively

$$k_1 = \exp\left(-\frac{\tau_1}{RC_S}\right) \qquad k_2 = \exp\left(-\frac{g_m}{C_S}\frac{\tau_2}{r_2(1+\lambda_1)}\right) \qquad (19)$$

and with the following additional parameters:

$$\lambda_1 = \frac{1}{r_1}\frac{g_m}{C_S}RC_S \qquad \lambda_2 = (1-f_{dc2})(1-r_2) \qquad (20)$$

The expressions for the factors r_1, r_2 and the DC-feedback factors are still given by (3) and (4), respectively.

Note that the transfer function of the integrator now exhibits two poles: one very close to unity that can be directly related to a pole error in the integrator transfer function, and one very close to zero, that increases even further the gain error of the integrator.

Once more, the parasitic input and load capacitance of the OTA enhances these effects even further. The exact equations are very cumbersome and therefore they are not presented here. The same reasoning as in section 3.2 can be applied and the most important consequence of the introduction of these capacitances is an increase of the exponential settling error of this structure. Therefore, taking C_P and C_L into account, the exponential settling error k_2 should be given by

$$k_2 = \exp\left(-p_{CL}\frac{\tau_2}{r_2(1+\lambda_1)}\right) \qquad (21)$$

where the closed loop pole is expressed as in (10) and (11). Note that the expression for λ_1 of (20) should also be adapted in a similar way to incorporate C_P and C_L:

$$\lambda_1 = \frac{p_{CL}}{r_1}RC_S \qquad (22)$$

where r_1, r_2 and the DC-feedback factors are now given by (3) and (5).

Compared to the equation for the settling error in the previous paragraph (9), which does not include the switch resistance, the exponential settling factor is degraded by $1+\lambda_1 = 1+\frac{p_{CL}}{r_1}RC_S$. This clearly shows the influence of the sampling time constant on the degradation of the closed loop pole of the integrator. Again, behavioral simulations with this transfer function are used to determine a design criterion for the switch resistance. Assuming a closed-loop pole of the

OTA five times higher than the sampling frequency and an OTA gain of 80 dB, this results in the following condition for the sampling network:

$$\frac{1}{2\pi \cdot R \cdot C_s} > 4 f_s \tag{23}$$

This means that the total switch resistance (switches S1 and S2 or switches S3 and S4 in Fig. 3) should be smaller than approximately 215 Ω if one takes into account the time loss to ensure non-overlapping clocks.

3.3.2 *Generation of distortion*

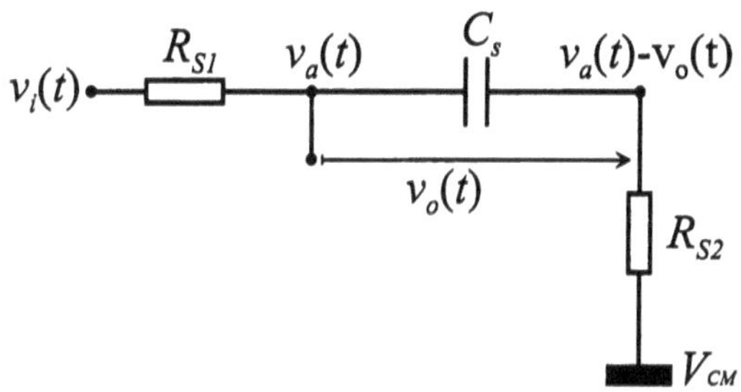

Fig. 8 Modeling of the sampling operation.

During the sampling phase, the input signal of the switched-capacitor integrator (Fig. 3) is sampled through switches S1 and S2 on the sampling capacitance. The resistance of switch S1 depends directly on the input signal, as illustrated in Fig. 6, and therefore causes harmonic distortion. Switch S2 on the other hand, has always one terminal connected to a fixed voltage. So at the end of the sampling phase, the voltages at the source and drain of that switch are about constant for each clock period. Therefore, the distortion generated by this switch can be neglected compared to the distortion introduced by switch S1. The same applies to switches S3 and S4, which at the end of the integration phase are connected to a reference voltage or to the virtual ground of the OTA, respectively. Furthermore, during the integration phase, no time varying input signal is driving the circuit. A constant charge, proportional to the input signal, is transferred from the sampling to the integration capacitance. Since this is like applying a DC signal during every clock phase, S3 and S4 generate considerably less distortion.

Therefore, to study the distortion introduced by the switches, only the sampling operation through switches S1 and S2 is considered. The model is shown in Fig. 8. The resistance of switch S1 is calculated by (13), (14) and (15). This resistance is a function of the switch size and voltages $v_i(t)$ and $v_a(t)$. The resistance of switch S2 is considered to be only dependent on the switch size, for the reasons mentioned above. Behavioral simulations with a model where both resistance's depend on drain and source terminals show that this simplification introduces a negligible error. The model of Fig. 8 can be described by the following two differential equations:

$$\frac{dv_o(t)}{dt} = \frac{v_i(t) - v_a(t)}{C_s \cdot R_{S1}\left(v_i(t), v_a(t), \frac{W}{L}\right)} \tag{24}$$

$$\frac{v_i(t) - v_a(t)}{R_{S1}\left(v_i(t), v_a(t), \frac{W}{L}\right)} = \frac{v_a(t) - v_o(t) - V_{CM}}{R_{S2}\left(\frac{W}{L}\right)} \tag{25}$$

This set of equations is extended for a differential configuration and it is numerically solved for each clock-period, always assuming an ideal reset of the capacitance as initial condition and a sinusoidal input signal with a fixed frequency and amplitude.

The simulations show that the distortion increases approximately proportional to the frequency of the input signal for a fixed sampling frequency. If one considers that the sampled signal is always lagging the input signal at all instants and that this delay causes a larger sampling error for larger signal frequencies, then the dependency on the signal frequency is expected.

Since the model assumes fully differential circuits, the second harmonic distortion is relatively small and the distortion is completely determined by the third harmonic. Therefore, the worst case occurs for a signal frequency of 366 kHz, since the third harmonic falls just inside the signal band of the converter.

The resulting distortion for this input signal is shown in Fig. 9 for various input amplitudes and W/L ratio's of the switches. It shows two ways of reducing the distortion: decreasing the signal amplitude or increasing the switch-size.

Since decreasing the input signal amplitude implies using a lower reference voltage, the capacitive load of the OTA's would increase (eq. 1) causing an increase of the power consumption. Therefore, the maximum input signal amplitude is maintained ($OL \cdot V_{ref} = 720mV$) and the switches are sized for this worst case signal amplitude. Fig. 9 shows that the minimal W/L ratio should be 240 in order to achieve a sampling-distortion better than 90 dB. The resulting switch-resistance is also low enough to ensure a good settling of the switched-capacitor integrator (eq. 23). Increasing the switches even further would increase the clock-feedthrough, the charge-injection and the capacitive load of the on-chip clock-drivers.

These results show that, by a detailed study of the effect of the switch-resistance, appropriate design considerations can be obtained to ensure optimal settling and distortion specifications, even for a 3.3 V voltage supply and without the use of special clock-boosting circuits.

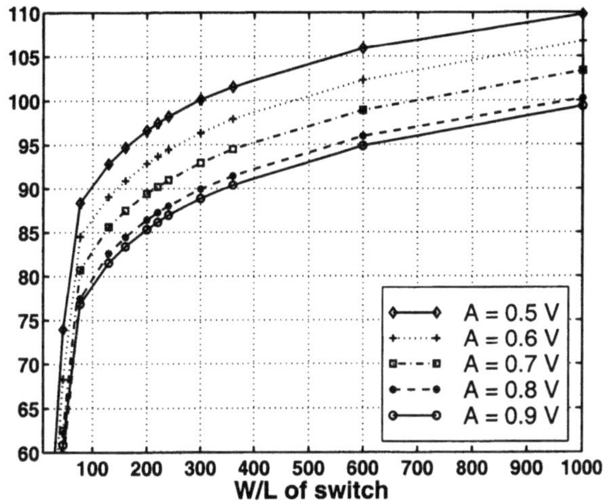

Fig. 9 Distortion of the sampling operation versus the size of the switches for various signal-amplitudes.

4 Design of the building blocks

4.1 Design of the OTA

The non-ideality study of the previous section revealed that the main requirements for the OTA are a high gain (>80 dB) and a large dominant closed-loop pole (>180 MHz). A fully differential folded-cascode OTA is well known for it's excellent frequency performance [8]. However, a gain-boosting stage needs to be added to obtain the high-gain requirement [10,11]. By a proper design of this gain-boosting stage, it is possible to maintain the excellent frequency behavior of the folded-cascode OTA.

The output swing of the OTA's determines the reference voltages, which set the capacitive load of the OTA and thereby it's power consumption. Therefore, it is very important to optimize the output swing of the OTA. In [11,12] a simple gain-boosting structure with just two transistors is employed. In this case, the output swing of the OTA is greatly reduced since it requires a V_{GS} over the top and bottom transistors of the folded-cascode output branch. For a voltage supply of 3.3 V, the resulting output swing is unacceptably low.

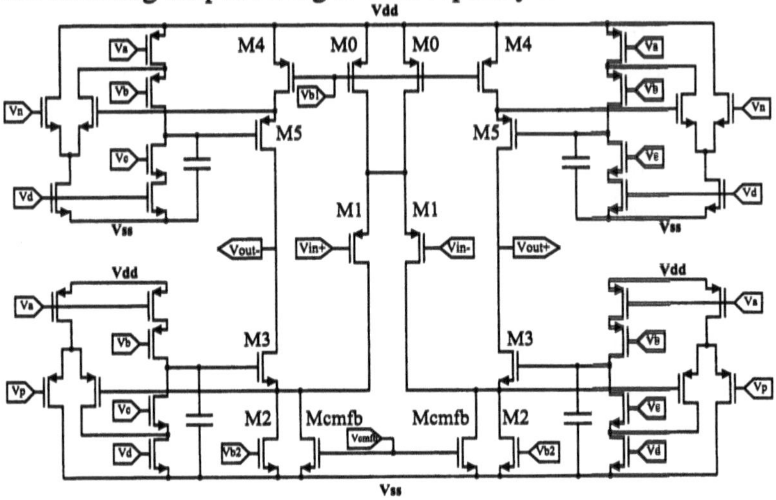

Fig. 10 Folded-cascode OTA with gain-boosting.

However, by employing a full OTA for the gain-boosting stages (Fig. 10), only a V_{DSsat} is required over these transistors and the resulting output swing is increased. In order to maintain a high-gain over the entire output swing of the OTA, the current source transistors should be kept in deep saturation (away from the linear region border). So, the V_{DS} of the transistors in the output branch is approximately 1.7 times V_{DSsat} in order to have some safety margin. In this design, the achieved output swing is 1.8 V with a gain of above 90 dB over the entire output range, which makes it possible to use symmetrical reference voltages as large as 900 mV. The closed-loop pole of the OTA is 300 MHz.

The dynamic common-mode feedback circuit [5,12] consists of a switched-capacitor circuit of Fig. 11. This kind of CMFB-circuit consumes almost no power. All switches are implemented as transmission gates.

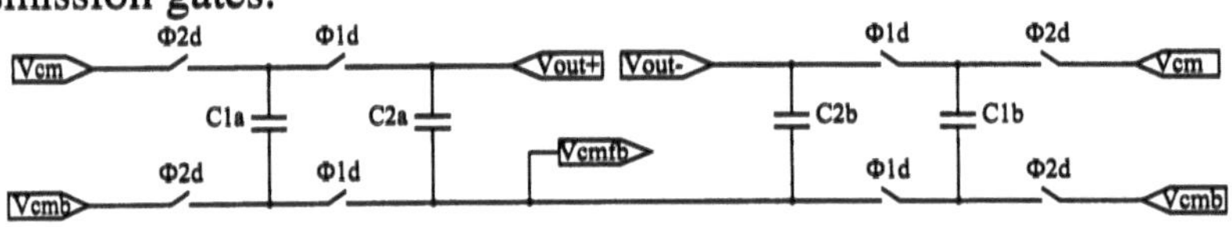

Fig. 11 Common Mode Feedback of the OTA.

4.2 Design of the comparator

Behavioral simulations show that the requirements for the comparator are very modest: an offset and an hysteresis smaller than respectively

100 mV and 40 mV are sufficient. The comparator [13], shown in Fig. 12, consists of a pMOS differential input pair and a top and bottom regeneration loop. The performance is much better than the required specifications. Some additional logic is also present to ensure the correct timing of the signals that control the feedback switches of the integrators.

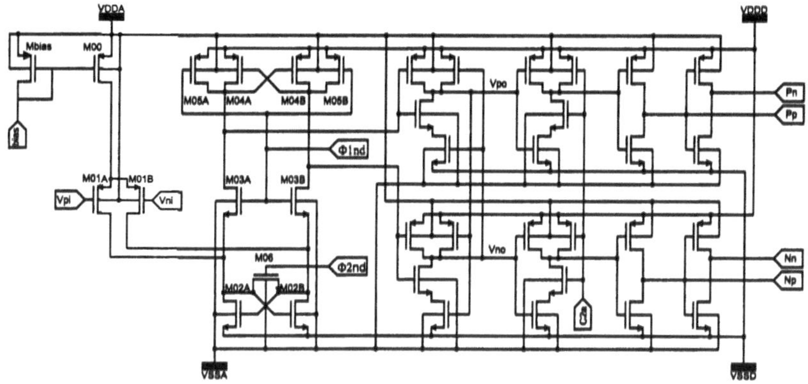

Fig. 12 Schematic of the comparator.

4.3 Design of the clock-driver

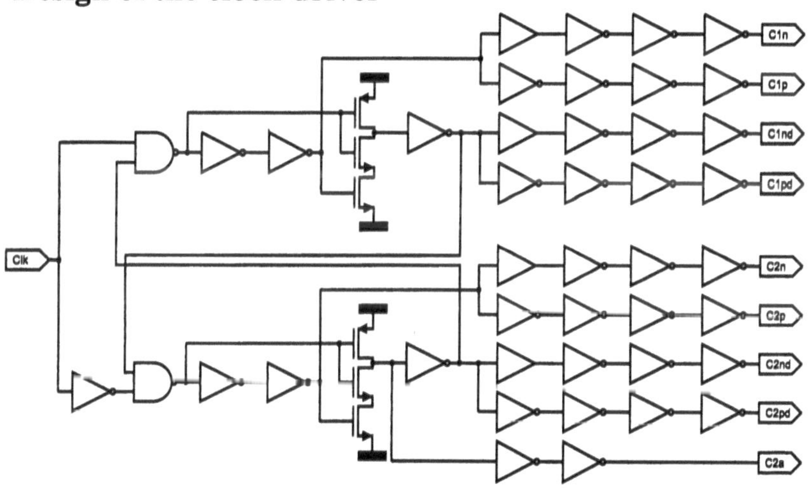

Fig. 13 Schematic of the clock-driver.

The switched-capacitor integrators require two non-overlapping clocks. In order to reduce the signal dependent charge injection [14], also a version of these signals with a delayed falling edge is needed. Since transmission gates are used for the switches, the inverse of these signals is also required. Finally, one more clock signal is necessary to ensure proper timing for the feedback signal of the comparator. The schematic shown in Fig. 13 [12] generates all these signals from one single clock input. Note that, in spite of the 3.3 V supply, no clock-boosters are used for the clock signals, as explained previously.

5 Measurement Results

The converter is implemented in a 0.5 μm n-well double-poly and triple-metal CMOS technology, operating from a 3.3 V supply. The die area, including the bonding pads and the decoupling capacitance, is 2.26 x 2.23 mm^2.

For measurement purposes, the chip is bounded on a thick-film substrate that is mounted in a copper-beryllium box to provide shielding from external noise sources. Great care has been taken to separate the analog and digital signals to avoid cross-talk problems. The external reference voltages for the DAC are passive RC-filtered and all the supplies and biasings are decoupled close to the chip. The converter is driven by a single 52.8 MHz clock. A transformer is used to convert the signal of the single ended signal source to a differential input signal. The three digital outputs of the converter are captured with a data-analyzer and transferred to a workstation, which performs the noise-cancellation logic and comb-filtering

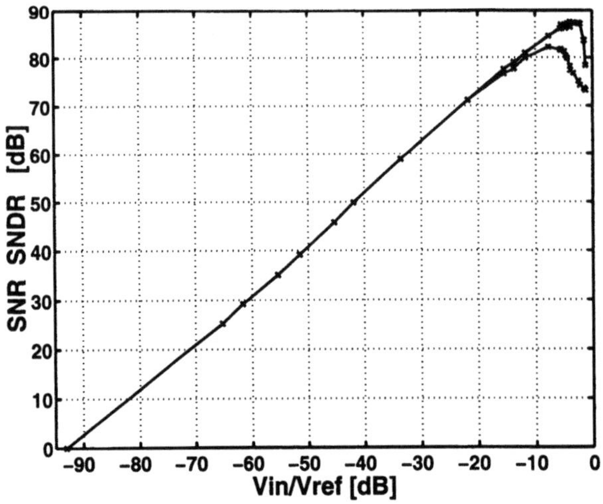

Fig. 14 SNR and SNDR vs. normalized input amplitude

The measured SNR and SNDR versus the relative input amplitude V_{in}/V_{ref} are shown in Fig. 14. The converter achieves a dynamic range of 92 dB, a peak SNR of 87 dB and a SNDR of 82 dB, after comb-filtering. The total power consumption from the 3.3 V voltage supply is 200 mW. The converter stays functional for a clock-frequency up to 125 MHz. The SNR is still 78 dB, however the harmonic distortion increases severely. Table 1 summarizes the most important specifications of the converter.

DR	92 dB
SNR	87 dB
SNDR	82 dB
Oversampling Ratio	24
Sampling Rate	52.8 MHz
Signal Bandwidth	1.1 MHz
Supply Voltage	3.3 V
Power Consumption	200 mW
Technology	0.5 μm CMOS TMDP
Die Area	2.2 x 2.3 mm^2

Table 1 Converter Performance

6 Conclusion

This paper addresses the feasibility of designing high-speed high-resolution ΔΣ ADC's for ADSL applications with a voltage supply of 3.3 V. The internal signal swings are optimized and several non-idealities are studied. A delta-sigma converter with a 3.3 V supply has been designed, exhibiting performances comparable to recently reported 5 V converters [5,9,12]. The converter achieves a resolution of 15 bits and a signal bandwidth of 1.1 MHz in a 0.5 μm CMOS technology.

7 References

[1] K. Maxwell, "Asymmetric Digital Subscriber Line: Interim Technology for the Next Forty Years", IEEE Communications magazine, October 1996, pp. 100-106.

[2] H. Casier, "Design of State-of-the-art Embeddable Delta-Sigma Converters for Mixed-Signal ASIC's ", Europractice course, January 1999.

[3] A. Marques, V. Peluso, M. Steyaert, W. Sansen, "Optimal Parameters for ΔΣ Modulator Topologies", IEEE Trans. on Circuits and System II, vol. 45, no. 9, pp. 1232-1241 , Sept 1998.

[4] B. H. Leung, S. Sutarja, "Multibit Σ-Δ A/D Converter Incorporating a Novel Class of Dynamic Element Matching Techniques", IEEE Trans. on Circuits and System II, vol. 39, no. 1, pp. 35-51, Jan. 1992.

[5] G. Yin, W. Sansen, "A High-Frequency and High-Resolution Fourth-Order Sigma-Delta A/D Converter in BiCMOS Technology", IEEE Journal of Solid State Circuits, Vol. 29, no. 8, pp. 857-865, Aug. 1994.

[6] D. B. Ribner, R. D. Baertsch, S.L. Garverick, D. T. McGrath, J.E. Krisciunas, T. Fujii, "A Third-Order Multistage Sigma-Delta Modulator with Reduced Sensitivity to Nonidealities", IEEE J. Solid-State Circuits, vol. 26, no. 12, pp. 1764-1773, Dec. 1991.

[7] G. C. Temes, "Finite Amplifier Gain and Bandwidth Effects in Switched-Capacitor Filters", IEEE J. Solid-State Circuits, vol. 15, no. 3, pp. 358-361, June 1980.

[8] K. Laker, W. Sansen, "Design of Analog Integrated Circuits and Systems", New York, McGraw-Hill, 1994.

[9] T. L. Brooks, D. H. Robertson, D. F. Kelly, A. Del Muro, S. W. Harston, "A Cascaded Sigma-Delta Pipeline A/D Converter with 1.25 MHz Signal Bandwidth and 89 dB SNR", IEEE Journal of Solid State Circuits, Vol.32, no.12, pp.1896-1906, Dec. 1997.

[10] K. Bult, G. Geelen, " A Fast-Settling CMOS Op Amp for SC Circuits with 90-dB DC Gain", IEEE Journal of Solid-State Circuits, vol. 25, no. 6, pp. 1379-1384, Dec. 1990.

[11] D. Flandre, A. Viviani, J. Eggermont, B. Gentine, P. Jespers, "Design Methodology for CMOS Gain-Boosted Folded-Cascode OTA with Application to SOI Technology", Proc. of the ESSCIRC'96, pp. 320-323, Sep. 1996.

[12] A. Marques, V. Peluso, M. Steyaert, W. Sansen, "A 15-b Resolution 2-MHz Nyquist Rate ΔΣ ADC in a 1-μm CMOS Technology", IEEE Journal of Solid State Circuits, Vol. 33, no.7, pp. 1065-1075, July 1998.

[13] G. Yin, F. Eynde, W. Sansen, "A High-Speed CMOS Comparator with 8-b Resolution", IEEE J. Solid-State Circuits, vol. 27, no. 2, pp. 208-211, Dec. 1990.

[14] D. Haigh, B. Singh, "A Switching Scheme for Switched Capacitor Filters which Reduces the Effects of Parasitic Capacitances Associated with Switch Control Terminals", Proc. of the ISCAS, pp. 586-589, 1983.

A 3.3 V 15-bit Delta-Sigma ADC with a Signal Bandwidth of 1.1 MHz for ADSL-Applications

Y. Geerts, M. Steyaert, W. Sansen

Katholieke Universiteit Leuven
ESAT - MICAS
Belgium

Overview

- Introduction
- Topology Selection
- System Level Optimization
- Study of Circuit Non-Idealities
- Design of Building Blocks
- Measurement Results
- Conclusion

Introduction (1)

Frequency spectrum of DMT-ADSL

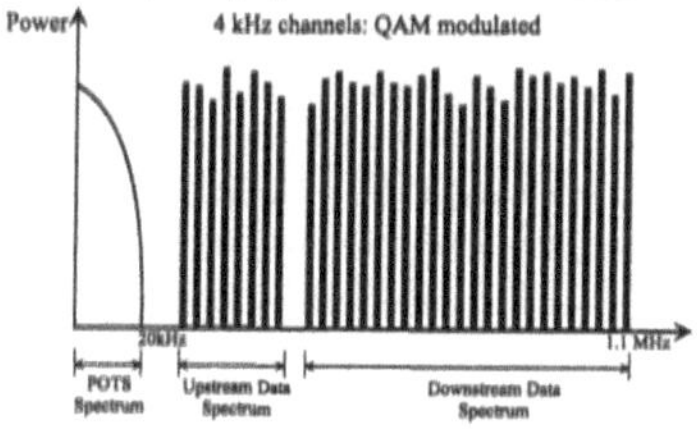

Introduction (2)

ADSL Receiver

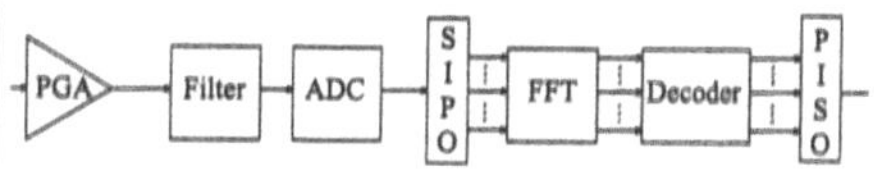

ADC specifications: [Ref:CAS99]

- SNR > 80 dB (13 bits)
- Signal Bandwidth of 1.1 MHz

Introduction (3)

High-Speed High-Resolution ΔΣ ADC's:
5 V Power Supply

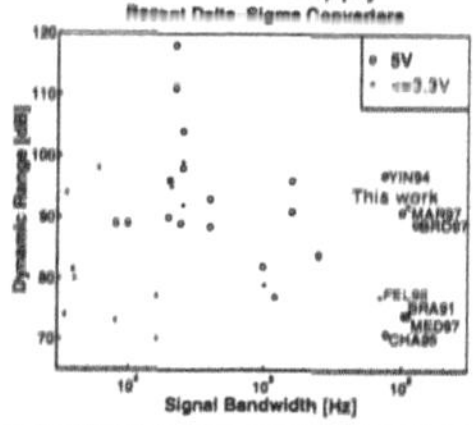

Introduction (4)

- High-Resolution High-Speed ΔΣ converters: 5 V Power Supply
- Design problems for a 3.3 V ΔΣ converter
  - reduced internal signal swings
    - system level: minimize required signal swing
    - OTA design: maximize output swing
  - reduced clock signals
    - less overdrive for switches: switch resistance
    - Clock boosters or Voltage Doublers are not used:
      - more difficult for high sampling frequencies
      - not suitable for all technologies (gate oxide)

Topology Selection (1)

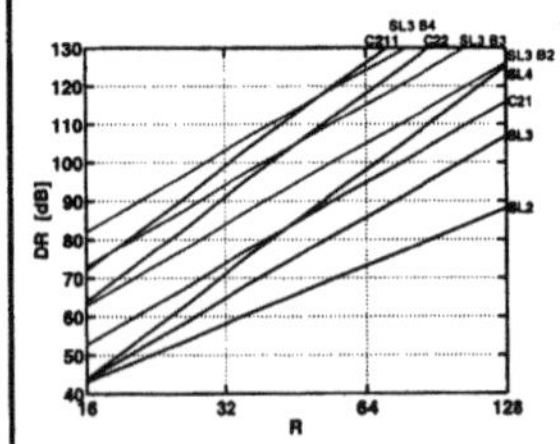

- Singe Loop:
  - ⊗ stability limits SNR_{peak}
- Multi-bit:
  - ☺ improves stability and SNR_{peak}
  - ⊗ DAC-linearity
- Cascade:
  - ☺ stability of 2nd order
  - ⊗ building block specifications

Topology Selection (2)

2-1-1 Cascade Architecture

- ☺ 4th order noise shaping
- ☺ stability of 2nd order topology
- ⊗ sensitive to building block specs (noise leakage)

- R = 24
- 15 bits
- BW = 1.1 MHz
- f_s = 52.8 MHz

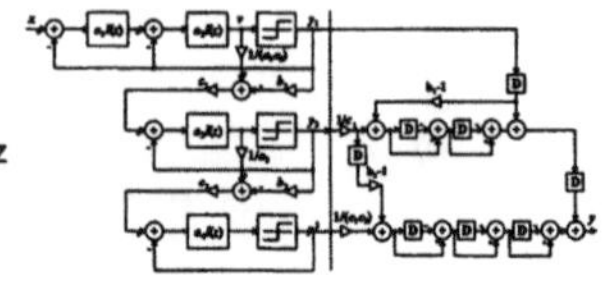

System Level Optimization (1)

- Key to High-Speed High-Resolution ΔΣ Design:
  - reduce capacitive loads of the OTA's
    - → $f_{d,cl}$ increases: larger f_s and signal bandwidths achievable
    - → power of OTA's decreases
  - kT/C noise determines the sampling capacitor in high-resolution converters

⇨ try to reduce kT/C noise

System Level Optimization (2)

- Reduce kT/C noise sources
  - Differential Implementation
    - 3 dB gain in signal to white noise ratio
    - charge and clock feedthrough cancels better
    - improves settling, HD, PSRR, ...
  - Double Reference Implementation
    - input and feedback signal share same capacitance
    - 3 dB gain in signal to white noise ratio
    - feedback factor of OTA increases (reduces load capacitance of the OTA)

$$\Rightarrow \quad SNR_{\frac{kT}{C}} = \frac{(2 \cdot OL \cdot V_{ref})^2}{2} \cdot \frac{C_s \cdot R}{4 \cdot k \cdot T}$$

System Level Optimization (3)

- Increase V_{REF} (increase input signal power)
  - decrease C_s quadraticly for equal $SNR_{kT/C}$
  - restriction: increase of internal signal swings
    - minimize required internal swing at the system level
    - maximize output swing of the OTA's (reduced power supply of 3.3V)
- Coefficients of converter chosen to
  - maximize SNR_{peak} but maintain stability
  - reduce required internal signal swings
  - share as many capacitors as possible (reduces capacitive load)

System Level Optimization (4)

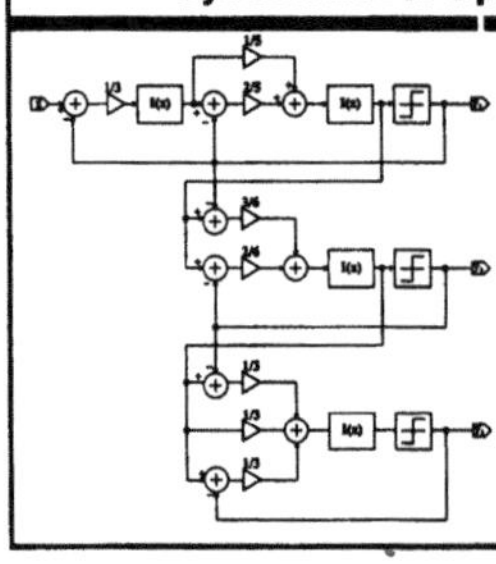

- V_{REF} = 900 mV
- Signal Swing < V_{REF}
- Sampling capacitors:
  - stage 1: kT/C noise (3pF)
  - stage 4: matching
- Scale integrators 1:0.5:0.35:0.35 (reduces power)

Study of non-idealities

- Switched - capacitor integrator

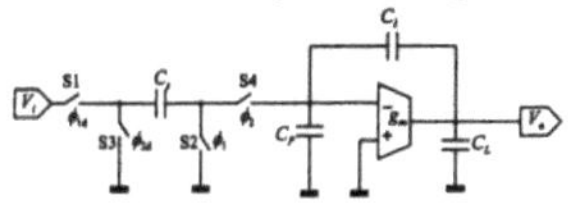

- OTA - Gain
- OTA - Dominant closed loop pole
- Switch Resistance
  - Influence settling of the integrator
  - Distortion of the sampling operation

Finite OTA Gain (1)

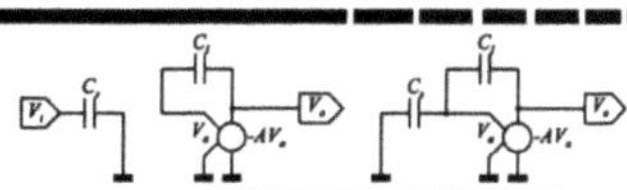

- Transferfunction: $I(z)=\frac{C_S}{C_I}\cdot\frac{\alpha z^{-1}}{1-\beta z^{-1}}=\frac{C_S}{C_I}\cdot\frac{r_2 z^{-1}}{1-\frac{r_2}{r_1}z^{-1}}$

with	$r_1=\frac{f_{dc1}A}{1+f_{dc1}A}$	$r_2=\frac{f_{dc2}A}{1+f_{dc2}A}$
feedback factors	$f_{dc1}=\frac{C_I}{C_I}=1$	$f_{dc2}=\frac{C_I}{C_s+C_I}$
Include C_P and C_L:	$f_{dc1}=\frac{C_I}{C_P+C_I}$	$f_{dc2}=\frac{C_I}{C_s+C_P+C_I}$

Finite OTA Gain (2)

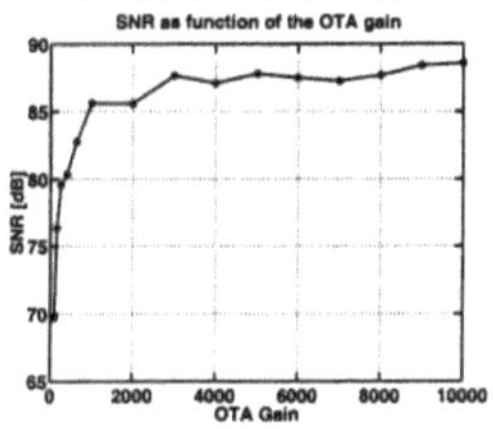

- Finite Gain effects:
  - Gain > 60 dB
- 80 dB selected
  - margin for degradations due to other non-idealities

Finite dominant closed loop pole (1)

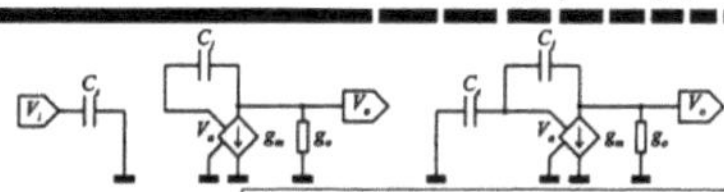

- Transferfunction: $I(z)=\frac{C_S}{C_I}\cdot\frac{\alpha z^{-1}}{1-\beta z^{-1}}=\frac{C_S}{C_I}\cdot\frac{r_2(1-k_2)z^{-1}}{1-\frac{r_2}{r_1}\left(1-k_2\left(1-\frac{r_1}{r_2}\right)\right)z^{-1}}$

with $k_2=\exp\left(-\frac{g_m}{C_S}\cdot\frac{\tau_2}{r_2}\right)$ and $A=\frac{g_m}{g_o}$

→ Gain and Pole errors of finite gain model are increased

Finite dominant closed loop pole (2)

- Include C_P and C_L
  → Complicated expressions
- Other approach:
  - change most inaccurately modeled factor
  → exponential term k2 (settling during integration phase)

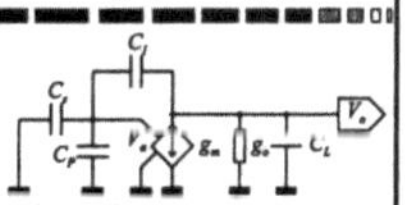

$$k_2=\exp\left(-p_{cl}\cdot\frac{\tau_2}{r_2}\right)=\exp\left(-\frac{g_m}{C_{eq,cl}}\cdot\frac{\tau_2}{r_2}\right)$$

with $C_{eq,cl}=\frac{C_{eq,ol}}{f_{dc2}}=C_S+C_P+\frac{(C_S+C_P+C_I)C_L}{C_I}$

Finite dominant closed loop pole (3)

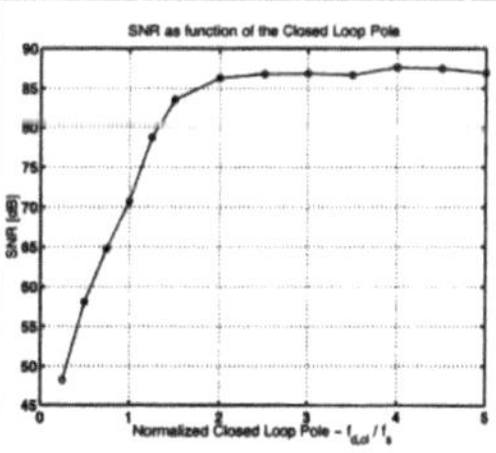

- Assumptions
  - Gain = 80 dB
- $f_{d,cl}>3\cdot f_s$
- Margin for
  - non-overlapping clocks
  - non-idealities (switches)
- OTA-design: 300 MHz
  - $f_{d,cl}=300\,MHz=5\cdot f_s$

Switch Resistance (1)

- Switch Resistance of nMOS

$$R_{sw} = \frac{1}{KP \cdot \frac{W}{L} \cdot \left(v_{GS} - v_T - \frac{v_{DS}}{2}\right)}$$

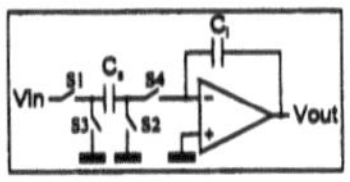

- Two problems
  - Resistance increases if power supply is reduced
    - ⇨ degrades settling performance of integrator
  - Resistance is signal dependent through V_d and V_s
    - ⇨ Switch S1 introduces harmonic distortion
- ⇨ Analyze both effect in detail to avoid the use of clock boosters

Switch Resistance (2)

- Transmission gates
  - reduces resistance
  - reduces clock feedthrough

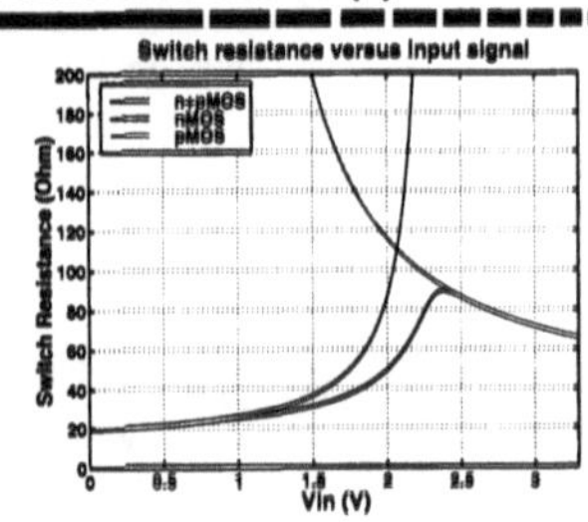

Switch Resistance: settling (1)

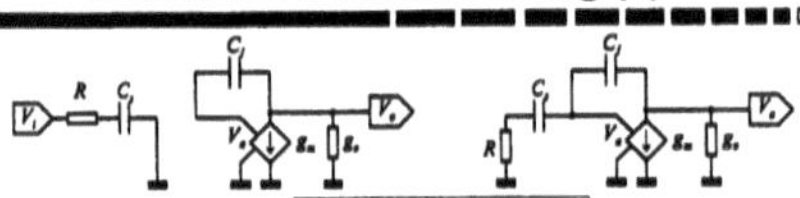

- Transferfunction: $I(z) = \frac{C_S}{C_I} \cdot \frac{r_2(1-k_1)(1-k_2)z^{-1}}{1 - b_1 z^{-1} + b_2 z^{-2}}$

with $k_1 = \exp\left(-\frac{\tau_1}{RC_S}\right)$ $\quad k_2 = \exp\left(-\frac{g_m}{C_S}\frac{\tau_2}{r_2(1+\lambda_1)}\right)$ $\quad \lambda_1 = \frac{1}{r_1}\frac{g_m}{C_S} RC_S$

- 2 poles
  - one close to unity → pole error
  - one close to zero → enhances gain error

Switch Resistance: settling (2)

- Include C_P and C_L: similar reasoning as $f_{d,cl}$

$k_2 = \exp\left(-p_{CL}\frac{\tau_2}{r_2(1+\lambda_1)}\right)$	$\lambda_1 = \frac{p_{CL}}{r_1} RC_S$	$p_{cl} = \frac{g_m}{C_{eq,cl}}$

- Dominant closed loop pole of the OTA is degraded by

$$1 + \lambda_1 = 1 + \frac{p_{CL}}{r_1} RC_S$$

→ RC_S time constant degrades settling of OTA

Switch-Resistance: settling (3)

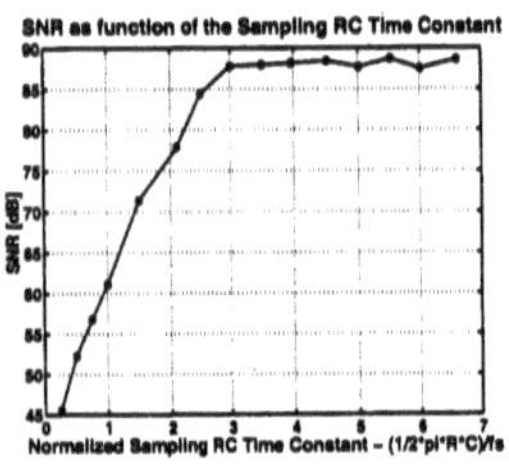

- Assumptions for OTA:
  - $\frac{g_m}{g_o} = 10000 = 80\ dB$
  - $f_{pole} = 5 \cdot f_s$

⇨ $\frac{1}{2\pi \cdot R_{switch} \cdot C_s} > 4 \cdot f_s$

e.g. Cs = 3pF and fs = 52.8 MHz

15 % non-overlap time

→ $R_{SWITCH} < 215\ \Omega$

Modeling of sampling operation

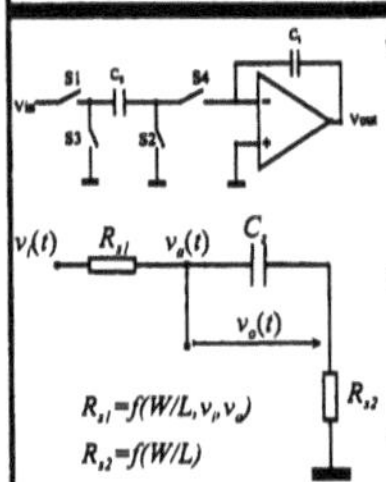

- Model solved numerically for each sampling phase
  - reset as initial condition
  - extended for differential implementation (reduces HD2)
- Resulting distortion:
  - increases with frequency of the input signal
  - dominated by HD3
- ⇨ Worst case: 366 kHz input signal (HD3 falls inside signal band)

Switch Resistance: Distortion of Sampling

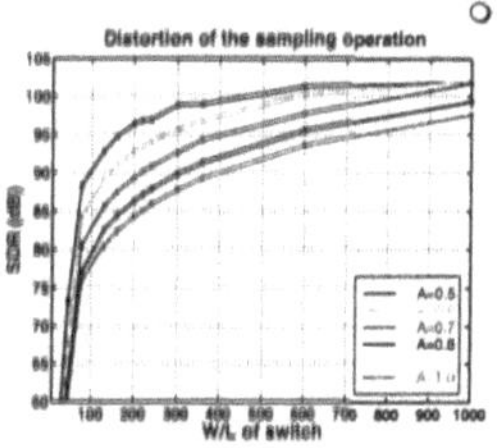

- Improve HD:
  - reduce signal amplitude
    - reduce the V_{REF}
    - increase C_S (due to kT/C noise)
    - bad for speed and power
    - ⇒ maintain $A_{MAX} = OL \cdot V_{REF}$
  - increase W/L of switch
    - W/L=240
    - SDR > 90 dB
    - R_{SWITCH} < 200 Ohm (settling of OTA)

OTA Design

- A_{dc} = 94dB
- GBW = 300 MHz
- I_{tot} = 20mA
- Output Swing 1.8V
  - V_{REF} = 900 mV
  - $V_{DS} > 1.7\ V_{DSsat}$

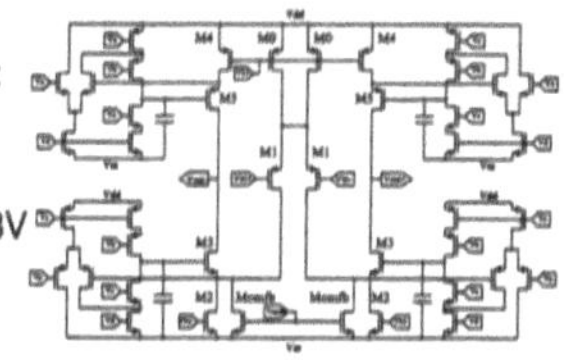

Comparator

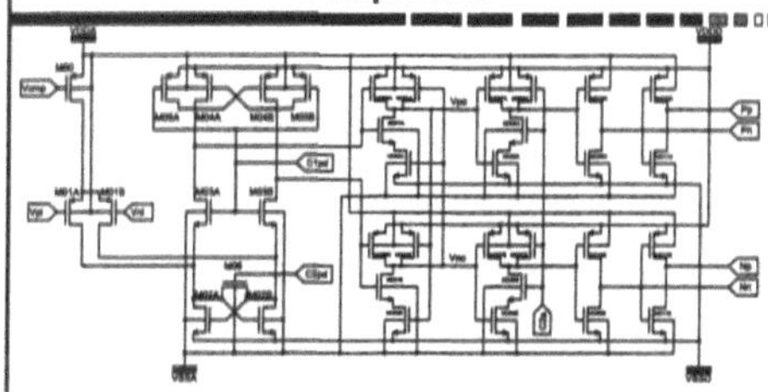

Φ_1: regeneration Φ_2: reset

Clock-circuit

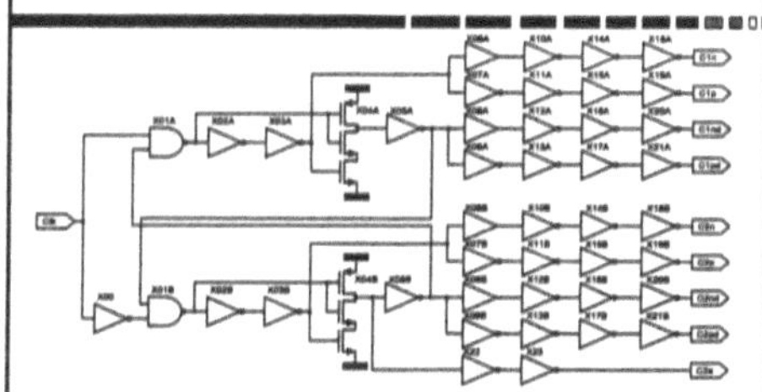

Only falling edges of the delayed clocks are delayed

Chip Photograph

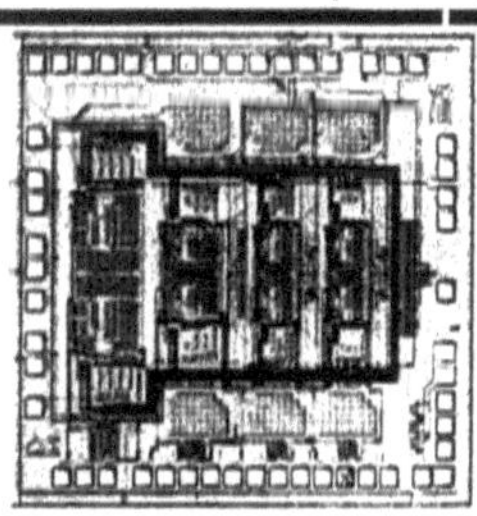

- Die Area: 2204 μm x 2230 μm
- Power Consumption
  - Analog: 170 mW
  - Digital: 30 mW
  - Total: 200 mW

Measurements Results (1)

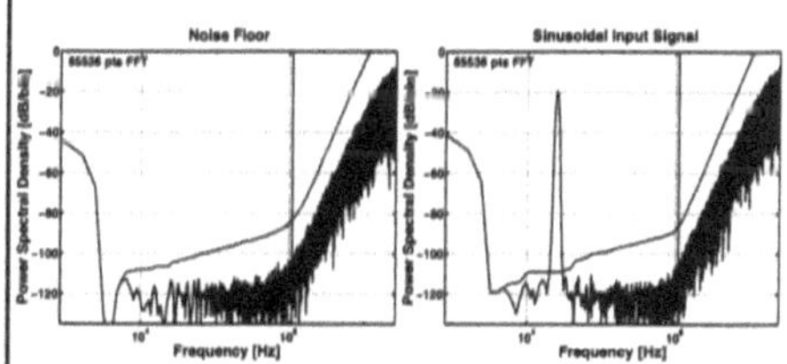

Measurement Results (2)

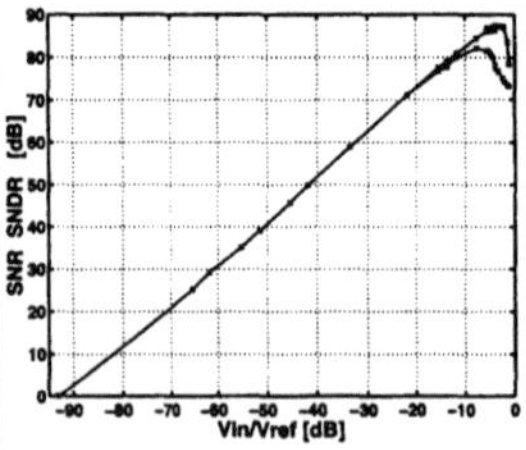

Summary	
DR	92 dB
SNR	87 dB
SNDR	82 dB
OSR	24
f_s	52.8 MHz
BW	1.1 MHz
Supply	3.3 V
Power	200 mW
Tech.	0.5 μm CMOS
Area	2.3×2.2 mm^2

Conclusion

- Design of 3.3 V CMOS ΔΣ for ADSL-applications
  - 1.1 MHz bandwidth and 15-bit dynamic range
  - standard 0.5 μm CMOS technology
  - 200 mW Power Consumption
  - →comparable to 5 V CMOS ΔΣ converters
- Performance obtained by
  - Optimization of system level (signal swing)
  - Detailed analysis of non-idealities (switch resistance)

TRANSMITTER & RECEIVER CLIPPING MITIGATION IN DMT TRANSCEIVERS

Denis J. G. Mestdagh * and Lennart B. Olsson **

* STMicroelectronics, France
** Telia Research AB, Sweden

- Abstract -

Two new methods for clipping mitigation in DMT transceivers are presented. The first method applies to the transmitter and spectrally shapes the clipping noise in order to increase the signal-to-noise ratio within the signal bandwidth. The second method applies to the receiver and can be implemented by design of pipeline A/D converters. The two methods are independent and can therefore be used separately or combined together without extra features. Both methods improve the overall system performance and relax the constraints of the implementation complexity of the analog front-end.

The Discrete Multi-Tone (DMT) modulation technique [1] has recently gained considerable interest from the communication engineering community for a wide range of applications in both wired and wireless networks. DMT is the international standard for Asymmetric Digital Subscriber Line (ADSL) [2] and its "light" version (G.Lite) [3], and is currently proposed for the VDSL standard under definition. It has also been standardized for wireless communication systems like Digital Audio Broadcast (DAB), Digital Video Broadcast Terrestrial (DVB-T) [4] as well as for the HIPERLAN wireless broadband LAN [5].

While DMT undoubtedly offers many advantages for high data rate communication over bandlimited channels having variations in gain and phase with frequency, a major drawback results from the fact that the instantaneous amplitude may exhibit large spikes that arise when several modulated carriers add in-phase. These spikes may have serious impact on the design complexity of the transceiver analog front-end; high resolution analog-to-digital and digital-to-analog converters [6], nonlinear effects causing intermodulation distortion and spectral spreading [7, 8, 9, 10].

The difficulty to maintain linearity and efficiency over a wide range of amplitudes and frequencies has motivated the use of clipping devices to limit the maximum signal excursions to a given level, i.e. $|A(t)|_{max} \leq A_{clip}$. This provides elements for the strategy to relax the implementation constraints of the analog front-end that includes, among others, analog-to-digital and digital-to-analog converters as well as line drivers.

However, by clipping the signal, one introduces a noise, the clipping noise, that degrades the performance of the communication system. Therefore, several methods have been proposed to avoid this performance degradation while compromising the implementation complexity by as much as possible. To the authors knowledge, all proposals to date are only applicable at the transmitter and no one can be applied at the receiver. While implementing clipping mitigation at the transmitter partly solve the implementation complexity of the analog front-end by relaxing the dynamic and linearity requirements of the digital-to-analog converter and amplifiers, overall system performance and implementation complexity are determined by the combination of transmitter with receiver.

In this paper, we present two new techniques for clipping mitigation; one is applied at the transmitter while the other is applied at the receiver. These techniques are independent and can therefore be applied separately at the transmitter or the receiver without prior knowledge of their use at the receiver or the transmitter, respectively. In other words, there is no need for a specific communication channel to tell whether clipping mitigation technique is applied at transmitter or receiver.

The paper is organized as follows. In Section II we briefly review the statistical characteristics of DMT signals and we discuss several methods for clipping mitigation that have been proposed so far. This is followed, in Section III by a description and analysis of the clipping mitigation method to be applied at the transmitter. This technique is borrowed from sigma-delta A/D - D/A converter theory and shapes the clipping noise by placing the noise outside of the signal spectrum in order to increase the signal-to-noise ratio within the useful bandwidth. Section IV is devoted to the description and analysis of the clipping mitigation technique at the receiver. The basic concept is first explained and a possible implementation using a pipeline A/D converter architecture is proposed. Finally, a summary is provided in Section V.

II. Statistical Characteristics of a DMT Signal

A DMT signal is the sum of N independently quadrature modulated (QAM) signals, referred to as subchannels, each being carried over distinct carrier frequency spaced by $1/T$ where T is the temporal length of a DMT symbol. These N modulated subchannels are efficiently synthesized by an inverse fast Fourier transform. Figure 1 shows a typical DMT signal: most of the signal is concentrated within σ while large spikes arise seldomly when several QAM modulated carriers add in phase.

For large values of N (say $N>10$), the ensemble average of the instantaneous amplitude $A(t)$ can be accurately modelled by a Gaussian random process (central limit theorem) with a zero mean and a variance σ^2 equal to the total signal power P_{tot}. That is, the probability that, at any given time, $A(t)$ takes the value x ($|x|<A_{max}$) is given by Eq.(1).

$$Prob\{A(t) = x\} = p(x) = \frac{1}{\sigma\sqrt{2\pi}}\exp\left\{-\frac{x^2}{2\sigma^2}\right\} \quad \text{with } |x|<A_{max} \qquad \text{Eq.(1)}$$

where A_{max} is the maximum amplitude obtained when all carriers are modulated with maximum amplitude and add in phase.

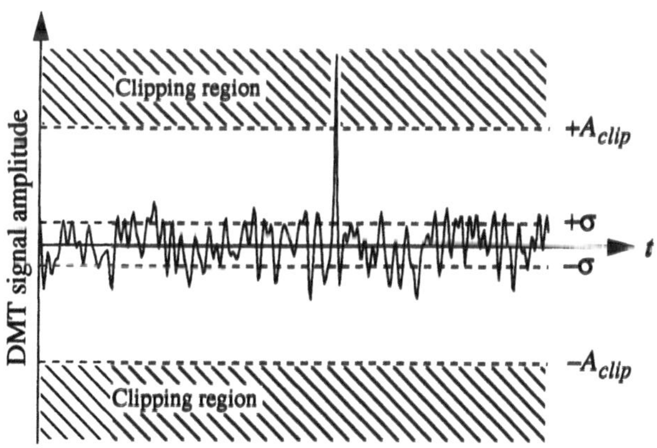

Figure 1. A typical DMT signal in the time domain representation. The clipping region is shown as shaded areas.

Therefore, the probability that a given sample in the DMT symbol is clipped is given by Eq.(2).

$$P(A_{clip}) = Prob\{|A(t)| \geq A_{clip}\} = 2\int_{A_{clip}}^{\infty} p(x)dx = 1 - erf\left(\frac{\mu}{\sqrt{2}}\right) \quad \text{Eq.(2)}$$

where $\mu = A_{clip}/\sigma$ is the crest-factor and $erf(t)$ is the error function defined by:

$$erf(t) = \frac{2}{\sqrt{\pi}}\int_0^t e^{-y^2} dy \cong 1 - \frac{e^{-t^2}}{t\sqrt{\pi}}\left[1 - \frac{1}{2t^2}\right] \quad \text{Eq.(3)}$$

Using the approximation of $erf(t)$ given by Eq.(3) gives an expression for $P(A_{clip})$ as a function of the crest-factor μ.

$$P(A_{clip}) \cong \sqrt{\frac{2}{\pi}} \cdot \frac{e^{-\mu^2/2}}{\mu}\left(1 - \frac{1}{\mu^2}\right) \quad \text{Eq.(4)}$$

It is clear from Eq.(4) that large amplitude spikes arise very rarely because of statistical averaging. For example, $\mu = 5.35$ implies $P(A_{clip}) = 10^{-7}$. Figures 2 (a) and (b) plot $P(A_{clip})$ and the power spectral density of the clipping noise as a function of the crest-factor μ, respectively.

In [11] it is shown that it is advantageous to accept some clipping in the DMT signal and to trade-off the resulting signal-to-noise ratio loss against a gain in signal-to-quantization noise ratio of the D/A or A/D converters, that have then to cover a reduced amplitude dynamic range. Results of this analysis reveal that a significant reduction of the D/A - A/D converter resolution can be obtained by clipping at an optimal level while keeping the same overall signal-to-noise ratio as when clipping is not applied. Figure 3 represents the gain $\Delta = b_2 - b_1$ in DAC/ADC resolution requirements as a function of μ for several values of b_1 (10, 12 and 14 bits).

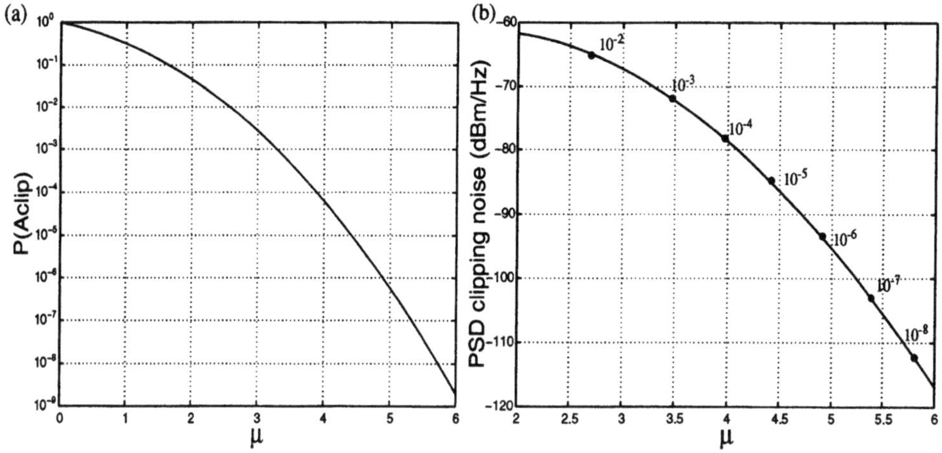

Figure 2. (a) Probability of sample clipping as a function of the crest-factor $\mu = A_{clip}/\sigma$, σ^2 being the total power of the DMT signal. (b) The power spectral density (flat) of the clipping noise as a function of μ. The dots in the curve indicate the corresponding clipping probability.

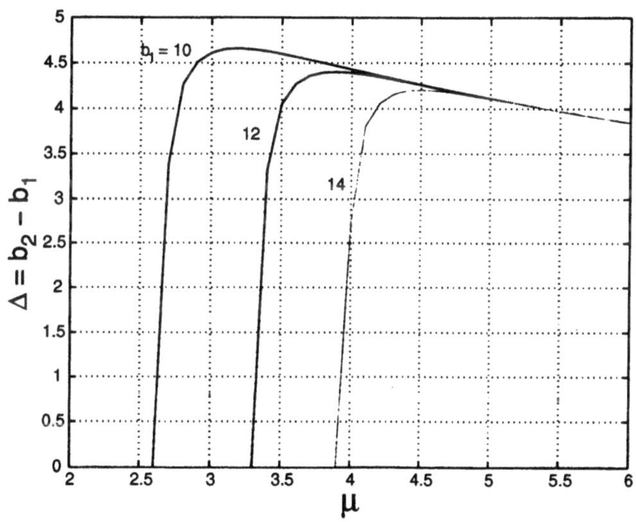

Figure 3. "Bit gain" DAC/ADC resolution gain requirements by "brute force" clipping of the DMT signal (N = 2048 carriers is assumed).

The curves correspond to 2048 carriers (recall that μ on the x-axis is independent of the number of carriers). It is seen that a gain of up to 4 bits is obtained by "brute force" clipping with $\mu = 5$.

Further improvements have been proposed in [12], [13]. These methods are based on input data encoding allowing to reduce the peak-to-average ratio of the DMT signal. For example, a 3-bit input data word is block coded and mapped onto a 4-bit code word such that the set of permissible words does not contain those that result in excessive peak envelope power. These coding methods, however, require an increase in data rate and hence a reduction of the energy per bit for the same total power, resulting in performance degradation in terms of information handling capacity of the communication system.

In [14] an alternative method based on the statistical properties of $A(t)$ is proposed. This method applies randomization (i.e. no coding) on DMT symbols whose instantaneous amplitude exceeds the clipping amplitude A_{clip}. The result is a decrease of the DMT symbol clipping probability by several orders of magnitude. Although the method doesn't use any kind of pre-coding and hence does not increase the actual transmission data rate, it requires a speed-up of the modulator which is certainly undesirable especially when power consumption must be minimized.

III Clipping Mitigation at the Transmitter

The basic idea behind the present proposal is to perform spectral shaping of the clipping noise such that the noise power is placed outside of the frequency band used by the signal. This is made possible at the transmitter because the noise introduced by clipping is known and can be processed digitally in a deterministic way.

Since the clipping-noise must be rejected outside of the useful bandwidth, the method is only applicable to oversampled signals. The higher the oversampling factor is, the more efficient the method is.

A schematic block diagram illustrating the method, referred to as the "Clipping Noise Shaper" is shown in Figure 4. The transfer function is given by:

$$S(z) = I(z) + N(z)[1 - z^{-1}F(z)] \qquad \text{Eq.(5)}$$

where $N(z)$ is the z-transform of the clipping noise.

The filter $F(z)$ is a finite impulse response linear filter that will shape $N(z)$. Output samples from $F(z)$ are superimposed onto the subsequent samples of the input signal. Since the clip-noise shaper is a closed loop, there is a delay introduced into the feed-back loop to ensure its stability. This delay is explicitly shown in Figure 1 as an external element of the filter $F(z)$.

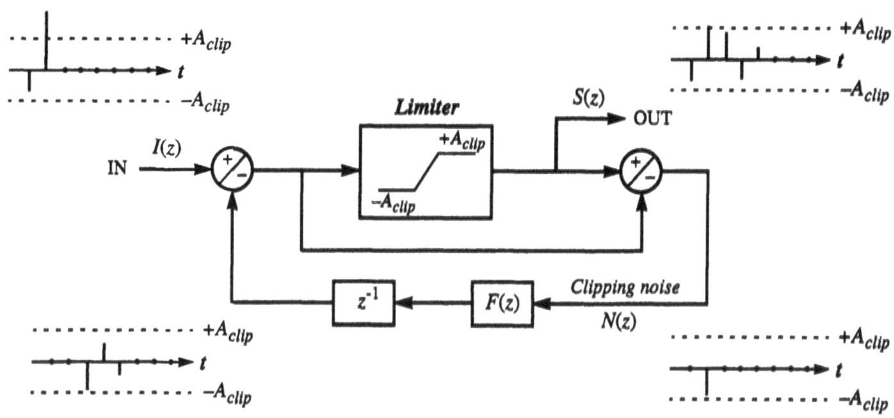

Figure 4. Schematic block diagram of the clip-noise shaper.

Case 1 - Band-pass filtering - One can choose $F(z)$ such that $[1-z^{-1}F(z)]$ is a band-pass filter centered around f_0. This can be for instance a damped sinusoid whose time function takes the form:

$$f(t) = \cos(2\pi f_0 t)e^{-\alpha t} \qquad \text{Eq.(6)}$$

As an example, Figure 5 shows the resulting spectrum of the shaped clipping noise with $\alpha=0.1$ and $f_0=17.6\ MHz$. The signal spectrum covers frequencies from dc up to 8.8 MHz, as in VDSL systems. In all subsequent examples we use a crest-factor $\mu=4.9$ in order to have a reasonable lerge amount of clipped samples and hence limit the computer simulation run time.

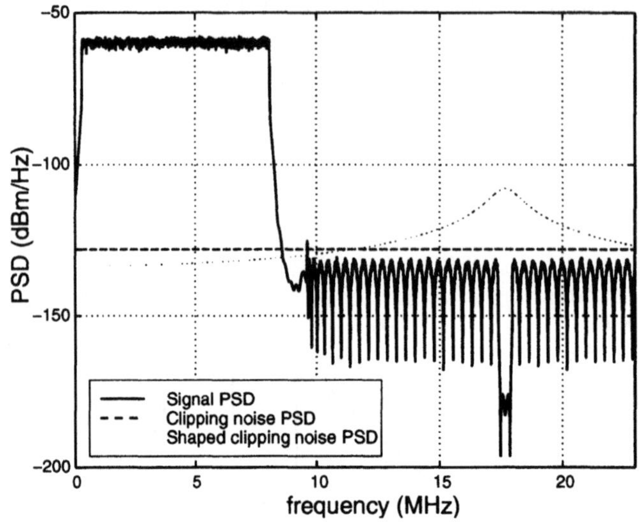

Figure 5. Power spectral density of the non-shaped clipping noise and the shaped clipping noise with bandpass filtering.

Case 2 - High-pass filtering - Setting $F(z) = 1$ (i.e. no feedback filter), the clipping noise is shaped by $[1-z^{-1}]$. Since $z = exp(j\omega t) = exp(j2\pi f/F_e)$ we obtain:

$$(1-z^{-1}) = 2j\sin\left(\frac{\pi f}{F_e}\right)e^{-j\pi\frac{f}{F_e}} \quad \text{or} \quad \left|1-z^{-1}\right| = 2\sin\left(\frac{\pi f}{F_e}\right) \qquad \text{Eq.(7)}$$

This function is depicted in Figure 6. This function is symmetric around $F_e/2$ with zero dc-component. Since the signal occupies a frequency band up to $F_e/4$, there is a significant increase of SNR at low frequency as can be seen from Figure 6. This effect can be further amplified by chosing an FIR filter such that:

$$[1-z^{-1}F(z)] = (1-z^{-1})^2 = 1-2z^{-1}+z^{-2} \qquad \text{Eq.(8)}$$

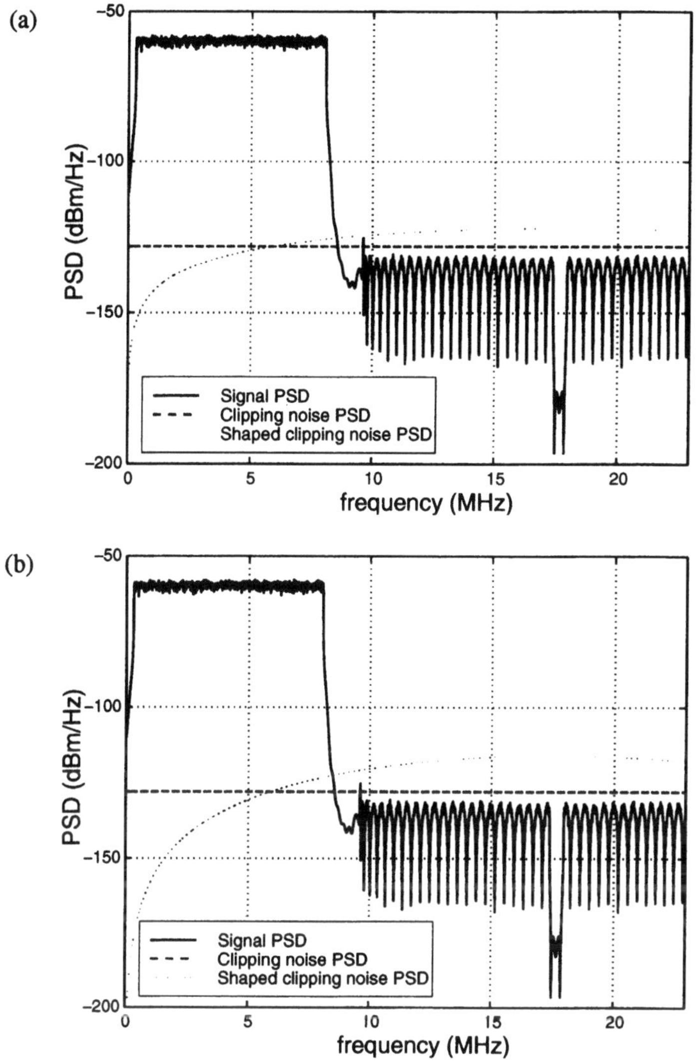

Figure 6. High-pass filtering. (a) 1st order filter ($F(z)$ = 1-tap FIR), (b) 2nd order filter ($F(z)$ = 2-tap FIR).

This is achieved with a 2 coefficient FIR filter whose z-transform is given by $F(z) = 2 - z^{-1}$. This function is also depicted in Figure 6 that shows larger attenuation at low frequencies and amplification at $F_e/2$ as compared to the previously discussed first-order high-pass filter.

Figure 7 shows the improvement of *SNR* within the 8.8 MHz signal bandwidth for the three filtering examples discussed above. It is seen that bandpass filtering provides *SNR* improvement over the entire signal bandwidth, while highpass filtering provides improvements only for low frequencies up to about 6 MHz. Above that point, the *SNR* degrades as compared with non-shaped clipping noise.

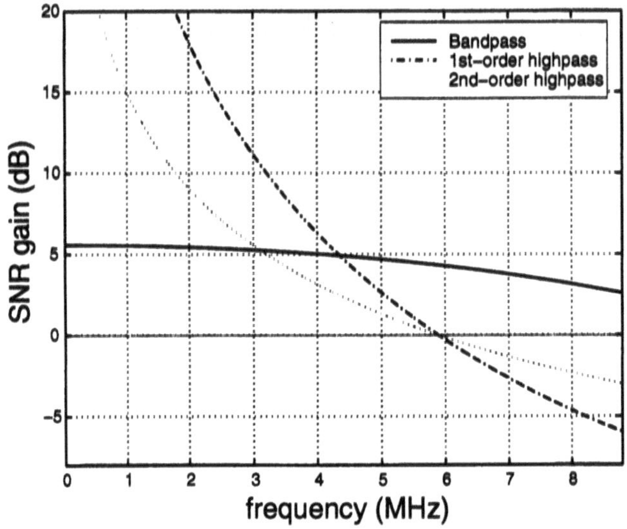

Figure 7. SNR improvement for different filters *F*(*z*).

The clip-noise shaper has been implemented within a 0.25 m CMOS IC for Zipper-VDSL applications. The size of the macroblock is 0.5 mm^2 and its power consumption is less than 50mW at 35MHz clock frequency.

IV Clipping Mitigation at the Receiver

The method proposed here for the receiver takes benefit from the statistical properties of the DMT signal that have been described in Section II. To explain the basic concept, let us assume that the receiver analog front-end uses a bank of *m* A/D converters, all with *n*-bit resolution, working in parallel on the same analog input signal. The input signal range of each A/D converter is twice as large as its lower range neighbor. At every sampling clock epoch, the output sample is selected from the lowest range A/D converter that is not indicating overflow. This way, small amplitude analog input will be sampled by the *n*-bit A/D converter that best fits the signal input range and hence provides the smallest quantization noise. Conversely, high amplitude analog input will be sampled by a higher input range n-bit A/D converter that will produce a higher quantization noise. However, since high amplitude analog inputs occur much less often than small ones, one can expect on average a smaller overall quantization noise, or equivalently a higher signal-to-noise ratio of the digitized signal.

This improvement in *SNR* can be quantified as follows. The quantization noise variance of the smallest input range n-bit A/D converter is given by [15]:

$$\sigma_q^2 = \left(2a_0/2^n\right)^2/12 \qquad \text{Eq.(9)}$$

where $\pm a_0$ is the input range of the A/D converter.

The larger input range A/D converters have higher quantization noise variance, because the same number of bits, n, is used for the digital representation of the samples. The average quantization noise variance over the m A/D converters is obtained by weighting the variance of each A/D converter according to the DMT signal amplitude probability. This gives:

$$\sigma_t^2 = \sigma_q^2 \cdot \left[P_0 + \sum_{k=1}^{m-1} (a_k/a_0)^2 (P_k - P_{k-1})\right] \qquad \text{Eq.(10)}$$

where a_k is the input range of the k:th A/D converter and $P_k = Prob\{|A(t)| \le a_k\} =$
$= Prob\{|x| \le a_0 2^k\}$ with $0 \le k \le m-1$.

Eq.(9) shows that since $a_k = a_0 2^k$, a single-range A/D converter with n bits and the range a_{m-1} has the quantization noise variance $2^{2(m-1)}\sigma_q^2$. Using this and Eq.(10), the improvement in *SNR* is directly obtained as:

$$SNR_m = 10log_{10}\left\{\frac{2^{2(m-1)}}{P_0 + \sum_{k=1}^{m-1} 2^{2k}(P_k - P_{k-1})}\right\} \qquad \text{Eq.(11)}$$

Figure 8 plots the *SNR* improvement SNR_m from Eq.(11). It is seen that the *SNR* improvement saturates for $m > 4$. This results simply from the fact that for $m > 4$ there is a very low probability that the input signal falls into the input range of the largest range A/D converters, so that their corresponding weighted factors in Eq.(10) become negligible.

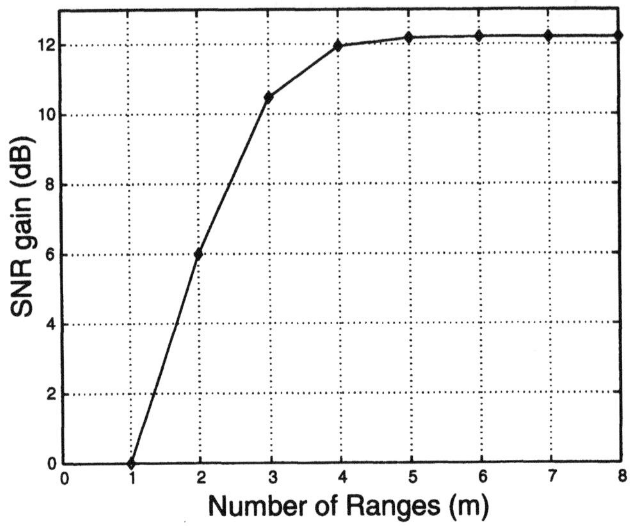

Figure 8. *SNR* improvement as a function of *m*.

The quality improvement can also be expressed as shown in Equation (11) as the number of extra bits that would be necessary for a single A/D converter having the largest input range.

$$\beta = \frac{SNR_m}{20\log_{10}(2)} \qquad \text{Eq.(12)}$$

A few examples have been analyzed using Eq.(12), assuming the maximum input signal excursion of $\pm 6\sigma$ (outside this range the signal is clipped). The results are summarized in Table 1.

Table 1:

m	a_0	β [bits]
2	3 σ	1
3	1.5 σ	1.7
4	0.75 σ	2

Straightforward implementation of the method would require a bank of *m* A/D converters, a range detection mechanism and a multiplexer for output selection. This is usually not feasible even for *m* as small as 4, which corresponds to the optimum. In order to better see how this is actually possible with current implementation of A/D converters, let us first consider Figure 9 that represents another view of the basic concept explained previously.

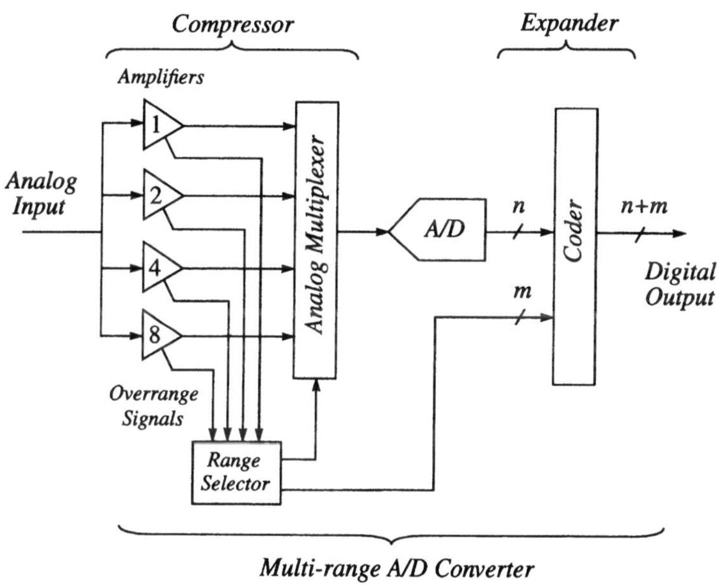

Figure 9. Schematic representation of a multi-range A/D converter.

There are 4 input amplifiers with gains 1, 2, 4 and 8 respectively. All amplifiers have the same input signal. The amplifiers are designed to saturate when the output signal exceeds the range of the A/D converter. This means that the higher the input signal amplitude is, more amplifier outputs are saturated. Saturation is indicated by a separate output from each amplifier.

The range selector is programmed to select the output of the highest gain amplifier that is not saturated and connect it to the A/D converter input. Thus, the smallest analog input signals are amplified by the gain = 8 amplifier before being digitized by the *n*-bit A/D converter. Medium range analog input signals are amplified by gain = 4 or gain = 2 amplifiers and the largest analog signals pass through the gain = 1 amplifier before being digitized by the *n*-bit converter. This way, analog signals at the input of the *n*-bit A/D converter cover the full input range of the converter.

Since the range selector provides information about the amplitude of the analog input, this information can be used by the coder at the digital output. The digital output can be seen as a floating-point representation with *n*-bit mantissa and an exponent determined by *m*. It can also be seen as a $(n+m)$-bit fixed-point representation with *n* true bits right or left shifted according to *m*.

As explained here, the basic concept is similar to the well-known PCM compander technique (*cf.* A-law and μ-law [15]) where compression is performed at the transmitter and expansion is performed at the receiver. The main difference here is that both compression and expansion are performed at the same location, *i.e.* at the receiver.

Modern principles for designing fast A/D converters open possibilities for actual implementation of the basic concept as explained above. As an example, fast A/D converters can be designed with a pipeline architecture that uses a number of similar pipelined stages labeled Stage 1, Stage 2, etc. as shown in Figure 10. All of the pipelined stages are similar in construction, consisting of a sample-and-hold amplifier (*SHA*), a digital-to-analog subconverter (*DASC*), an analog-to-digital subconverter (*ADSC*), a subtractor, and a multiply by 2^{mi} amplifier. The symbol m_i denotes the number of bits the *i*th stage of the pipeline resolves.

The input V_{IN} is first sampled-and-held and then digitized by the *ADSC* to provide the first m_0 most significant bits (MSB's), represented by d_0. This digital code d_0 is applied to the *DASC* to produce an analog voltage which is subtracted from the sampled-and-held input. The difference represents the residue and is amplified by 2^{m0} to scale it back to the full scale. This amplified residue, V_{RES1}, is passed to Stage 2 as an input.

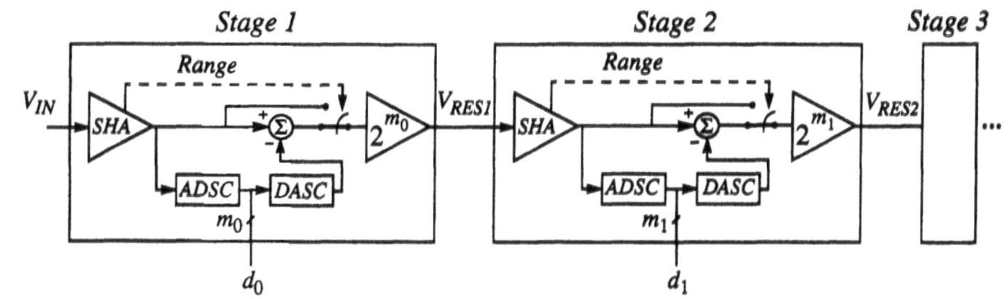

Figure 10. A pipeline A/D converter.

After performing a similar set of operations as described for Stage1, Stage 2 resolves the next m_1 MSB's. In this manner, as the input signal is processed by Stage 1, Stage 2 concurrently processes the residue signal V_{RES1} from the previous sample. The concurrency of operation by each stage in the pipeline allows the pipelined A/D converter to achieve high throughput, suitable for broadband communication applications.

In a pipelined architecture, the growth in hardware complexity is linear with the number of bits resolved. This linear dependency is in contrast with an exponential (2^N, where N is the number of resolved bits) dependency for a full flash architecture. The number of bits that are meaningful to resolve in a pipelined architecture depends mainly on the precision of the analog processing in each stage. The subtraction of the *DASC* outputs introduce errors that altogether must be smaller than the quantization step of the LSB.

With this technique it is possible to implement the presented receiver clipping mitigation method using a single A/D converter. In each stage the analog input signal level is checked. If this range check shows that the signal amplitude is small enough to be in the range of the next stage, the subtraction of the *DASC* output is bypassed, as shown in Figure 10, to avoid introducing any error. The following stages then work as if they were only preceded by amplifiers. The precision of the stages allows a specific number of bits to be extracted from the signal. For different signal amplitudes, this range of bits is thus shifted along the pipeline structure and a function similar to the converter shown in Figure 9 is achieved.

With advanced sub-micron CMOS process and power-optimized designs, the resulting increase in hardware size and power consumption can be kept at a reasonable level. As a last example, DMT-based VDSL transceivers would require

A/D converters with 12-bit resolution over the full analog input range from $-A_{clip}$ to $+A_{clip}$ with $\mu = 5$. With the presented technique one would need a 12 stage pipe-line A/D converter (assuming 1 bit per stage) with only 10-bit resolution technology.

V Summary

The Gaussian characteristics of the time-domain signals encountered in multicarrier communication systems, calls for methods to mitigate the effects of the rare high peak amplitudes that occur in both the transmitter and the receiver. It is very important for the system performance that the quantization noise level is not increased while a design for low clipping probability is made.

The clip-noise shaper, presented in Section III, reduces the noise caused by the clipping, by moving most of the noise power outside the spectrum of interest. This allows for the use of lower values of the crest-factor μ and thereby better utilizing the resolution of the transmitter D/A converter. The clip-noise shaper has the potential to provide 1 to 2 bits reduction of D/A converter requirements. The requirements on the line-driver are also relaxed by this, provided that the full bandwidth, including all the shaped clip-noise, is used. Limiting the bandwidth to the band of interest before the line-driver might create new peak amplitudes.

The method, presented in Section IV, enables the receiver to handle large peak amplitudes without sacrificing the performance for small signal amplitudes. The small signal amplitudes are still digitized using the smallest possible quantization steps that available technology allows, while larger amplitudes are quantized using larger steps. The signal probability density for this kind of signals, however, makes the total quantization error variance stay on a low level. This technique has the potential to increase the allowed input signal dynamic range by up to approximately 12 dB. This corresponds to 2 bits reduction of A/D converter resolution requirements.

References

[1] A. Riuz, J.M. Cioffi, and S. Kasturia, "Discrete multitone modulation with coset coding for the spectrally shaped channel", IEEE Transactions on Communications, 1991.

[2] G.dmt: Draft G.dmt Recommendation, Stephen Palm, Editor, ITU SG15/Q4 Contribution PO-07, Portland, OR, January 18, 1999. (ADSL)

[3] G.lite: Draft G.lite Recommendation, Chris Hansen, Editor, ITU SG15/Q4 Contribution PO-08, Portland, OR, January 18, 1999.

[4] ETS 300 744 ed.1, Digital Video Broadcasting (DVB); Framing structure, channel coding and modulation for digital terrestrial television (DVB-T), ETSI, Feb 1997.

[5] M. Aldinger, "Multicarrier COFDM scheme in high bit rate radio local area networks", ICCC Conf. on Wireless Computer Networks, September 1994.

[6] D. J. G. Mestdagh, "Calculation of ADC resolution for DMT modulation", IEE Electronics Letters, vol. 31, n. 16, pp. 1315-1316, 3rd August 1995.

[7] R. Gross et al., "SNR and spectral properties for a clipped DMT ADSL signal", Proceedings IEEE International Conference on Communications, ICC'94, pp. 843-847, May 1-5, 1994.

[8] J. H. Van Vleck and D. Middleton, "The spectrum of clipped noise", Proceedings of the IEEE, vol. 54, n. 1, pp. 2-19, January 1996.

[9] F. E. Bond and H. F. Meyer, "Intermodulation effects in limiter amplifier repeaters", IEEE Transactions on Communication Technology, vol. COM-18, n. 2, pp. 127-135, April 1970.

[10] J. E. Mazo, "Asymptotic distortion spectrum of clipped, dc-biased, Gaussian noise", IEEE Transactions on Communications, vol. 40, n. 8, pp. 1339-1344, August 1992.

[11] D. J. G. Mestdagh, P. Spruyt and B. Biran, "Effect of amplitude clipping in DMT-ADSL transceivers", IEE Electronics Letters, vol. 29, n. 15, pp. 1354-1355, July 1993. *See also* D. J. G. Mestdagh et al., "Analysis of clipping

effect in DMT-based ADSL systems", Proceedings IEEE International Conference on Communications, ICC'94, pp. 293-300, May 1-5, 1994.

[12] A. E. Jones, T. A. Wilkinson and S. K. Barton, "Block coding scheme for reduction of peak to mean envelope power ratio of multicarrier transmission schemes", IEE Electronics Letters, vol. 30, n. 25, pp. 2098-2099, December 1994, *See also* T. A. Wilkinson and A. E. Jones, "Minimization of the peak to envelope power ratio of multicarrier transmission schemes by block coding", IEEE 45th Vehicular Technology Conference, pp. 825-829, Chicago 1995.

[13] S. J. Sheperd, P. W. J. Van Eetvelt, C. W. Wyatt-Millington and S. K. Barton, "Simple coding scheme to reduce peak factor in QPSK multicarrier modulation", IEE Electronics Letters, vol. 31, n. 14, pp. 1131-1132, July 1995.

[14] D.J.G. Mestdagh and P. M. Spruyt, "A method to reduce the probability of clipping in DMT-based transceivers", IEEE Transactions on Communications, vol. 44, n. 10, pp. 1-5, October 1996.

[15] B. Sklar, Digital Communications: Fundamentals and Applications, Prentice-Hall, 1988.

Acknowledgement

The authors would like to thank Olivier Isson from STMicroelectronics and Freddy Balestro from CNET France Telecom for their support in the study of the clip-noise shaper.

SURVEY OF THE STATE OF THE ART ANALOG FRONT END CIRCUIT TECHNIQUES FOR ADSL

C. Dru Cabler
Senior Member of Technical Staff
Advanced Micro Devices, Communication Products Division,
Dresden, Germany

ABSTRACT

In the last several years, there has been increasing demand by consumers and small businesses for higher speed connectivity to the Internet. ADSL promises to fulfill the demand for faster data rates. However, in order to achieve the highest data rates for a given line, careful consideration must be given to the analog front end design such that it does not become the limiting factor. Based on recent publications, the fundamental techniques for building analog front ends are explored. Possibilities for future advancements in analog front ends are addressed in the conclusion.

1. INTRODUCTION

Like many higher speed communication systems using more complex modulation schemes, a typical analog front end for an ADSL system may consist of the following building blocks:

- D/A converter for the transmit path and reconstruction filter.
- A/D converter for the receive path and anti-alias filter.
- Transmit driver capability with programmable gain
- Receiver amplifier with programmable gain
- Auxiliary analog modules

This paper will first examine some of the original techniques that were used to build each of these building blocks. It will then discuss some recent techniques that are being employed in ADSL analog front end circuits. Finally, it will discuss some possible directions that future designs may take in order to reduce power and die size.

2. FIRST ATTEMPTS AT AN ANALOG FRONT END FOR ADSL

Before the various governing bodies had completed the standardization of full rate ADSL, at least one attempt was made at building an analog front end for DMT based ADSL. This first generation chip [1] has all of the building blocks previously described in Section 1.

Beginning with the receive path, the signal is first amplified by a low noise, variable gain preamplifier in order to maximize SNR. Then the signal is passed through two filters. An active-RC filter using a tunable capacitor is employed to reduce the echo signal [2]. According to [1], this is done not only to reduce the echo before A/D conversion, but to also prevent saturation of the subsequent bandpass filters.

Bandpass filtering is implemented using a Gm-C architecture to realize a 14th order Chebyshev filter. In order to improve the linearity of the transconductor, high gain with feedback along with linear resistors are used. Simulations indicate that 3rd order distortion is better than –65dB. An on-chip continuously operated tuning circuit is used to reduce filter frequency response variation due to process drift, power supply and temperature extremes. As compared more recent publications cited later in this paper, this is an extremely high order filter with more complexity than may be needed for ADSL. However, there will always be tradeoffs between A/D converter performance, amount of on-chip filtering, amount of off-chip filtering and analog echo canceling.

Finally, the A/D converter used is a 4th order delta-sigma modulator which is realized as a cascade of a single bit, 2nd order modulator with two single bit, 1st order modulators implemented using switched capacitor techniques, as described in [3,4]. An oversampling ratio of 32 is used in order to achieve 12 bit performance. However, there is a penalty with this technique. The opamp requires 10mA of bias current in order to realize acceptable performance levels in the A/D converter. A 5th order comb filter is used in the decimator to downsample the digital signal to the desired sample rate.

In the transmit path, on-chip interpolation filters are used to increase the input signal sample rate by a factor of 32. A single bit, 6th order digital delta-sigma modulator using multiple feedback is used to shape the quantization noise in-band to obtain more than 14 bits of theoretical SNR, as described in [5,6]. The actual SNR is limited to at least 12 bits by the kT/C noise of the following 1st order switched capacitor low pass filter. Next is a 2nd order continuous time, active-RC low pass Sallen & Key filter, which is used to further reduce out of band noise. It should be noted that using such a high order delta sigma converter results in an extremely large amount of out of band noise. Additional bandpass filtering using a similar Gm-C technique to that used in the transmitter path follows the Sallen & Key filter.

This early attempt at an analog front end was fabricated in a 0.7 micron CMOS process and required $65mm^2$. Power consumption was 1.9W with a single 5V power supply. Though the reported power consumption was rather high, reasonable performance was obtained. This chip represents a "proof of concept" -- it is possible to fully integrate an analog front end for ADSL, including decimation and interpolation, and achieve reasonable performance. As will be shown in a review of subsequent published works, substantial progress has been made toward reducing the die size and power consumption in the second generation chips.

3. RECENT PUBLICATIONS DESCRIBING ANALOG FRONT ENDS FOR ADSL

Recently, a number of publications have described some second generation analog front end circuits for ADSL. Common to many of these more recent attempts are the techniques for transmitter D/A conversion and receiver A/D conversion.

In [7], an analog front end chip is described. Included in this chip is a 12 bit current source output D/A converter which can operate at a 20 Msps rate. Typical performance for the entire transmit path is 72dB THD and 70dB SNR. An interesting feature of this D/A converter is that the full scale current is set by an external resistor. This allows reduced power consumption to be achieved at the expense of reduced THD performance.

Also in [7], is a 12 bit 10 Msps pipelined A/D converter. As in the transmitter path, typical performance for the entire receiver path is 72dB THD and 70dB SNR. No additional details are given in this reference with respect to the actual architecture of the A/D.

Other auxiliary circuits are also included in this device. They include a programmable transmit filter with two nominal 3dB frequencies such that it can operate in either the CO or CPE. Likewise, there exists a receive filter with nominal 3dB frequency of 4Mhz. There is also a low noise, programmable gain amplifier with 30dB program gain range and input referred noise of 12nV/√Hz. Additionally, there is a 7 bit D/A converter for driving a VCXO for timing recovery. Typical power consumption of 485mW is claimed in normal operating mode.

In [8], a BiCMOS analog front end for ADSL is described. In this approach, a 14 bit D/A converter is employed. The approach used is similar to that in [9]. Here, the 4 MSBs are thermometer coded and use unit weighted current sources. The 10 LSBs are implemented as an R-2R ladder. Reconstruction filtering is performed with a 3rd order active R-C Chebyshev filter. The filter is trimmed using NiCr links to trim the resistive portions of the filter. Multi-tone Power Ratio (MTPR) of 65dB is achieved for the transmit path.

The receiver section of this chip includes two input programmable gain amplifiers (PGA). The first PGA has 24dB of programmable range and input referred noise of 12nV/√Hz. The second PGA is programmable from –9 to +18dB. The two PGAs can be cascaded together directly or external filtering can be inserted between them to insure that clipping does not occur.

This particular solution is interesting for several reasons. It is in a BiCMOS process utilizing +/-5V supplies and thin film resistors. Nominal power consumption is 750mW and the die size is 25.8mm^2. No A/D converter is included on this implementation and no auxiliary circuits exist.

1999 brought to light three new ADSL analog front end circuits which were published at ISSCC. The first, [10], uses a D/A converter with a 12bit architecture with unit weighted current sources for the 8 MSBs and weighted current sources for the 4 MSBs. A unique switching scheme is employed to reduce the effects of 1st and 2nd order gradients. Measured DNL and INL are 0.75LSB and 1LSB respectively when operating at 8.832Msps.

The A/D converter in this reference is a pipelined architecture with 6 stages, each stage having 3 bits. Redundancy and digital error correction allows the A/D to achieve DNL and INL of 0.75LSB and 1.5LSB, respectively. Nominal power consumption of the A/D is 120mW when operating from a 3.3V supply.

The receive path contains an LNA with input referred noise of 12nV/√Hz and 31dB programmable gain range in order to compensate for channel losses up to 70dB. Power consumption of the LNA is 10mW. Two continuous time filters are used with cutoff frequencies of 138kHa and 1.104Mhz. These filters are continuously tuned R-MOSFET-C leapfrog topologies.

Auxiliary circuits include an 8 bit monotonic DAC for VCXO control and an on chip crystal driver. The part is fabricated in a 0.5um 2P3M CMOS process with nominal power consumption of 400mW. No laser trimming or fuses are utilized.

The next paper [11] is interesting for its performance. The D/A converter has 16 bit resolution and uses a switched-capacitor digitally-calibrated multistage architecture. Two separate transmit paths exist: one for the main transmit path and one for analog echo canceling with a single, shared D/A converter for both paths. The transmit path and echo path filters are 4th order Butterworth implementations with programmable 3dB frequencies. A programmable gain amplifier with 18dB of programmable gain range is the final stage in the transmit path.

The receiver path includes two stages of programmable gain amplification which results in a total programmable capability of –6dB to +38dB. Like the transmitter path, the receiver path also includes 4th order lowpass Butterworth filters with a programmable 3dB frequency. The A/D is a switched capacitor, digitally calibrated, multistage pipelined architecture with 16 bits of resolution. MTPR in excess of 81dB is achieved with the A/D in a loopback mode with the D/A.

Auxiliary circuits include a VCXO D/A converter and an energy detector to allow for ADSL wakeup. Filter calibration is performed at powerup. The part consumes 675mW with a single 5V supply in an 0.5um, 2P3M process. Total die size is 5.63mmx5.83mm.

The last DMT based ADSL chip is described in [12]. As in the first analog front end referenced [1], this chip is self-contained, including both interpolation and decimation. In the transmitter path, the input signal is upsampled to 8.832Msps (CO mode) or 4.416Msps (RT mode) before being presented to the D/A converter. The 14 bit D/A converters employ continuous calibration in order to achieve 0.5LSB static INL and 0.2LSB static DNL. An active R-C filter is used to remove signal images. A programmable gain amplifier with 24dB of programmable gain is the last stage in the transmitter path. A signal chain identical to the transmitter path also exists for echo canceling.

The receiver utilizes two digitally programmable gain amplifiers with an Equalizer when configured for RT mode. The 14 bit A/D operates at 4.416Msps, and is a fully differential 1.5 bit/stage pipelined architecture described in [13]. In order to achieve required performance, the first 5 stages are trimmed using fused polysilicon links.

Power consumption in the CO is 600mW while RT mode is 760mW with a 3.3V power supply. The part was fabricated in a CMOS process.

4. POSSIBLE FUTURE TRENDS IN ANALOG FRONT ENDS FOR ADSL

Perhaps one of the most interesting aspects for near term deployment of ADSL is the interest in less expensive, lower data rate, so-called splitterless DSL. This leads to considerable interest in lowering the overall power and cost of analog front ends that support lower bandwidth ADSL. As can be seen in the previous section, the analog architectures being implemented for D/A conversion, A/D conversion and filtering are much more alike than they are different. Typical D/A conversion techniques involve current sources that may be a mixture of thermometer coding and binary weighting, with only a small amount of oversampling, typically on the order of 4X. Likewise, A/D conversion techniques in recent publications typically employ pipelined methods with the same amount of oversampling as the D/A path. On chip filtering typically consists of continuous time, active filters employing some type of tuning technique in either a continuous manner [1,10], automatic one-time tuning at powerup [11], or fuse link trimming [12].

Future circuits may employ methods that are currently implemented in the voiceband and audio technologies. As seen in the first generation paper [1], early attempts employed single bit delta sigma conversion schemes. The penalty is that since the signal can change from positive to negative full scale in a single sample, extreme slew rates and, consequently, high power consumption are required. Additionally, because of the higher order modulators employed in this system, extreme amounts of filtering are required to remove the out of band quantization noise generated by the delta sigma process. These issues have been addressed in voiceband and audio circuits with multibit systems.

Much attention has recently focused on multibit delta sigma techniques using data weighted averaging techniques to improve the effective linearity of the D/A converters that are employed in both A/D and D/A conversion [14]. With these techniques, a reasonable tradeoff can be made with delta sigma order, number of bits, oversampling ratio and, ultimately, area and power. It is not unreasonable to expect oversampling ratios of 16x or even 32x to be used in an ADSL analog front end with a lower order, multibit delta sigma modulator. The use of higher oversampling ratios and larger number of bits would may also ease both the on and off chip filtering requirements which would result in both a power and cost savings. By having the flexibility to increase the number of effective bits in the A/D converter, one can trade off on-chip filtering, off chip filtering, and number of bits used analog echo cancellation in order to optimize the overall cost and power consumption.

Additional power consumption reduction in oversampled delta sigma A/D conversion may be also be achieved by the use of continuous time delta sigma techniques. Combining this technique with a multibit quantizer and D/A converter and including data weighted averaging may prove to be a viable option for ADSL, assuming that jitter problems can be avoided.

One example of a candidate for further study is given in [15]. In this paper, a continuous time, 5 bit delta-sigma A/D conversion with 10X oversampling for data acquisition is described. Quantization noise in a 500kHz bandwidth is –64dB and a third harmonic distortion of –70dB is achieved. Of particular interest is the fact that the analog portion of this chip consumes 7.2mW in a standard 0.5 micron digital CMOS process.

A second example of a DSL-type implementation is given in [16]. Here, a 6 bit plus sign D/A converter employing data weighted averaging is employed along with a 2nd order digital sigma delta noise shaper. This is implemented in an 0.5um CMOS technology that includes two layers of poly.

5. SUMMARY

Recently, a number of publications have appeared describing second generation analog front end devices for ADSL. Consumer demand for increasing data rates when connecting to the Internet will drive silicon suppliers to continue to lower costs for the physical front end devices. Cost of supplying power and its accompanying cooling requirements will drive silicon suppliers to continue to lower power for these same sorts of devices located in the Central Office end of the network. It is clear that there is much additional work that can be done to lower the cost and power consumption of these devices. The knowledge gained by this exercise will undoubtedly be employed as newer technologies become available such as VDSL.

REFERENCES

[1] Chang, Z.Y., et al. "A CMOS Analog Front-End Circuit for an FDM-Based ADSL System," *IEEE J. Solid-State Circuits*, vol. 30, no. 12, pp.1449-1456, Dec. 1995.

[2] Durham, A.M. and Redham-White, W., "Integrated continuous-time balanced filter for 16-bit DSP interfaces," *IEEE J. Solid-State Circuits,* vol. 28, pp. 835-839, Jul. 1993.

[3] Yin, G. and Sansen, W., "A high frequency and high resolution fourth order Sigma Delta A/D converter in BiCMOS technology," *IEEE J. Solid-State Circuits,* vol. 29, pp 857-865, Aug. 1994.

[4] Cabler, C.D, "Fourth-Order Cascaded Sigma-Delta Modulator", US Patent 5414424.

[5] Ritoniemi, T., et al., "The design of stable high order 1-bit sigma-delta modulators," *Proc. IEEE ISCAS,* 1990, pp. 3267-3270.

[6] Lee, W.L., and Sodini, C.G., "A topology for high order interpolative coders," *Proc. IEEE ICCAS,* 1987, pp. 459-462.

[7] Analog Devices, Inc. "AD6437, Analog Front End for ADSL".

[8] Lanford, D.S, et al., "A BiCMOS Analog Front-End Circuit for an FDM-Based ADSL System," *IEEE J. Solid State Circuits,* vol. 33, no. 9, Sep. 1998, pp. 1383-1393.

[9] Tesch, B.J. and Garcia, J.C., "A low glitch 14 bit 100Mhz D/A converter," *IEEE J. Solid State Circuits,* vol. 32, Sep. 1997, pp. 1465-1469.

[10] Cornil, J.P., et al., "CODEC for Echo-Cancelling, Full-Rate ADSL Modems," *ISSCC Digest of Technical Papers*, San Francisco, CA, Feb. 1999, pp. 242-243.

[11] Conroy, C., et al., "A CMOS Analog Front-End IC for DMT ADSL," *ISSCC Digest of Technical Papers,* San Francisco, CA, Feb. 1999, pp. 240-241.

[12] Hester, R., et al., "CODEC for Echo-Canceling, Full-Rate ADSL Modems," *ISSCC Digest of Technical Papers*, San Francisco, CA, Feb. 1999, pp. 242-243.

[13] Lewis, S., et al., "A 10-b 20Msample/s Analog-to-Digital Converter," *IEEE J. Solid State Circuits,* vol. 27, No. 3, March, 1992, pp. 351-358.

[14] Baird, R.T. and Fiez, T., "Linearity Enhancement of Mutability SD A/D and D/A Converters Using DWA," *IEEE Transactions on Circuits and Systems,* Vol. 42, No. 12, Dec. 1995, pp. 753-762.

[15] van der Zwan, E.J., et al., "A 13mW 500kHz Data Acquisition IC with a 4.5 Digit DC and 0.02% Accurate True-RMS Extraction,", *ISSCC Digest of Technical Papers,* San Francisco, CA, Feb. 1999, pp. 398-399.

[16] Moyal, M. et al., "A 25kft 768kb/s CMOS Transceiver for Multiple Bit-Rate DSL", *ISSCC Digest of Technical Papers,* San Francisco, CA, Feb. 1999, pp. 398-399.

RF CMOS Modelling

Introduction

Modeling MOST transistors has once more gained importance because of the inaccuracy of the models which include all three regions i.e. the weak- and strong-inversion region and the velocity saturation region. Moreover the increasing success of MOST technologies for RF applications , has evidenced the need for more accurate models at very high frequencies, in the GHz region. This session provides the state-of-the-art in RF MOST models. It compares the existing BSIM3v3, Philips' Model 9 and the EKV model with respect to their capability to accurately predict GHz performance with submicron CMOST technologies.

The first presentation, given by Luuk Tiemeijer of Philips, Eindhoven, gives an overview of all effects that come in at high frequencies. The non-quasi-static effect is detailed in addition to model parameters such as gate resistance. etc. The resultant noise and distortion are specified as well.

It is followed by a presentation by Kristin De Meyer of IMEC, Leuven, on the extraction of MOST parameters which play an important role at high frequencies. They are mainly the effective channel length and the source resistance. Data is given for channel lengths down to 0.08 μm.

Jeffrey Ou of UCBerkeley then the recent additions to the BSIM3v3 model to enhance its accuracy at GHz frequencies. Again a series gate resistance has been added to improve performance up to 3 GHz. The noise characteristics have been verified as well.

The fourth paper is on the EKV model as explained by Christian Enz of CSEM, Neuchatel. This model is known for its very smooth transition from the weak-inversion region to the strong-inversion region. Now parasitic capacitances and resistances have been added to provide excellent fits of the Y parameters up to 10 GHz and noise up to 6 GHz.

A purely microwave approach is offered by Renaud Gillon of Alcatel Microelectronics, Oudenaarde, A new s-parameter de-embedding strategy up to 20 GHz is applied to both bulk and SOI MOSFET modelling.

Finally Tajinder Manku of University Waterloo, provides a new model to explain the noise behavior of submicron MOSTs. Scaling rules are added, leading to optimum finger widths for ultra-low-noise applications such as LNA's, etc.

It can be concluded from these presentations that it has become quite feasible to model a MOST at very high frequencies, giving rise to an increased use of MOST technologies in RF applications.

Willy Sansen

RF CMOS Modelling

L.F. Tiemeijer, L.M.F. de Maaijer[1], R. van Langevelde
A.J. Scholten, and D.B.M. Klaassen
Philips Research Laboratories, Eindhoven, The Netherlands

ABSTRACT

RF application of CMOS requires accurate modelling of transistor properties, such as impedance levels, power gain, noise, distortion, substrate parasitics and non-quasi-static effects. Apart from noise, until now relative little attention has been paid to these properties in compact models for circuit simulation. Here clear experimental results will be used to illustrate the accuracy and limitations of state-of-the-art compact models. Wherever possible, improvements or macro models that can be used for circuit simulation will be presented.

1 Introduction

The application of CMOS in circuits such as RF low noise amplifiers, mixers, local oscillators, and IF circuitry is becoming more and more attractive due to the ongoing increase in the RF performance of mainstream deep-submicron CMOS processes.

These processes are provided by foundries, which commonly supply the parameters for public-domain compact MOS models, which are suitable for analog circuit design: the BSIM3v3 model from UC-Berkeley and the Philips compact MOS model, MOS Model 9 [1, 2, 3]. The latter has been introduced within Philips in 1990 and became available in the public domain in 1993. Many of its features and capabilities have been elucidated in publications (see e.g. [4] to [6]), while the derivation of many equations and the underlying physical mechanisms are described in [7]. In contrast to the well known BSIM3v3 model, the charge model of MOS Model 9 has a

[1]L.M.F. de Maaijer is presently with NOKIA Telecommunications Düsseldorf

bias-dependent charge partitioning between source and drain. All intrinsic MOSFET capacitances are derived from this charge model.
In this contribution the specific requirements put forward for adequate RF design will be discussed based on a few key performance figures which are essential to evaluate the building blocks of any RF-CMOS receiver architecture. These figures are power gain, noise figure, and intermodulation distortion. This will be completed by a discussion of $1/f$ noise, which for CMOS provides the dominating contribution to local oscillator phase noise, and the transistor input and output impedances, which must be accurately modelled in order to allow proper interconnection between the different building blocks. The design of the receiver building blocks is not the topic of this paper, instead the focus is on the modelling of the individual MOSFETs which make up these building blocks.
In the following sections first impedance levels and power gain will be discussed based on high-frequency S-parameter measurements of deep-submicron MOSFETs. The admittance or Y-parameters are introduced, to allow for an easy comparison of conductances and capacitances to simulations using a compact model. In this comparison the importance of including various transistor parasitics will be discussed in detail. Furthermore it will be shown that non-quasi-static modelling is required in order to accurately calculate power gain and input impedance levels. In the next section the modelling of the thermal noise and the $1/f$ noise is discussed and compared to experimental data. In a final section static linearity and RF distortion are discussed, based upon a recently published model, which provides a very accurate description of the higher order derivatives of the transconductance and output conductance. Throughout this paper experimental data will be denoted by symbols and lines will represent simulations.

2 Impedance levels and power gain

To evaluate the device input and output impedance levels and power gain at RF the small-signal scattering (S) parameters are used. These two-port S-parameters have been measured with an HP8510 Network Analyser using air coplanar high-frequency probes in ground-signal-ground configuration. Special RF test structures have been designed with the MOSFETs in common source-substrate configuration. In some of these structures the substrate was contacted directly beneath the source area. The S-parameter measurements have been corrected for the interconnect parasitics using

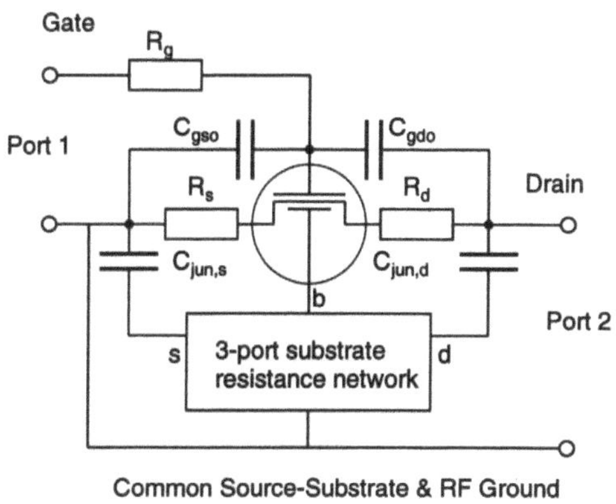

Figure 1: *Equivalent circuit for the parasitic resistances and capacitances of a typical RF MOSFET in common source bulk configuration.*

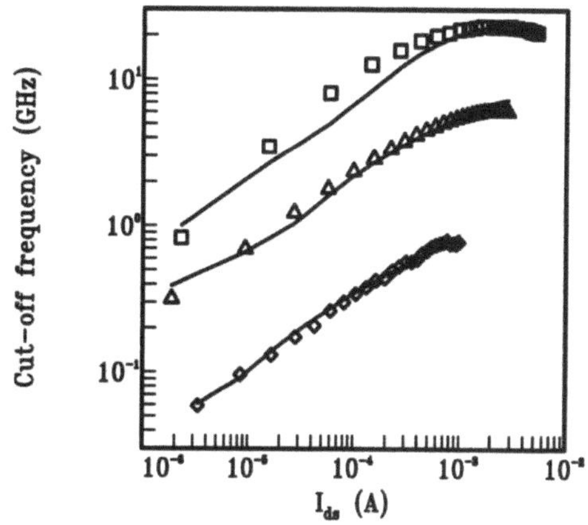

Figure 2: *Extrapolated cut-off frequency versus drain current for a 10/4 (diamonds), a 10/1 (triangles) and a 10/0.35 (squares) NMOS device*[2]. *Lines: Simulations using MOS Model 9.*

"open" and "short" dummy structures [8] and converted to two-port admittance (Y) parameters. After this correction the MOSFET device is best described by the equivalent circuit shown in fig. 1. The source and drain resistances R_s and R_d are usually accounted for in the compact models for the DC currents, for instance by the mobility reduction parameters. Since they are not part of the intrinsic device, they will be seen in the input impedance, and correct RF modelling is only possible when these resistances are explicitly accounted for externally. For the simulations we therefore used MOS Model 9 with parameters describing the device with external series resistances. As an example of what this can do for RF, measured and simulated cut-off frequencies are compared in fig. 2 for three different device lengths versus drain current.

2.1 Substrate resistance and the output impedance

At high frequencies the MOSFET output impedance is determined by junction capacitances and the substrate loss resistance network. For the junction capacitances the Philips JUNCAP model has been used [2], while for the substrate resistance network a description based on the three-port substrate impedance (Zs_{ij}) parameter matrix is introduced. The need for a non-zero Zs_{dd} is illustrated in fig. 3. This substrate resistance component

[2]We use the notation 10/4 to indicate $W = 10\ \mu$m and $L = 4\ \mu$m.

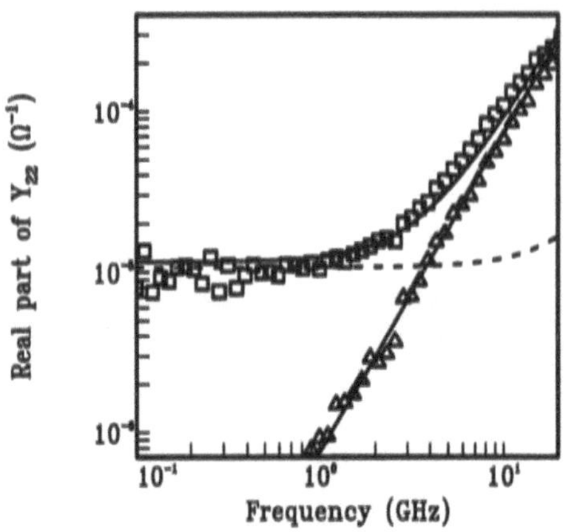

Figure 3: *Real part of the output admittance Y_{22} at V_{gs} = 0V (triangles) and V_{gs} = 1.5V (squares) for a 10/1 NMOS device. Solid line: $Zs_{dd} = 320\Omega$, Dashed line: $Zs_{dd} = 0\Omega$.*

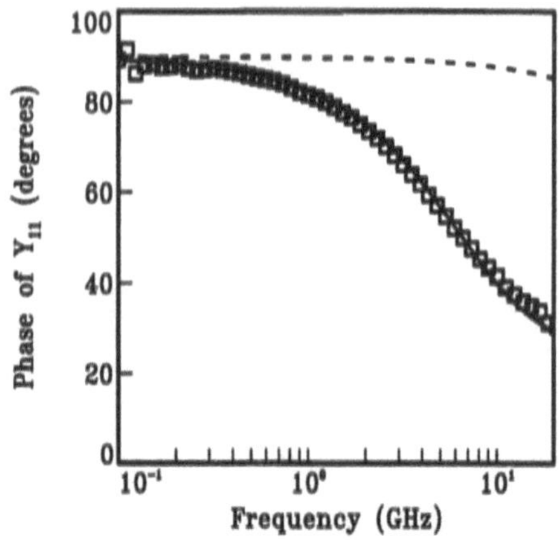

Figure 4: *Phase of the input admittance Y_{11} at V_{gs} = 0V for a 10/10 NMOS device. Solid line: $Zs_{bb} = 200\Omega$, Dashed line: $Zs_{bb} = 0\Omega$.*

is required to explain the increase in the real part of the output admittance with frequency, which will limit power gain at high frequencies. In the off-state there is no inversion layer and the input capacitance is dominated by the gate-bulk capacitance. Therefore the input impedance is also influenced by the substrate loss resistance network. This is illustrated in fig. 4, which shows the phase of the input admittance y_{11}. A non-zero substrate resistance component Zs_{bb} is needed to correctly explain the dissipative behaviour at high frequency. As shown in fig. 5 the substrate resistance network also causes an increase in the feedback at high frequencies due to a non-zero Zs_{bd} and Zs_{db}. When the Zs_{dd} values extracted for devices where the p-well and the substrate are contacted directly beneath the source area are compared to those reported previously for a standard device [9] it is found that the substrate resistance has been significantly reduced as shown in fig. 6. It thus shows that when the device cannot be used in a common source-substrate configuration, one has to accept a somewhat reduced device performance due to larger parasitic substrate resistances.

2.2 Gate resistance and input impedance

At high frequencies the MOSFET input impedance can no longer be regarded as purely capacitive, and instead the effective gate, source and substrate resistances must be accounted for. The effective gate resistance R_g is calculated from the sheet resistance R_{poly} of the silicidized poly silicon

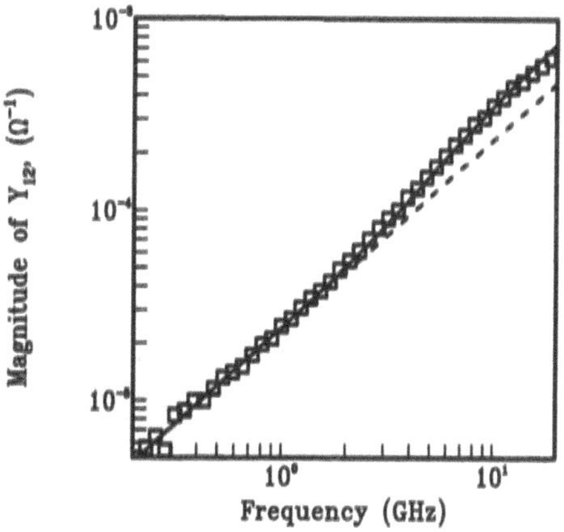

Figure 5: *Magnitude of the feedback admittance Y_{12} at V_{gs} = 0V for a 10/10 NMOS device. Solid line: $Zs_{db} = Zs_{bd} = 60\Omega$, Dashed line: $Zs_{db} = Zs_{bd} = 0\Omega$.*

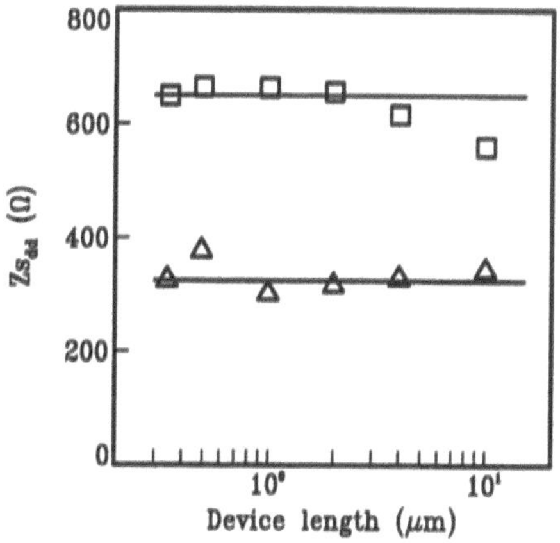

Figure 6: *Summary of Zs_{dd} values extracted for standardW = 20μm NMOS devices [9] (squares) and W = 10μm NMOS devices with an extra p-well contact directly beneath the source (triangles).*

gate and the device layout using:

$$R_g = \frac{1}{N_{gc}FM}\left[\frac{W_{\mathrm{E}}}{3N_{gc}F} + W_{ext}\right]\frac{R_{poly}}{L_{\mathrm{E}}} \tag{1}$$

Here N_{gc} equals either one or two depending on whether the gate is single-sided or double-sided contacted, F and M account for folding and multiplying operations, W_{E} and L_{E} are the effective device width and length respectively, and W_{ext} is the distance from where the connection to the gate-poly is made to the beginning of the active area.

With the parasitic capacitances and the gate and substrate resistances properly accounted for, one might expect that accurate modelling of impedance levels and power gain is within reach. However there are still some unsolved issues, as shown in figs. 7 and 8. The first figure shows the input series resistances which was extracted for various channel lengths ranging from 0.35μm to 10μm versus transconductance g_m. Here it appears that the gate and source resistances cannot account for the observed input series resistances. The second figure shows the maximum available power gain measured for 3 devices biased in saturation. As can be seen the simulation is way off, and the maximum oscillation frequency f_{max} is grossly overestimated. These deviations are known limitations of the quasi-static model used here, which can be overcome by using non-quasi-static modelling. This is the topic of the next subsection.

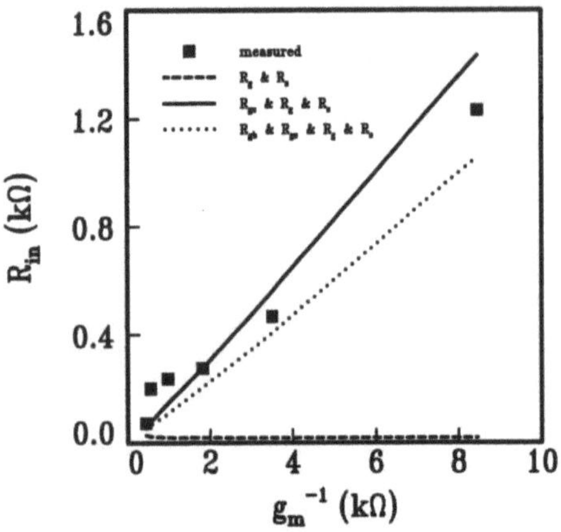

Figure 7: *Extracted input series resistances for NMOS devices with channel lengths ranging from 10/10 down to 10/0.35. Solid line: non-quasi-static model, Dashed line: quasi-static model, Dotted line: non-quasi-static model with R_{gb}.*

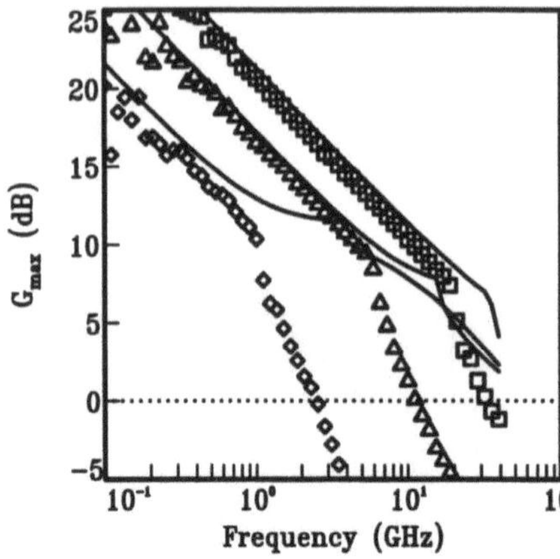

Figure 8: *Maximum available power gain versus frequency for a 10/4 (diamonds), a 10/1 (triangles) and a 10/0.35 (squares) NMOS device, biased into saturation at V_{gs} = 1.5V and V_{ds} = 3.3V. Solid line: quasi-static model.*

2.3 Non-Quasi-Static modelling

Popular MOS models like MOS Model 9 and BSIM3v3 show their limitations when it comes to RF power gain modelling [10, 11, 12]. These limitations stem from the underlying quasi-static assumption where the charges associated with the various device terminals are assumed to be identical to those in the corresponding steady state. However when the terminal voltages change on a time scale which is comparable to the channel transit time Non-Quasi-Static (NQS) effects can be observed. A straightforward approach to deal with these effects is to divide the MOSFET in several short sections such that the transit time for each section remains sufficiently short. The non-quasi-static models proposed in the literature [13, 14] take a more general analytical approach by solving the transmission line equations for a distributed MOSFET and result in exact but very complicated expressions for the admittance parameters. However, with suitable simplifications they can be used to construct computationally highly efficient and accurate models for RF simulations [15].

To illustrate this we note that in a quasi-static model the four-port MOSFET admittance parameters are obtained from the DC current and terminal charges by:

$$y_{ij} = \frac{\partial I_i}{\partial V_j} + \jmath\omega \frac{\partial Q_i}{\partial V_j} \tag{2}$$

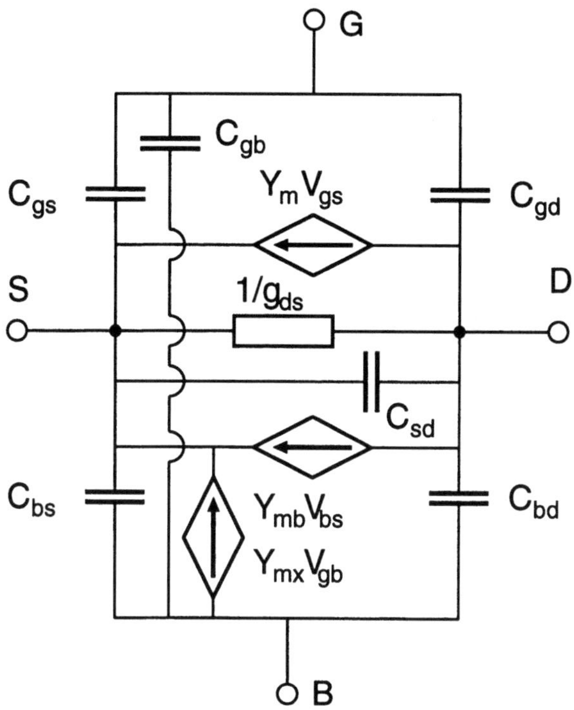

Figure 9: *Equivalent circuit of the MOS Model 9 quasi-static model in its most general form. In the representation chosen here the number of controlled current sources is reduced to the minimum.*

where the first term represents the (trans)conductances g_m, g_{mb}, and g_{ds}, and the second term the (trans)capacitances C_{ij}. Only 9 out of the 16 elements of the admittance matrix can be defined independently. This redundancy allows the definition of several different equivalent circuits, of which the one displayed in fig. 9 is the most suitable to illustrate the need for non-quasi-static modelling. In this circuit the three complex voltage controlled current sources are:

$$y_m = y_{dg} - y_{gd} = g_m + \jmath\omega(C_{gd} - C_{dg}) \quad (3)$$

$$y_{mb} = y_{db} - y_{bd} = g_{mb} + \jmath\omega(C_{bd} - C_{db}) \quad (4)$$

$$y_{mx} = y_{bg} - y_{gb} = \jmath\omega(C_{gb} - C_{bg}) \quad (5)$$

Examining the equivalent circuit of the quasi-static model reveals three serious problems which result in unphysical behaviour:

- The input impedance at the gate and the bulk terminals is purely capacitive. Therefore no power is dissipated in driving the device, yielding an infinite power gain.

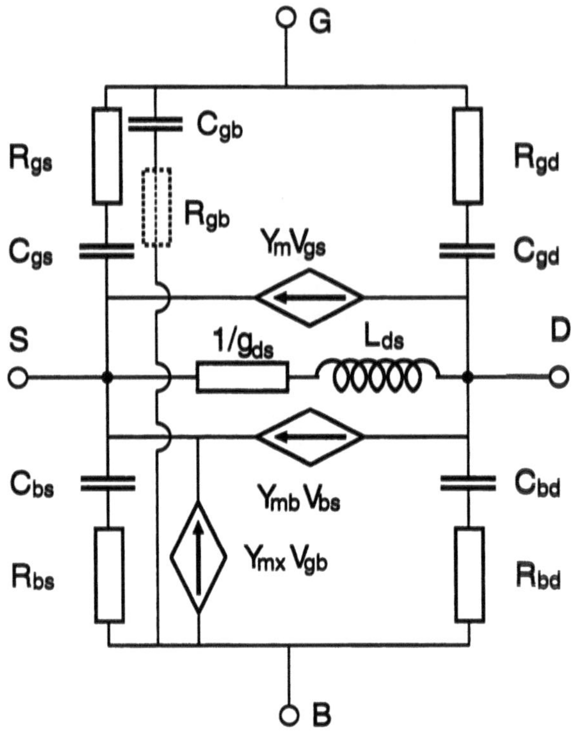

Figure 10: *Equivalent circuit of the Tsividis first order non-quasi-static model [15].*

- The transconductances increase with frequency, which is highly unlikely to be based on reality, in view of the finite time taken by the carriers to travel through the active channel;
- The drain to source impedance decreases with frequency, which for similar reasons is highly unlikely to occur in reality;

These problems are solved by the first order non-quasi-static model described by Tsividis [15], and which is shown in fig. 10. Here two complex voltage controlled current sources are modified into:

$$y_m = \frac{g_m}{1 + j\omega\tau_1} \tag{6}$$

$$y_{mb} = \frac{g_{mb}}{1 + j\omega\tau_2} \tag{7}$$

Comparing this model to the quasi-static case it can be observed that:

- The branch capacitances to gate and bulk are replaced by a series connection of a capacitor and a resistor, thus solving the input power dissipation issue;

- The effective transconductances contain terms that result in a decrease with frequency. The time constants τ reflect the channel transit time;
- The drain to source impedance now becomes a resistance in series with an inductance;

By calculating the values of the 5 branch resistances from the momentary values of g_m, g_{ds}, and g_{mb} following Tsividis [15] and requesting the low frequency limit of the non-quasi-static admittance parameters to be identical to the quasi-static case, yielding:

$$Y_m = \frac{g_m}{1 + \jmath\omega(C_{dg} - C_{gd})/g_m} \quad (8)$$

$$Y_{mb} = \frac{g_{mb}}{1 + \jmath\omega(C_{db} - C_{bd})/g_{mb}} \quad (9)$$

such a non-quasi-static model can be implemented without the need for additional parameters. The expressions for the bias dependent internal resistances are fairly lengthy in their general form and will not presented here. We will restrict ourselves to a short discussion of their impact on the input series resistance of fig. 7 given by:

$$R_i = Re(Y_{11}^{-1}) \quad (10)$$

In deep saturation the main contributions to this resistance come from R_{gs}, which in saturation is equal to:

$$R_{gs} = \frac{1}{5g_m} \quad (11)$$

and from R_{gb}, which in saturation has a negative value equal to:

$$R_{gb} = -\frac{4(g_m + g_{mb})}{5g_m g_{mb}} \quad (12)$$

At the gate terminal these resistances transform into the internal resistance shown in fig. 11. Including all intrinsic non-quasi-static and external parasitic resistances the dotted line of fig. 7 is obtained which slightly underestimates the actual input resistances. Including them all except for the negative R_{gb} the solid line of fig. 7 is obtained which appears to slightly overestimate the input resistances. However, on both cases the situation has improved considerably compared to the quasi-static case represented by the dashed line. In the off-state the transconductances become zero and the non-quasi-static resistances approach infinity. To avoid erroneous off-state input impedances R_{gb} is put to zero at all bias settings in the final non-quasi-static model.

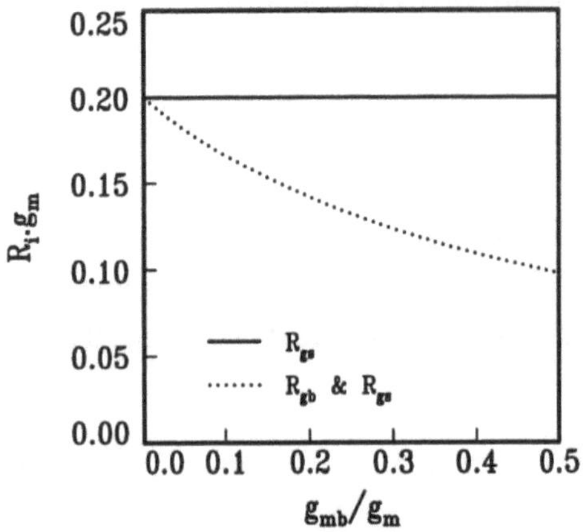

Figure 11: *Normalised internal resistance R_i in saturation calculated using R_{gs} (solid line), and R_{gs} and R_{gb} (dotted line). R_{gd} is not seen in saturation. In the linear regime one finds $R_i = R_{gs} \parallel R_{gd} = 1/12g_{ds}$. R_{gb} is not seen in this regime.*

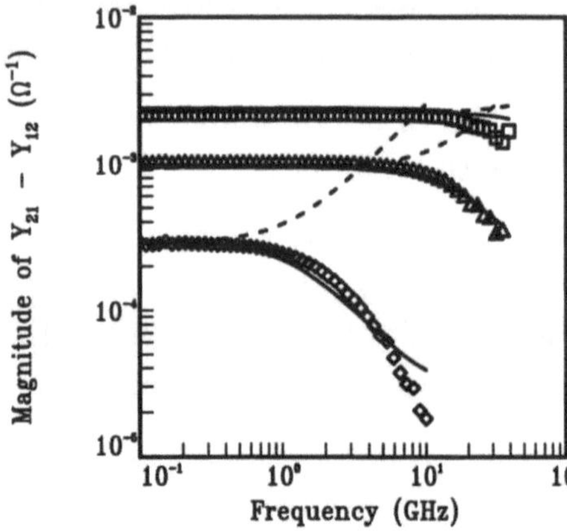

Figure 12: *Decrease in magnitude of the transconductance with frequency for a 10/4 (diamonds), a 10/1 (triangles) and a 10/0.35 (squares) NMOS device. Solid line: non-quasi-static model, Dashed line: quasi-static model.*

2.4 Power gain and f_{max}

As indicated in the previous sections a quasi-static model cannot predict f_{max}, and therefore a comparison of measured and simulated maximum oscillation frequencies f_{max} is a very relevant benchmark for any non-quasi-static compact RF MOSFET model. The same comparison also provides a good benchmark test for the accuracy of the S-parameter measurements. Recalling that f_{max} is the frequency where the maximum available power gain given by

$$G_{max} = \frac{|y_{21}|^2}{4(Re(y_{11})Re(y_{22}) - 0.5Re(y_{21}y_{12}))}, \tag{13}$$

becomes unity, it is immediately seen that f_{max} can only be obtained correctly when both the decrease in magnitude of the transconductance with frequency, and the real parts of the input and output conductances are accurately modelled.

We will first review these individual contributions. The measured decrease in magnitude of the transconductance with frequency for various channel lengths is shown in fig. 12. Clearly this is only modelled correctly by the NQS model. For the real part of the input admittance we recall that only for the non-quasi-static case, where both the internal resistance R_i and the effective gate and source resistances are accounted for, the simulated input series resistance agrees with the value which can be extracted from the in-

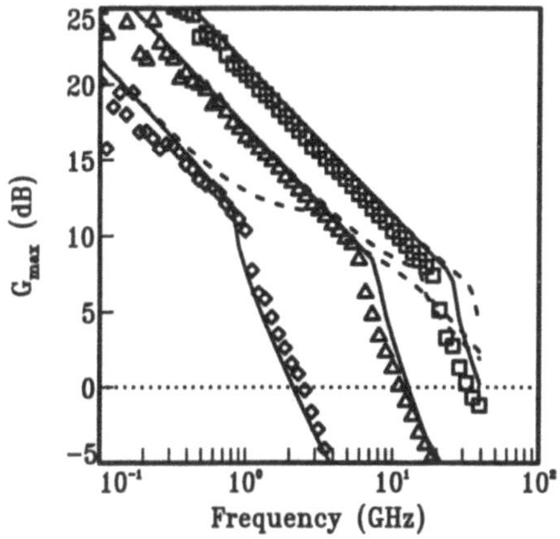

Figure 13: *Maximum available power gain versus frequency for a 10/4 (diamonds), a 10/1 (triangles) and a 10/0.35 (squares) NMOS device. Solid line: non-quasi-static model, Dashed line: quasi-static model.*

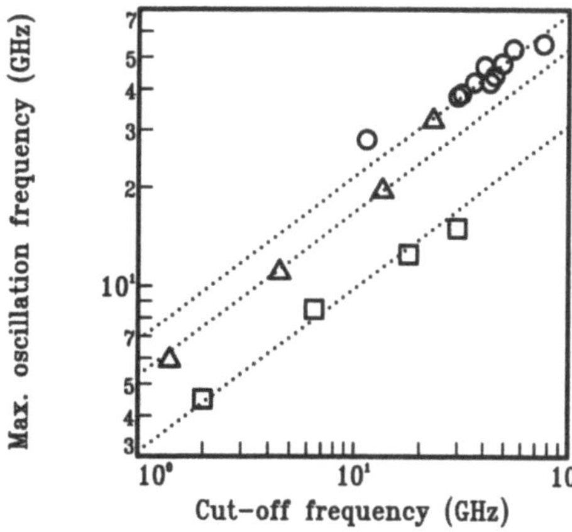

Figure 14: *Maximum oscillation frequency versus cut-off frequency for various layouts and processes (see text). The dashed lines represent a square-root relation ship between* f_{max} *and* f_t.

put admittance parameters(fig. 7). The real part of the output admittance is dominated by the substrate parasitics. As shown before in fig. 3 accurate modelling requires a correct description of the junction capacitance and Zs_{dd}.

With all these ingredients in place, accurate modelling of the maximum stable gain and maximum available power gain is now possible as illustrated in fig. 13, which redisplays the results of fig. 8 but now with the results of the non-quasi-static model included. As can be seen power gain and the maximum oscillation frequency f_{max} are modelled correctly.

Although non-quasi-static effects have to be included, for the minimum length devices f_{max} is still to a large extent limited by the parasitics [9]:

$$f_{max} \approx \sqrt{\frac{f_t}{4\pi C_{jun,d}\sqrt{(R_g + R_i + R_s)Zs_{dd}}}} \tag{14}$$

with the cut-off frequency f_t given by:

$$f_t \approx \frac{g_m}{2\pi C_{in}} \tag{15}$$

Using these formulae as a guideline the advantages of layouts which reduce the substrate resistance Zs_{dd} and drain junction capacitance $C_{jun,d}$ with respect to high f_{max} are clear, although they should preferably be realized without increasing the total input capacitance C_{in}. This is illustrated in fig. 14, which shows a comparison of maximum oscillation frequency versus cut-off frequency for standard NMOS devices (squares) and devices with improved gate and substrate contact (triangles), for gate lengths of 2, 1, 0.5 and 0.35 μm, respectively, and results reported for NMOS multi-finger structures (circles) with gate lengths down to 0.13 μm. The dashed lines illustrate the square-root relation ship between f_{max} and f_t, which is obtained when the power gain is fully limited by the parasitics.

3 Noise

Having discussed the AC modelling in the previous section we now focus on the noise figure. Starting with the drain current thermal noise [16], in MOS Model 9 this is further refined by including the gate induced noise and accounting for the effects of velocity saturation and hot carriers [2, 16]. Omitting the short channel effects for a moment, and including the parasitic resistances, the minimum noise figure, i.e the noise figure obtained at the optimum source impedance, can be approximated as [17]:

$$F_{min} \approx 1 + 1.9\frac{f}{f_t}\sqrt{0.4 + g_m R_g + \frac{(g_m + g_{mb})^2}{g_m} R_s + \frac{g_{mb}^2}{g_m} Z s_{bb}} \qquad (16)$$

Although this equation shows that for low frequencies the MOSFET noise figure approaches 0 dB, it must be realized that it is accompanied by a highly reflective optimum source impedance, and that losses in the matching circuitry will provide practical limits. Nevertheless, with the continuously increasing cut-off frequencies very attractive noise figures become available at RF.

Recent work has cast a doubt on whether the present compact noise models would still be applicable to sub-micron CMOS devices [18, 19] due to the high electric field encountered in these devices. To clarify these matters, in the next subsection the noise figures measured on a single 0.25 μm CMOS transistor and on a 0.5 μm CMOS based low noise amplifier are reviewed [20] and compared to the noise level predicted by MOS Model 9.

3.1 RF noise figure

The noise figure measurements were performed on-wafer using a Maury Microwave automated tuner system and a noise figure test-set, with the input tuned for minimum noise figure and the output conjugately matched for optimal power extraction. Fig. 15 shows minimum noise figures measured on a 0.25 μm CMOS transistor. The solid lines represent simulated

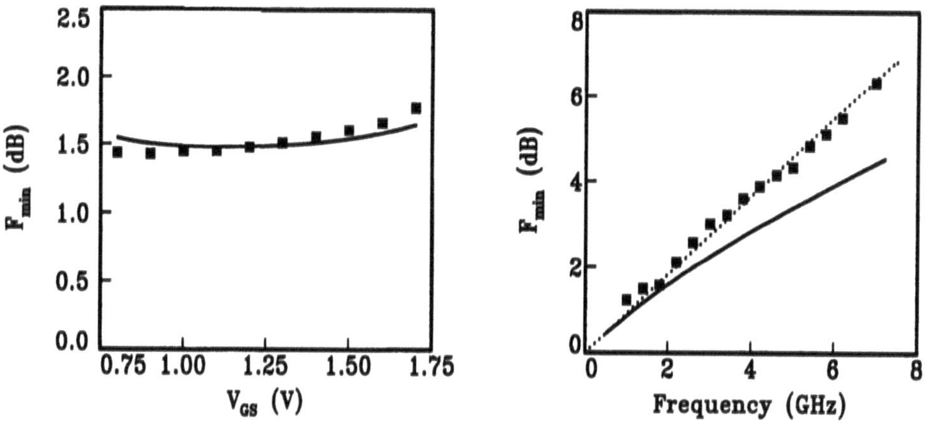

Figure 15: *Minimum noise figure versus gate voltage at a frequency of 1.8 GHz and versus frequency at* $V_{gs} = 1.1V$*, of a 20/0.25 NMOS device at* $V_{ds} = 1.2V$*. Solid line: MOS Model 9, Dotted line: Fit according to ref. [21].*

LNA @ 900 MHz	F_{min} dB	G_a dB	Re Z_{opt} Ω	Im Z_{opt} Ω
Measured	2.2	16.9	112	354
Simulated	2.3	16.4	64	345

Table 1: *LNA noise performance at 900 MHz.* F_{min} *is the minimum noise figure and* G_a *the associated available gain, both at the optimum source impedance* Z_{opt}.

values using MOS Model 9. The agreement is good although it must be noted that sophisticated models offering a better description of the frequency dependence exist [21]. To verify noise figures at the circuit level, a cascode LNA operating at 900 Mhz, was designed and realized in an industrial 0.5 μm IC process [20]. The LNA was designed for off-chip matching and therefore the noise parameters were measured on the tuner system. A minimum noise figure of 2.2 dB and an associated available gain of 16.9 dB were obtained. These values are in excellent agreement with the simulations (table 1). For the real part of the optimum source impedance a small deviation from the simulated value was found, which is attributed to the fact that non-quasi-static effects were not included in these circuit simulations. Considering that the excess noise figures were observed in devices with poor output characteristics due to avalanche [18, 22], it is thus concluded that for modern submicron CMOS processes where the electric field distribution is properly optimised, the noise figures predicted by MOS Model 9 are fairly accurate down to the minimum device length.

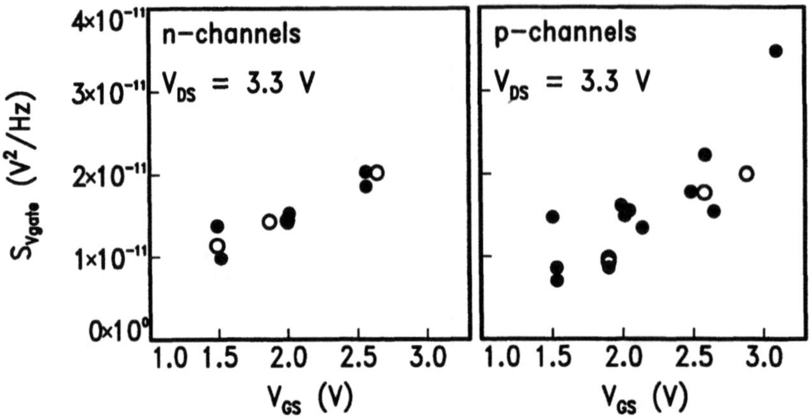

Figure 16: *Input referred voltage noise at 1 Hz as a function of gate-source voltage, of 10/5 devices with (solid symbols) and without (open symbols) gate protection. The 1/f noise level is not deteriorated by the absence of the gate protection.*

3.2 1/f noise

Low-frequency noise, and $1/f$ noise in particular, is becoming increasingly important. Evidently, $1/f$ noise limits the performance of low-frequency circuits. However, due to up-conversion, it will also contribute to local oscillator phase noise, and thus have an impact on RF performance. In most textbooks [7, 15] the following empirical modelling formula for $1/f$ noise is found:

$$S_{V_{\text{gate}}} = \frac{B}{W_{\text{E}} L_{\text{E}} f} \tag{17}$$

where $S_{V_{\text{gate}}}$ is the input-referred $1/f$ noise spectral density, and B is a technology-dependent parameter. This model can also be found in MOS Model 9, level 902. In practice, however, the input-referred $1/f$ noise shows a gate- and drain-bias dependence which is not captured by eq. (17).

Recently, Hung et al. [23, 24] published a $1/f$ noise model that excellently describes this bias-dependence throughout all operating regimes of the MOSFET. The model assumes trapping and de-trapping of charge carriers in the gate oxide to be responsible for the observed $1/f$ noise. The trapping and de-trapping processes cause both the number of carriers and their mobility to fluctuate in a correlated fashion. The distribution of distances between a gate oxide trap and the conducting channel explains the $1/f$ spectrum of the current fluctuations. This model, which is also, in a slightly different form, implemented in BSIM3v3 [3], is now available in MOS Model 9, level 903.

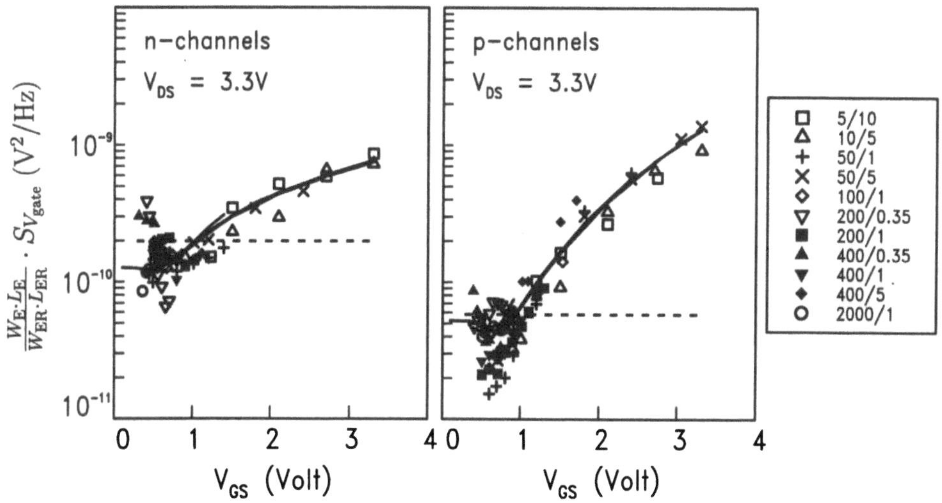

Figure 17: *Input referred voltage noise at 1 Hz in saturation (V_{DS} = 3.3 V), multiplied by the ratio of device area to reference device area, as a function of gate-source voltage, for a number of geometries. The MOS Model 903* $1/f$ *noise model is represented by the solid line, and the MOS Model 902* $1/f$ *noise model by the dashed line.*

We have evaluated the model on measurements performed on the Philips 0.35 μm CMOS technology. To test whether the number of intrinsic oxide traps and thus the $1/f$ noise level might have been increased during process steps such as plasma etching or packaging, first a comparison was made between the input-referred noise of 10/5 devices with and without gate protection. By AC-short-circuiting the gate to the ground any $1/f$ noise generated in the gate protection itself will not influence the measured $1/f$ noise level. The results are shown in fig. 16. Evidently, an increase in $1/f$ noise for devices without gate protection is not seen.
The subsequent measurements were performed on a selection of 11 geometries, including "conventional" device sizes like 10/5 and 5/10, as well as a number of wide, short devices, like 2000/1 and 400/0.35, which determine the $1/f$ noise in many practical circuits. When the measured and simulated $S_{V_{\text{gate}}}$ results are multiplied by the device area $L_E \cdot W_E$ and divided by a reference device area $L_{ER} \cdot W_{ER}$, as shown in fig. 17, all results merge nicely onto a single curve. No systematic deviations for short-channel devices are observed, and the deviations between different geometries fall within the sample-to-sample spread for one geometry [25]. Compared to eq. (17) the modelling of the bias dependence of the $1/f$ noise has improved significantly using the new $1/f$ noise model.

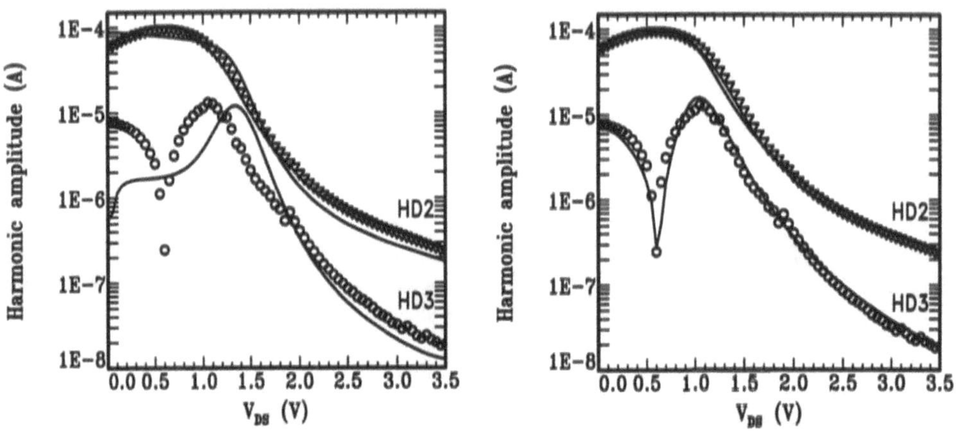

Figure 18: *Measured second-order ($HD2$) and third-order ($HD3$) harmonic amplitude of the drain current when a sinusoidal voltage ($\hat{V} \cdot \sin 2\pi f t$) is applied to the drain terminal ($\hat{V}$ = 0.5 V and f = 1 kHz) as a function of drain bias in comparison to: (left) MOS Model 9, and (right) the new MOS model. 10/1 NMOS device, 0.8 μm CMOS technology, V_{GS} = 2.5 V.*

4 Distortion

In RF receivers amplifier linearity is very important to prevent intermodulation distortion and increased noise floors. In general, owing to the use of balanced circuits, second order distortion can be cancelled out and third-order distortion forms a principal limit to the circuit linearity. In this respect MOS transistors, with their quadratic transfer characteristics are much more linear than bipolar transistors which have an exponential transfer characteristics. To accurately model the residual distortion in a balanced circuit, a compact MOSFET model suitable for RF modelling should give an accurate description of the second-order derivatives of the transconductance and output conductance.

Contemporary compact MOS models such as BSIM3v3 and MOS Model 9 were developed aiming at an accurate and continuous description of current, charges and (trans)conductances, and little attention was paid to the description of the higher-order derivatives. These models are therefore less accurate in predicting distortion effects in analog circuit design. As an example the measured and modelled distortion of a single MOSFET for different DC bias conditions can be seen in fig. 18 (left), where modelled results were obtained using MOS Model 9. It is clear that there is a deviation between model and measurements, especially for low drain bias values (i.e. the ohmic region). In the above case the MOSFET was driven by a purely sinusoidal input voltage on the drain terminal, similar inaccurate

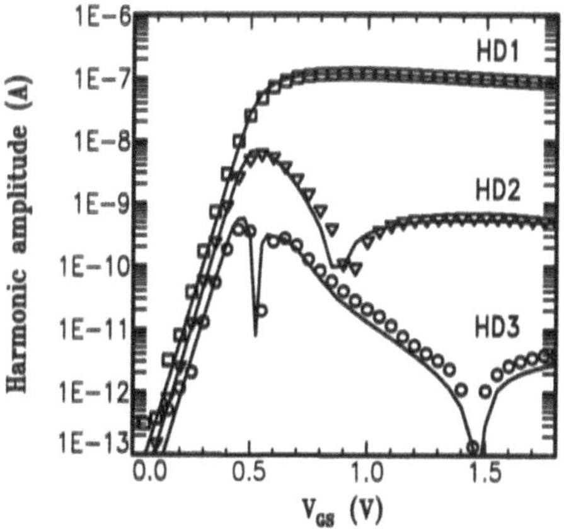

Figure 19: *First-order ($HD1$), second-order ($HD2$) and third-order ($HD3$) drain current harmonic amplitude modelling as a function of gate bias using the new MOS model. A sinusoidal voltage ($\hat{V} \cdot \sin 2\pi f t$) is applied to the gate terminal ($\hat{V} = 0.5$ V and $f = 1$ kHz). 10/10 PMOS device, 0.25 μm CMOS technology, $V_{DS} = -50$ mV.*

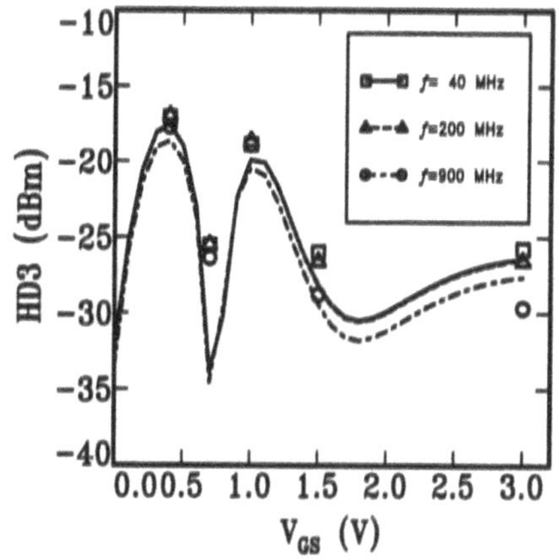

Figure 20: *Third-order RF harmonic distortion ($HD3$) modelling for different values of gate bias using the new MOS model when a 0 dBm sinusoidal input signal is applied to a MOSFET operating in common source-substrate configuration with 50Ω source and load impedances. 16×10/0.35 NMOS device, 0.35 μm CMOS technology, $V_{DS} = 3.3$ V.*

results, however, were obtained for gate terminal driven transistors. The signal frequency used in fig. 18 is very low (i.e. $f = 1$ kHz), so that in this case the influence of capacitances can be neglected and the deviation in modelled distortion is completely due to an imprecise expression of the DC drain current. A more accurate description of the MOS transistor's static behaviour is thus essential before attention can be shifted to the distortion behaviour at higher frequencies.

4.1 Static linearity

Over the past years a new model has been developed which is particularly suitable for distortion analysis in circuit design [26]. For the development of the above model special attention has been paid to the modelling of several physical mechanisms that play a role in a MOSFET. For an accurate modelling of transconductance well above threshold a new description of the mobility reduction effect and the voltage dependence of series resistance has been introduced, as reported in [27], [28]. On the other hand an accurate modelling of output conductance has been established by a new description of the velocity saturation effect in the ohmic region, and

new descriptions of channel length modulation, static feedback and self-heating in the saturation region [29]. In analog circuit design an accurate description of the weak, the moderate and the strong inversion region is essential. In particular the moderate inversion region becomes more and more important with the continuous down-scaling of supply voltage. The above-mentioned model is based on a charge sheet model, including both drift and diffusion current, and consequently comprises all inversion regions. To establish this, a new explicit relation for the surface potential has been developed [30].
The resulting model gives an improved description of the output conductance as becomes clear from the comparison in fig 18,and a highly accurate description of the transconductance as depicted in fig 19. It is thus evident that the new model gives a precise prediction of distortion in all inversion regions, even in the moderate inversion region. The model has been tested on different CMOS technologies (0.8 μm, 0.35 μm and 0.25 μm), and has been found to give good results on a large range of gate, drain and bulk bias values both in single transistors and in real circuits.

4.2 RF distortion

For a correct description of RF distortion the intrinsic capacitances of the MOS transistor have to be taken into account as well. In order to do so the charge model of MOS Model 9 has been adapted for implementation in the new MOS model. The distortion at different signal frequencies of a MOSFET operating in common source-substrate configuration with 50Ω source and load impedances is shown in fig. 20. It is clear that the modelled results are accurate even up to high frequency values. The distortion behaviour at high frequencies is still dominated by the description of static behaviour, and is only slightly affected by the influence of the intrinsic capacitances. Of course at even higher frequencies Non-Quasi-Static effects will come into play, and the above quasi-static model will no longer give satisfying results. The effect of the latter on distortion is still subject to examination. A straightforward approach though, which has been found to give promising results, is to replace the single MOSFET by a series-circuit of a number of MOS transistors.

5 Summary and outlook

An overview has been given of the capabilities of the present compact models to describe the actual MOSFET behaviour at RF. Starting from the MOSFET input and output impedance levels and the power gain it is shown that parasitic resistances must be included for an accurate description. The source, drain and gate resistances can be calculated from parameters usually provided by the CMOS foundries, and their inclusion in circuit simulation packages is probably not too far away. A further improvement in simulation accuracy is possible by accounting for the substrate resistance network. Including this in a circuit simulator is one of the challenges for the near future, as this will involve the development of complicated geometrical scaling rules for the substrate resistances, and involve extensive S-parameter characterisation efforts. MOSFET devices which are optimised for RF performance will have their parasitics reduced to the maximum extend possible within a given technology. As a consequence their behaviour can only be described accurately when non-quasi-static effects are included. Having demonstrated that the non-quasi-static admittance parameters can be obtained based on the existing DC and charge models, i.e. without the need for additional parameters, implementation of suitable algorithms in circuit simulation packages can be foreseen in the near future. With respect to thermal noise figure it has been shown that the present MOS Model 9 formulation is sufficiently accurate, even for the modern deep-submicron processes, where fairly high electrical fields are encountered. In view of its impact on local oscillator phase noise, the $1/f$ noise formulation in MOS Model 9 has been improved. The bias dependence of the input referred $1/f$ noise level is now explicitly accounted for. Regarding distortion a new model has been discussed which gives a more accurate description of the distortion behaviour of a MOSFET in all inversion regions, even in the moderate inversion region. The model has been tested on different CMOS technologies both at DC and RF, and has been found to give good results on a large range of gate, drain and bulk bias values.

Acknowledgements

The authors wish to acknowledge Johan Knol and Peter de Vreede for the design of test structures, Rob Verhaar and Andre Montree for providing silicon, and Adrie Zegers for conducting $1/f$ noise measurements.

References

[1] F. Najm, *Simulation and Modelling, A New Beginning*, IEEE Circuits & Devices, Vol.12, No.1, 8-10, 1996.

[2] *MOS MODEL 9*, complete model description and documentation for implementation available at: http://www.semiconductors.philips.com/Philips_Models.

[3] Weidong Liu *et al.*, *BSIM3v3.2 MOSFET Model*, 1997-1998, available from http://rely.eecs.berkeley.edu:8080/bsim3www.

[4] R.R.J. Vanoppen, J.A.M. Geelen and D.B.M. Klaassen, *The High-Frequency Analogue Performance of MOSFETs*, Proceedings IEDM, 173-176, 1994.

[5] R.M.D.A. Velghe, D.B.M. Klaassen and F.M. Klaassen, *Compact MOS Modeling for Analog Circuit Simulation*, Proceedings IEDM, 485-488, 1993; ibid. Proceedings ESSDERC, 833-836, 1994.

[6] M.J. van Dort and D.B.M. Klaassen, *Circuit Sensitivity Analysis in Terms of Process Parameters*, Proceedings IEDM, 941-944, 1995.

[7] H.C. de Graaff and F.M. Klaassen, *Compact Transistor Modelling for Circuit Design*, Springer-Verlag, Wien, 1990.

[8] M.A.C. Koolen, J.A.M. Geelen, and M.P.J.G. Versleijen, *An improved de-embedding technique for on-wafer high frequency characterization*, Proceedings BCTM 1991, pp. 188-191.

[9] L.F. Tiemeijer and D.B.M. Klaassen, *Geometry scaling of the substrate loss of RF MOSFETs*, Proceedings ESSDERC 1998, 480-483, 1998.

[10] R. Singh, A. Juge, R. Joly and G. Morin, *An Investigation into the Nonquasi-Static Effects in MOS Devices with On-Wafer S-Parameter Techniques*, Proceedings IEEE IC-MTS, Vol.6, 21-25, 1993.

[11] W. Liu, *R.F. MOSFET Modeling Accounting for Distributed Substrate and Channel Resistances with Emphasis on the BSIM3v3 SPICE Model*, Proceedings IEDM 1997, pp. 309-312.

[12] M. Chan, K.Y. Hui, C. Hu, and P.K. Ko, *A Robust and Physical BSIM3 Non-Quasi-Static Transient and AC Small-signal Model for Circuit Simulation*, IEEE Trans. Electron. Devices, Vol. 45, No.4, pp.834-841, 1998.

[13] M. Bagheri and Y.P. Tsividis, *A Small-signal DC-to-High-Frequency Non-quasi-static Model for the Four-Terminal MOSFET Valid in All Regions of Operation* IEEE Trans. on Electron Devices, Vol. ED-32, No. 11, pp. 3283-2391, 1985.

[14] T. Smedes and F.M. Klaassen, *An Analytical Model for the Non-Quasi-Static Small-Signal Behaviour of Submicron MOSFETs*, Solid-State Electronics, Vol.38, No.1, 121-130, 1995.

[15] Y.P. Tsividis, *Operation and Modelling of the MOS transistor*, Mc Graw-Hill Book Co., New York, 1987.

[16] A. van der Ziel, *Noise in Solid State devices and Circuits*, Chapter 5, J. Wiley & Sons, New York, 1986.

[17] R.A. Pucel, H.A. Haus, and H. Statz, *Signal and noise properties of GaAs microwave FET*, in Advances in Electronics and Electron Physics, vol. 38, L. Morton, Ed. New York: Academic Press, 1975.

[18] A.A. Abidi, *High-Frequency Noise measurements on FET's with small dimensions*, IEEE Trans. on Electron Devices, vol. ED-33, No. 11, pp. 1801-1805, Nov. 1986.

[19] P. Klein, *An analytical thermal noise model of dep submicron MOSFETs for circuit simulation with emphasis on the BSIM3v3 SPICE model*, Proceedings ESSDERC 1998, pp. 460-463, 1998.

[20] R.R.J. Vanoppen, L.M.F. de Maaijer, D.B.M. Klaassen, and L.F. Tiemeijer, *RF noise modelling of 0.25 μm CMOS and low power LNA's*, Proceedings IEDM, pp. 317-320, 1998.

[21] J.B. Lee, H.S. Min and Y.J. Park, *Reinvestigation and extension of the steady-state Nyquist theorem for multi-terminal semiconductor devices and its application to minimum noise figure in microwave field effect transistors*, J. Appl. Phys. Vol. 79, No. 1 pp. 228-241, 1996.

[22] L.M. Franca-Neto, E. Mao, J.S. Harris Jr., *Low noise FET design for wireless communications*, Proceedings IEDM, pp. 305-308, 1998.

[23] Kwok K. Hung *et al.*, *A unified model for the flicker noise in metal-oxide-semiconductor field-effect transistors*, IEEE Trans El. Dev. Vol. 37, No. 3, March 1990.

[24] Kwok K. Hung *et al.*, *A physics-based MOSFET noise model for circuit simulators*, IEEE Trans El. Dev. Vol. 37, No. 5, May 1990.

[25] G. Ghibaudo and O. Roux-dit-Buisson, *Low-frequency fluctuations in scaled down silicon CMOS devices status and trends*, Proc. ESSDERC '94, pp.693-700 (1994).

[26] R. van Langevelde, *A Compact MOSFET Model for Distortion Analysis in Analog Circuit Design*, *PhD Thesis*, TU Eindhoven, Eindhoven 1998.

[27] R. van Langevelde and F.M. Klaassen, *Influence of Mobility Degradation on Distortion Analysis in MOSFETs*, *in Proceedings ESSDERC 1996*, Bologna, Italy, pp. 667-670, 1996.

[28] R. van Langevelde and F.M. Klaassen, *Effect of Gate-Field Dependent Mobility Degradation on Distortion Analysis in MOSFET's*, *IEEE Trans. Electron Devices*, Vol. ED-44, No. 11, pp. 2044-2052, 1997.

[29] R. van Langevelde and F.M. Klaassen, *Accurate Drain Conductance Modeling for Distortion Analysis in MOSFETs*, *IEDM 1997 Tech. Digest*, pp. 313-316, 1997.

[30] R. van Langevelde and F.M. Klaassen, *An Explicit Surface-Potential Based MOSFET Model for Circuit Simulation*, submitted for publication in Solid-State Electronics.

PHYSICS BASED ACCURATE EXTRACTION OF LEFF AND RS FOR DEEP SUBMICRON MOSFETS

Kristin M. De Meyer [1,2] and Serge Biesemans[3]

1 IMEC, Kapeldreef 75, B-3001, Leuven, Belgium
2 KULeuven, ESAT-INSYS, Kardinaal Mercierlaan 94, B-3001, Leuven, Belgium
3 currently at IBM, Hopewell Junction, NY 12533, USA

ABSTRACT

In leading edge VLSI technologies the exact knowledge of both the effective gate length Leff and the parasitic series resistance Rs are of utmost importance for process engineers to evaluate technological issues and for circuit designers to accurately predict circuit performance especially in analog applications. In this paper first the most important existing techniques and their error analysis will be presented. Then a more accurate technique especially for deep submicron technologies will be explained. Its advantages will be clearly illustrated.

1. INTRODUCTION

In scaling down MOSFETs into the deep submicron regime the values of the effective channel length and the ΔL arising from lithography, etching and underdiffusion are comparable and therefore the requirements for the determination of Leff are becoming much more stringent. Also the length of the channel region and that of the underdiffused extensions are close to each other. This means that the series resistance arising from these extensions which are not extremely highly doped strongly affects the drive current and has to be taken into account in models predicting circuit behaviour.

In this paper we will first illustrate that classical extraction methods for Leff and Rs fail for a 0.1μm technology. An overview of the existing methods and their error analysis will explain the reasons for their failure. Next the shift and ratio method [1] will be presented. Also this method has its weaknesses and therefore a new "modified " shift and ratio method was developed to overcome these. The advantages and improved accuracy of the latter will be illustrated.

2. MOTIVATION

The motivation for the accuracy analysis of the conventional L-array methods and for the development of a new method are the results of ΔL coming from an experimental 0.1 μm nMOS run [2,3] as shown in Fig.1.

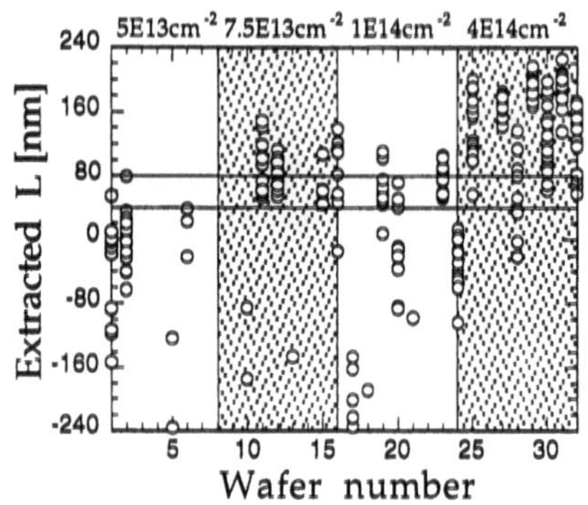

Fig.1 Extracted ΔL values from an experimental 0.1 μm technology showing a large spread and unphysical results. The expected ΔL should be between 40-80 nm. Four extension conditions are present and indicated above the graph (As implanted at 20 KeV).

The experiment counted 32 wafers with various splits in the channel, extension and spacer conditions. The standard extraction method used, is the one of Laux [4]. The following problems can clearly be noticed.

- 16 wafers out of 32 have no result
- the results are clearly not physical. The real ΔL is expected to be 40-80 nm as indicated. This is concluded from the fact that devices with a poly length down to 80 nm (as measured by TEM) are present and functioning [2].

The results of Fig.1 will be explained in section 3.4 in terms of the limited resolution of the method, expressed by $\sigma\Delta L$ and σR_S, and its robustness against different noise sources present. It will be shown that this limited accuracy or resolution directly depends on the process variation σL_g. The results for ΔL should therefore be written as $\Delta\bar{L}\pm\sigma\Delta L$. Fig.1 provides only $\Delta\bar{L}$. In practical terms this means that if in the L-array one transistor has an $L_g = L_{gtarget} + \delta L_g$, where δL_g is an error due to process uncertainties, the extracted $L_{eff} + \delta L_{eff}$ and $R_S + \delta R_S$ will deviate from the real value. The robustness of the extraction method can be analyzed by performing a statistical sensitivity analysis and calculating $\sigma\Delta L$ and σR_S as a function of process variation δLg. A method which has a large σ is considered to be ill-conditioned and can give large erroneous results for ΔL and R_S. Sources of noise considered here are

- uncertainty of the poly lengths in the L-array (σL_g)
- the assumption of a V_{gs} independent L_{eff} and R_S
- extrapolation of the linear regression line
- non-linearity of R_{tot} as a function of L
- small angle effect
- outliers taken into the L-array

For larger technologies as used in the past, e.g. $L_g \in [0.75, 5\ \mu m]$, the ΔL was of the order of [400-1000 nm]. An uncertainty of for instance 20 nm is of no consequence because its relative error ($\delta\Delta L/\Delta L<1\%$) is insignificant. However, in today's extreme deep submicron technologies with a $\Delta L\sim$[40-80 nm], the L_g uncertainty will contribute to errors in ΔL larger than 100-300%.

3. OVERVIEW OF EXISTING METHODS AND THEIR ERROR ANALYSIS

In what follows the most commonly used extraction methods will be discussed. These are all based on L-arrays.

Without being complete, most of the conventional L-array methods can be classified into three groups

- the single V_{gs}-method
- the double V_{gs}-method
- the multiple V_{gs}-method

The single V_{gs}-method is known in literature as the $1/\beta$ method [5]. Most methods belong to the double V_{gs}-method as for instance the Terada and Muta [6], Chern [7] and Hu [8] methods. The multiple V_{gs}-method is known as the Laux method [4]. It is found that the robustness and the accuracy increases as more values for the V_{gs} parameter are taken into account.

3.1 The single Vgs-method

The single V_{gs} method assumes $R_S=0$. The g_m ($=dI_{ds}/dV_{gs}$) or R_{tot} ($=I_{ds}/V_{gs} = R_S+R'ch(L_g-\Delta L)$) is plotted as a function of L_g for a fixed V_{gs}. Then, a linear regression is performed as shown in Fig.2.

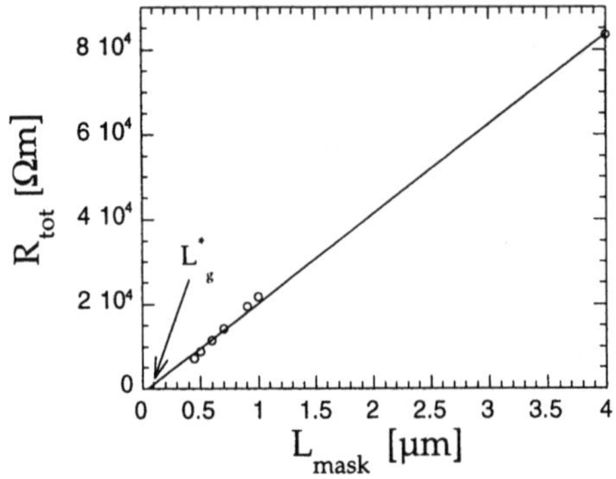

Fig.2 Principle of the single V_{gs} method. The intersection of the regression line with the x-axis gives L_g^* which is interpreted as ΔL

The special point in this algorithm is the intersection of the regression line with the L_p-axis (i.e. $R_{tot}(L_g)=0$).
According to standard theory, this results in ($L_{eff}=L_g-\Delta L$)

$$0 = \frac{L_g^* - \Delta L}{W C_{ox} \mu (V_{gs} - V_T - V_{ds}/2)} \quad => \quad \Delta L = L_g^* \tag{1}$$

This provides the required result.

3.2 The double V_{gs}-method

The main idea in this method is to avoid the assumption $R_S=0$. Because two unknowns appear, ΔL and R_S, two equations are required. By performing a linear regression on two sets of data, each at a different V_{gs}, one can find the intersection (L_g^*, R_s^*) of the two lines $R_{tot1}=R_{S1}+R_{ch1}*L_{eff1}$ and $R_{tot2}=R_{S2}+R_{ch2}*L_{eff2}$ as shown in Fig.3.

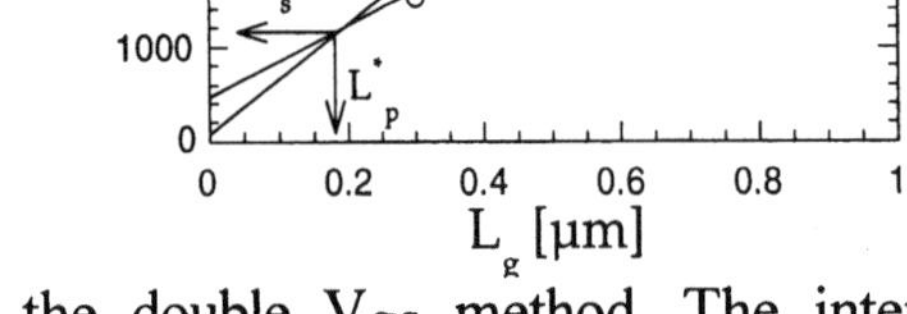

Fig.3 Principle of the double V_{gs} method. The intersection of the two regression lines gives the couple (L_g^*, R_s^*) which is interpreted as $(\Delta L, R_S)$

This results in ($f=1/WC_{ox}\mu_{eff}(V_{gs}-V_T-V_{ds}/2)$)

$$L_g^* = \Delta L_1 + \frac{R_{s1} - R_{s2} + f_2(\Delta L_2 - \Delta L_1)}{f_2 - f_1} \quad (2a)$$

$$R_{tot1}^* = R_{s1} + \frac{f_1}{f_2 - f_1}(R_{s1} - R_{s2} + f_2(\Delta L_2 - \Delta L_1)) \quad (2b)$$

Under the assumption that as the difference in V_{gs}, i.e. $V_{gs2}-V_{gs1}$, is taken small enough but larger than zero to have an intersection point, R_{s1} equals R_{s2} and ΔL_1 equals ΔL_2, (2) reduces to

$$L_g^* = \Delta L_1 \quad (3a)$$

$$R_{tot1}^* = R_{s1} \quad (3b)$$

Which again provides the required result. By repeating the algorithm for different V_{gs} pairs, one can scan in principle the V_{gs} dependence of L_{eff} and R_S.

3.3 The multiple V_{gs}-method

If one cannot neglect the second term of (2), the special point provides only two equations for four unknowns (ΔL_1, ΔL_2, R_{s1}, R_{s2}). The assumption made above leads to a systematic error in (3). This systematic error can be removed by replotting the data. Without going into detail the idea to remove the systematic error is to go one step further and to perform a second linear regression (C+Dx) on the set of data obtained from the double V_{gs}-method at several V_{gs} values [4].

The multiple V_{gs}-method removes the systematic error, but not the random errors due to the various noise sources mentioned before, e.g. the non-linearity of R_{tot} versus L_g and σL_g.

3.4 Error analysis

In this section, the accuracy σ is calculated. As mentioned earlier, σ gives the deviation of the extracted ΔL and R_S from the real one in case that noise sources are present. For instance, in the derivation of the L-array methods, one assumes to know exactly the poly length L_g. The accuracy σ will quantify how much the ΔL and R_S will change if one or more devices have an $L_g+\delta L_g$ instead of L_g. This δL_g is considered to be part of the measurement noise. Other noise sources have been mentioned above.

3.4.1 The single Vgs-method

In this section we only will develop an intuitive picture of the error introduced by the uncertainty on L_p introduced due to process variation. Suppose one has a perfect etching process which results in an L-array with well known values of Lg. Hence the (Lg,Rtot) points will be on a straight line which is represented in Fig.4 a by the thick line.

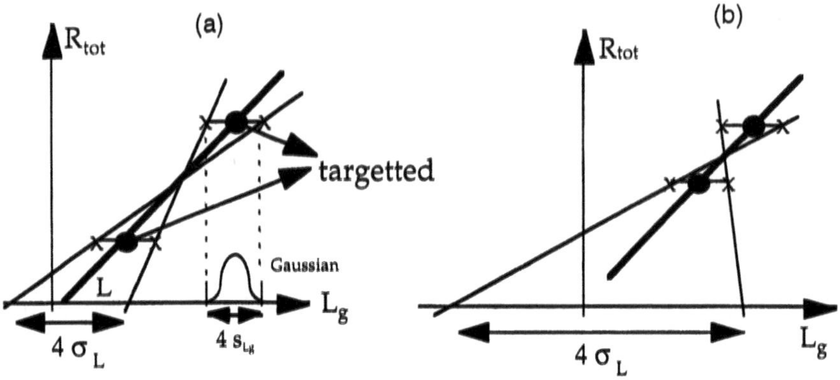

Fig.4 Interpretation of the uncertainty in the extracted ΔL due to uncertainties in Lg. Case (a) is for an L-array including small devices while case (b) only has larger devices resulting in a larger ΔΔL

From the data set {Lg,Rtot} the regression model is given as

$$R_{tot} = \alpha + \beta L_g \tag{4}$$

where α and β are the least square fits of the data set, found from least square minimization to be (x->Lg, y->Rtot)

$$\beta = \frac{\sum (x-\bar{x})(y-\bar{y})}{\sum (x-\bar{x})^2} \tag{5a}$$

$$\alpha = \bar{y} - \beta \bar{x} \tag{5b}$$

The next step is to relate α and β with the required ΔL. From the model Rtot=Rs+R'ch(Lg-ΔL), we identify

$$\alpha = R_s - f\Delta L \qquad \beta = f \tag{6}$$

By putting Rs=0, ΔL is immediately seen to be ΔL=-α/β.

Putting Rs=0, however, introduces a systematic absolute error Δ(ΔL). We can find this error to be

$$\Delta L = \frac{R_s}{f} - \frac{\alpha}{\beta} \quad => \quad \Delta(\Delta L) = \frac{R_s}{f} \tag{7}$$

and the absolute systematic error is proportional to Rs.

Besides the systematic error, one can calculate the random error σ. So far, perfect knowledge of the poly length Lg constituting the L-array was assumed. However, process variations in lithography and poly etching will introduce a certain amount of uncertainty in Lg (=σLg). Because of these process variations, the regression model for $\overline{R}_{tot}$

$$\overline{R}_{tot} = a + b\,L_g \tag{8}$$

will differ from the exact one. As a result, the extracted value $\overline{\Delta L}$ will differ from the real ΔL.

The origin and the geometrical meaning of σ is pictured in Fig.4 a. The thick dots represent the presumed L_g value, while the two 'x' marks represent the outer boundaries of the interval in which the real L_g will be. A Gaussian distribution is assumed with a standard variation σ_{Lg}. The thick line represents the regression line), while the two thinner lines indicate the worst case deviation (8) in case the real L_g would differ from the assumed one (the big dot).

To estimate the total uncertainty $\sigma_{\Delta L}$ in the presence of some σ_{Lg}, one has to use the full expression including the R_S, even if it is put artificially equal to zero. Applying the σ-operator to ΔL (7) results in (we replace L_g by x)

$$\sigma_{\Delta L}^2 = \frac{\sigma_{R_s}^2}{f^2} + \frac{\sigma_a^2}{b^2} + \left(\frac{a}{b^2}\right)^2 \sigma_b^2 \tag{9}$$

with

$$\sigma_a^2 = \frac{S_R}{(n-2)}\left(\frac{1}{n} + \frac{(x_0 - \overline{x})^2}{\sum (x - \overline{x})^2}\right) \qquad @ \; x_0 = 0 \tag{10}$$

$$\sigma_b^2 = \frac{S_R}{(n-2)\sum (x - \overline{x})^2} \tag{11}$$

$$S_R = \sum (y - \overline{y})^2 - b^2 \sum (x - \overline{x})^2 \tag{12}$$

For the uncertainty σ_{Rs}, one might take a value of 10%. The function 'f' itself is explicitly dependent on V_{gs} and contains the mobility μ_{eff}. This mobility can be found from a large reference device. The factor n-2 comes from the fact that one loses two degrees of freedom in estimating a and b.

As shown in Fig.6a, the presence of a σ_{Lg} is the origin of $\sigma_{\Delta L}$. However, there are other factors involved which will determine the final value of $\sigma_{\Delta L}$. Depending on what L-array one takes, different results might be obtained. This is coded as the 'which L-array'-problem and can be further broken down in the extrapolation and the non-linearity problem. Without going into the mathematical details these are illustrated below.

Extrapolation problem

Fig.4 a and b compare two cases : an L-array including small values of L_g and an L-array with only large values of L_g. It is clearly seen that the second case is worse than the first one in the sense that it has a larger $\sigma_{\Delta L}$ for the same amount of σ_{Lp}. Extrapolation of a linear regression model is in principle not allowed..

Non-linearity problem

Fig.5 shows the experimental data. Due to the non-linearity of R_{tot} as a function of L_p, ΔL depends very much on which sub-array one takes. From the extrapolation problem it could be concluded that one should take an array with mainly small devices, however, one should make sure that they are still exhibiting a linear behavior. In fig.5 it can be seen that by taking the extra 0.25 μm device, a shift in ΔL of 85 nm is observed. The non-linearity is quantified by the fact that the $\sigma_{\Delta L}$ increases from 22.5 nm for the array without the 0.25 μm device to 50 nm for the array with this device.

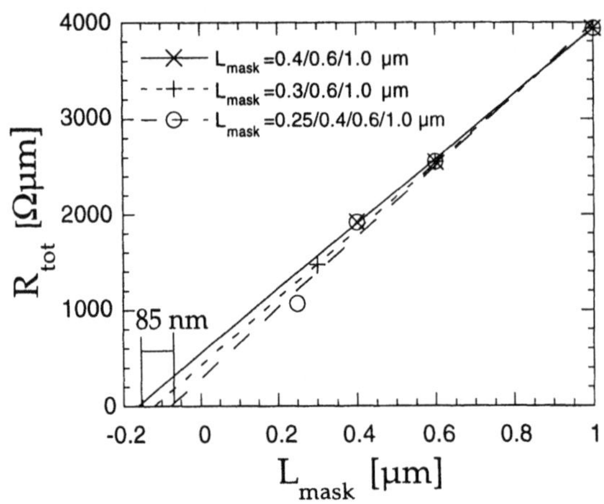

Fig.5 Graphical representation of the 'non-linearity' problem. Taking different sub-arrays results in different regression lines

3.4.2 The double V_{gs}-method

Much of the work developed in the previous section can directly be applied to the double V_{gs}-method. The same sources of error are present being the uncertainty in L_g, the extrapolation and the non-linearity

The special point from which ΔL and R_s could be extracted is the intersection of two regression lines each at different V_{gs} values.Only an intuitive picture of the uncertainties is given. Again, due to the uncertainty in L_g, the extracted values will differ from the physical ones. As a result, the cross section will not be clearly defined and this results in an uncertainty area as shown in Fig.6a.

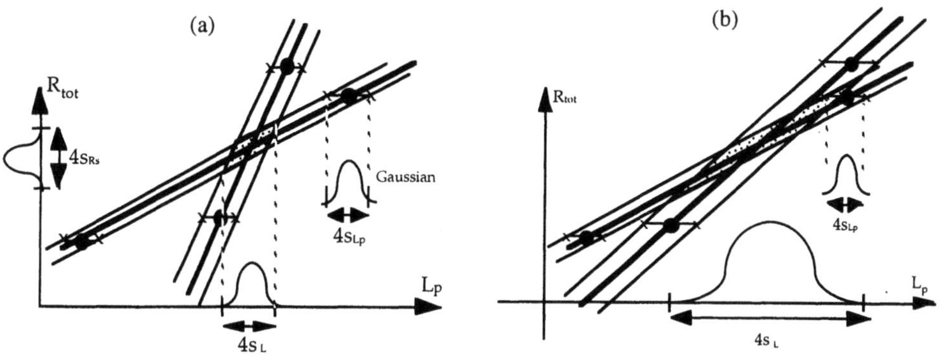

Fig.6 (a) The uncertainty in L_g gives rise to an uncertainty in the interception point (ΔL,R_S). (b) If the angle under which the two regression lines meet decreases, the uncertainty in the intersection point (ΔL,R_S) increases

A new element, typical for the double V_{gs} method and relevant for the accuracy of the extraction, is the angle between the two lines. Fig.6 b shows for a center point of V_{gs}, which is higher than in Fig.6 a, the R_{tot} as a function of L_g. The angle between the two lines is smaller for the larger V_{gs} value. From numerical mathematics it follows that finding the intersection of two lines which meet under a small angle is an ill-conditioned problem. This means that the uncertainty in the intersection $\sigma_{\Delta L}$ increases as the angle reduces.

The problems inherent to the double Vgs method are found back in the multiple Vgs method.

In the next section we propose a possible way to avoid the problems inherently to L-array methods. Small ΔL-technologies and experimental runs with less process control need a very sensitive and robust tool to monitor the L_{eff} and R_S. It will be shown next that single transistor algorithms have more resolution and are more robust due to their lower σ [9] and their ability to look at each single device separately. On top of that, single transistor methods are able to extract L_{eff} and R_S accurately at low V_{gs}, near V_T, which is expected to give a value closer to the metallurgical channel length L_{met}. Existing examples of single transistor algorithms are the method of Whitfield [10] and Taur [1]. We have elaborated on the method in [1] to improve its accuracy by using this σ analysis [3].

4.THE ORIGINAL AND MODIFIED SHIFT AND RATIO METHOD

Whereas the original S&R-method [1] assumes L_{eff} and R_S to be independent of V_{gs}, we removed that assumption and achieved an accuracy σLeff of 15-20 nm [15]. This simplifies the interpretation of the extracted Leff compared to the original S&R because in the latter case the extracted value Leff cannot directly be related to L_{met}. In this modified S&R method, which will be presented in the second part of this section, a measure of L_{met} can be determined as the saturated part at low V_{gs} (e.g. 0.175 μm in Fig.7).

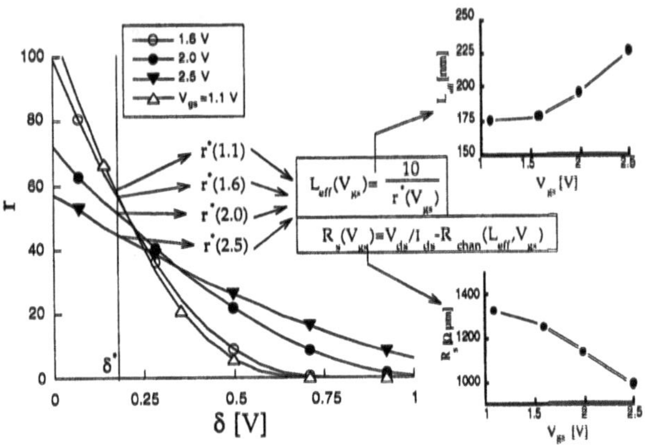

Fig.7 Summary of the modified S&R-method. The intersection of r(δ) with the vertical line $\delta=\delta^*$ is recorded and transformed into Leff(Vgs) and Rs(Vgs)

Fig.7 is the graphical representation of the m-S&R method and will be explained in detail in the coming sections.
The algorithm starts from the expression for the total MOSFET resistance in the linear (low drain bias) regime

$$R_{tot} = \frac{V_{ds}}{I_{ds}} = R_s + R_{ch} \tag{13}$$

with R_{ch} is the intrinsic channel resistance. For an ideal MOSFET, the channel resistance can be expanded into

$$R_{ch} = \frac{V'_{ds}}{I_{ds}} \quad with\ I_{ds} \cong \frac{W}{L_{effi}} \mu_{eff} C_{ox} (V'_{gs} - V_T - V'_{ds}/2) V'_{ds} \tag{14}$$

where Leff,i is the effective channel length of the specific i-th device, μ_{eff} is the effective channel mobility, C_{ox} the gate oxide capacitance per unit area, V'_{gs} and V'_{ds} the internal gate and drain voltages. The Leff,i differs from L_i defined in [1] in that sense that it can have a Vgs-dependence and a non-zero derivative with respect to V_{gs}. Because experimental data are based on external terminal voltages (Vgs, Vds), (14) has to be reworked by substituting

$$V'_{gs} = V_{gs} - I_{ds} \frac{R_s}{2} \tag{15a}$$

$$V_{ds}' = V_{ds} - I_{ds}R_s \tag{15b}$$

with R_S the sum of the source and drain parasitic resistances. Combining (14) and (15) results in

$$I_{ds} = \frac{W}{L^*}\mu_{eff}C_{ox}(V_{gs} - V_T - V_{ds}/2)V_{ds} \tag{16}$$

with

$$L^* = L_{eff,i} + W\mu_{eff}C_{ox}R_s(V_{gs} - V_T - V_{ds}/2) \tag{17}$$

We have made the approximation that $\mu_{eff}(V_{gs}-I_{ds}R_s/2)=\mu_{eff}(V_{gs})$ which is sufficiently accurate under all conditions. Expression (16) which is referenced to the external terminal voltages has the same functional form as the expression referenced to the internal voltages (14) except for the value of L^*.

Both the S&R and m-S&R algorithms start as follows. First, the total resistance is written as

$$R_{tot} = \frac{V_{ds}}{I_{ds}} = \frac{L^*}{W\mu_{eff}C_{ox}(V_{gs} - V_T - V_{ds}/2)} = L^* f(V_{gs} - V_T) \tag{18}$$

Next, the function $S(V_{gs})$ is defined as

$$S(V_{gs}) = \frac{dR_{tot}}{dV_{gs}} = f L^{*'} + L^* f' \tag{19a}$$

$$\cong R_s' + L_{eff,i}' f - L_{eff,i} f/(V_{gs} - V_T - V_{ds}/2) \tag{19b}$$

where the prime means derivative with respect to V_{gs} and 'f' is defined in (18). It is assumed that 'f' reduces to

$$f' = \frac{-f}{(V_{gs} - V_T - V_{ds}/2)} \tag{20}$$

which means that the mobility μ_{eff} has a weak V_{gs}-dependence compared to the linear function ($V_{gs}-V_T-V_{ds}/2$). In order to find the $L_{eff,i}$ of a particular device, its curve $S_i(V_{gs})$ is shifted to lower V_{gs} values by a varying amount, δ, and the ratio 'r'

$$r(\delta) = \frac{S_o(V_{gs})}{S_i(V_{gs} - \delta)} \tag{21}$$

between two devices is computed as a function of δ with V_{gs} as a parameter. Here, $S_o(V_{gs})$ represents the curve from a big reference device (e.g. L_o=10 μm). The reason this L_o has to be as big as possible shall become clear in the next paragraph.

4.1 The original S&R method

In the original S&R algorithm [1], it was assumed that L_{eff} and R_S have a weak V_{gs} dependence and consequently have a vanishing derivative. Combining (19b) and (21), the ratio r(δ) becomes under these assumptions

$$r(\delta)=\frac{L_o}{L_{effi}}\frac{f(V_{gs}-V_{To})}{f(V_{gs}-V_{Ti}-\delta)} \tag{22}$$

if δ equals $\delta^*=V_{To}-V_{Ti}$, the ratio becomes independent of V_{gs}

$$r(\delta^*)=\frac{L_o}{L_{effi}}\frac{f(V_{gs}-V_{To})}{f(V_{gs}-V_{Ti}-(V_{To}-V_{Ti}))}=\frac{L_o}{L_{effi}} \tag{23}$$

This is plotted in Fig.8.

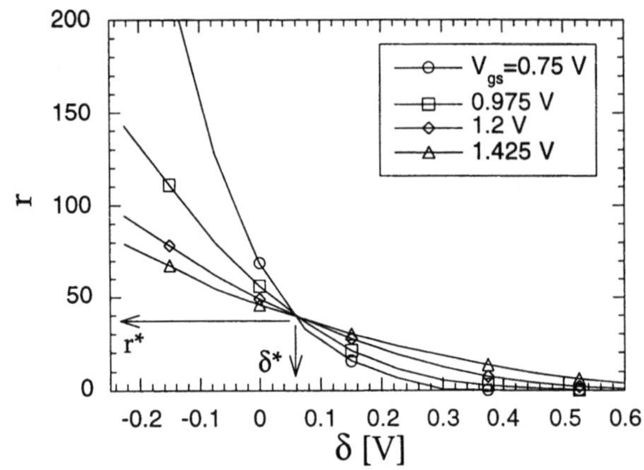

Fig.8 The ratio 'r' as a function of δ with V_{gs} as a parameter. The intersection point of interest is $(\delta^*, r(\delta^*))$

The different curves correspond to different values for V_{gs} as indicated. A single point of intersection is observed and $L_{eff,i}$ can be calculated as

$$L_{eff,i}=\frac{L_o}{r(\delta^*)} \tag{24}$$

Now it becomes clear why L_o has to big. If one would take for L_o a small transistor, than this value L_o would not be known with sufficient accuracy and hence cannot serve as a reference. For the big transistor, on the other hand, L_o-ΔL L_o because ΔL is very small compared to e.g. L_o=10 μm.

If, however, the V_{gs} spans an interval which is of practical interest (e.g. [0V,2.5V]), as done in Fig.9, multiple intersection points are observed and it is unclear which one to take.

Since multiple δ^* can be found in this way, one and the same device seems to have many V_T's ($=\delta^*-V_{To}$) which is physically meaningless. This apparent contradiction can be traced back to the assumption of V_{gs}-independent L_{eff}. In order to cope with the problem of multiple intersection points, the assumption of the weak V_{gs}-dependence of L_{eff} and R_s in the derivation has been removed.

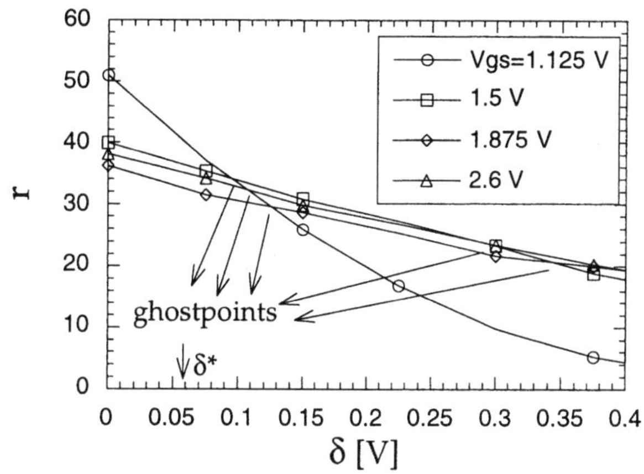

Fig.9 As the V_{gs} range is increased, more intersection points appear. The indicated δ^* is the δ where all curves are expected to intersect according to the original S&R method

4.2 The modified S&R algorithm

In the modified 'Shift & Ratio' method, the r(δ) is reinterpreted without neglecting the terms R'_s and fL'_{eff} resulting in

$$r(\delta)=\frac{R'_{s,o}(V_{gs})+f\,L'_o(V_{gs})-\dfrac{L_o f(V_{gs}-V_{To})}{(V_{gs}-V_{To}-V_{ds}/2)}}{R'_{s,i}(V_{gs})+f\,L'_{eff,i}(V_{gs})-\dfrac{L_{eff,i}(V_{gs})f(V_{gs}-V_T-\delta)}{(V_{gs}-V_{Ti}-V_{ds}/2-\delta)}} \tag{25}$$

It can simply be shown using any example that R'_s and fL'_{eff} are of the same order of magnitude but opposite in sign. This results in the fact that the value of the sum of both terms, $R'_s + fL'_{eff}$, will be negligible with respect to the value of the third term in (25). The ratio r(δ) then becomes

$$r(\delta)=\frac{L_o}{L_{effi}(V_{gs})}\frac{f(V_{gs}-V_{To})}{f(V_{gs}-V_{Ti}-\delta)}\frac{V_{gs}-V_{Ti}-V_{ds}/2-\delta}{V_{gs}-V_{To}-V_{ds}/2} \tag{26}$$

In the point $\delta^*=V_{To}-V_{Ti}$, this reduces to

$$r(\delta^*)=\frac{L_o}{L_{effi}(V_{gs})} \tag{27}$$

which looks very alike (23) but with the constant $L_{eff,i}$ substituted by the Vgs dependent quantity $L_{eff,i}(V_{gs})$. At first sight, apart from the derivation, no difference is observed between the modified and the original S&R method [1]. However, the present derivation allows to solve the problem of the multiple intersection points and to remove the requirement of a single intersection point to extract $L_{eff}(V_{gs})$. Additionally, it is able to find the

correct V_{gs}-dependence of both L_{eff} and R_S. However, the 'shift' δ^* cannot be found easily anymore.
The extraction of L_{eff} proceeds as follows. The V_T is a constant of the device and can be determined best by using the maximum slope method (see next section). The intersection of the different $r(\delta,V_{gs})$ curves with the vertical line $\delta=\delta^*$ is recorded as $r(\delta^*,V_{gs})$, shown in Fig.7.
The $L_{eff,i}(V_{gs})$ can then be found as

$$L_{eff,i}(V_{gs}) = \frac{L_o}{r(\delta^*, V_{gs})} \tag{28}$$

In this way, the reduction of V_T due to the short channel effect or the V_T increase due to the reverse short channel effect can be taken into account explicitly by the position of the value δ^*. A correct and consistent determination of the V_T is thus of utmost importance because it directly influences the value of $r(\delta^*)$ and consequently the value of the extracted L_{eff}. In the next paragraph, the algorithm to extract the correct V_T of a device self-consistently will be given.
To proceed, in order to find $R_S(V_{gs})$, a combination of (16) and (17) results in

$$R_s = \frac{W\mu_{eff} C_{ox} V_{ds}(V_{gs} - V_T - V_{ds}/2) - I_{ds} L_{eff,i}}{I_{ds} W\mu_{eff} C_{ox}(V_{gs} - V_T - V_{ds}/2)} \tag{29}$$

The right hand side of (29) can easily be evaluated using the experimental data for I_{ds} and the $L_{eff,i}$ values obtained from the above algorithm. The μ_{eff} was left unspecified so far because it cancels out in the extraction of $L_{eff,i}$. However, it plays a major role in the determination of R_S. Some authors take it to be a constant while others use some analytical expression for its degradation with V_{gs} [1]. Our approach is such that μ_{eff} is identified with the universal mobility [11] and is extracted from the reference (e.g. L_o=10 μm) device. In the formulas above this experimentally determined $\mu_{eff}(V_{gs})$ was used.

4.3 Determination of V_T

There exist several ways to determine the V_T of MOSFETs [17]. The most often used method is the maximum slope method at low V_{ds}, shown in Fig.10.

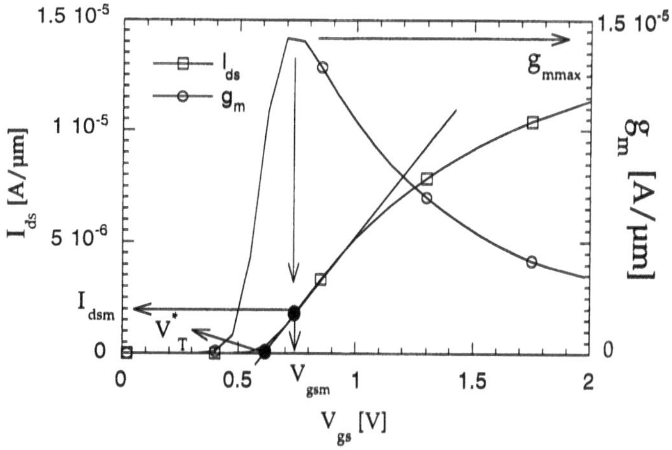

Fig.10 The maximum slope method. In the point where the slope or g_m reaches its maximum, the tangent to the I_{ds} curve is drawn. The intersection with the coordinate axis is V_T^*

At the gate voltage, V_{gsm}, where the slope reaches a maximum, one draws the tangent line at the I_{ds} curve. The intersection of this line with the V_{gs}-axis is the threshold voltage V_T^*.
However, if the V_{ds} is not small compared to V_T^* nor the R_S is negligible with respect to R_{ch}, the real threshold voltage V_T of the device will differ from the extracted value V_T^*. This can be seen in Fig.11 where identical devices are taken except for the R_S value. Applying the maximum gm-method results in a different extrapolated V_T^*.

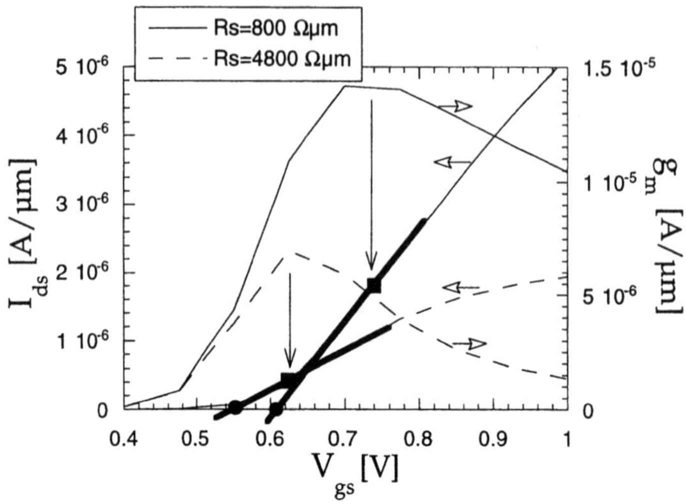

Fig.11 Plot showing that R_S can influence the V_T extraction with the maximum g_m-method

It can be shown that the V_T can be derived from the extracted value V_T^* with the following formula:

$$V_T = V_{gsm} - \frac{V_{ds}}{2} + \frac{1 - \sqrt{1 + 4\frac{W}{L_{eff,i}} R_s \mu_{eff} C_{ox}(V_{gsm} - V_T^*)}}{2\frac{W}{L_{eff,i}} R_s \mu_{eff} C_{ox}} \tag{30}$$

which is the real V_T corrected for R_S and V_{ds}. Typical values for R_S in nMOS devices are below 1000 ohmμm. For sub-quarter micron devices, the error can become as large as 60 mV.

4.4 Discussion and results

Two possible routes can be taken at this point to determine L_{eff}. Either one neglects the R_S in the determination of V_T, i.e. putting $V_T = V_T^*$, or one uses (39). The second case is a bit more complicated because one needs to know L_{eff} and R_S to determine V_T from the measured V_T^*. We have chosen for the second option. This, however, calls for an iterative procedure as shown in Fig.12.

The V_T^* is extracted using the conventional maximum slope method. Then, the two starting values are taken to be R_S=1000 ohmμm and L_{eff}=100 nm. Next the real V_T is calculated using (39). Applying the m-S&R method provides two new values for R_S and L_{eff} which are used as a new estimation in the determination of V_T. This procedure converges very fast. After 3 to 5 iterations, a steady state value is obtained .

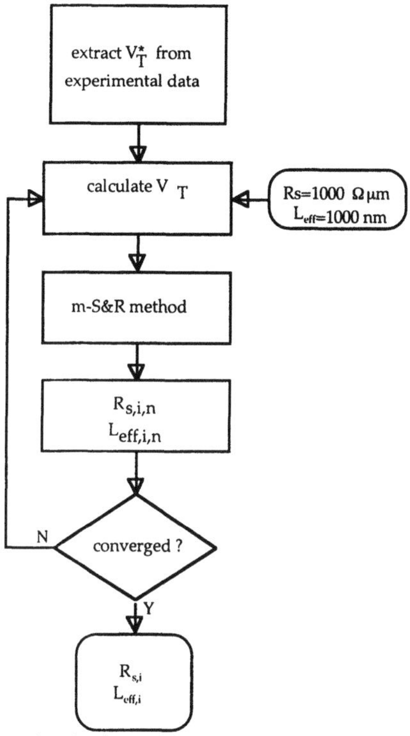

Fig.12 Self-consistent calculation of V_T, L_{eff} and R_s

The extracted results, L_{eff} and R_s, for our 0.1 µm devices as a function of V_{gs} are shown in Fig.3 or different poly-Si gate lengths L_g.

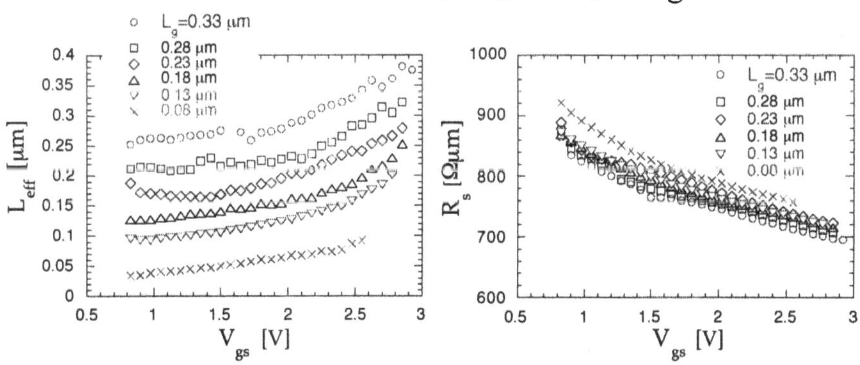

Fig.13 Results of the m-S&R method for (a) $L_{eff}(V_{gs})$ and (b) $R_s(V_{gs})$

A similar behavior as function of V_{gs} among the devices with different poly-Si gate lengths is observed as expected. Notice that devices have been made with gate lengths down to 0.08 µm resulting in a metallurgical channel length of only 30 nm. The quality of the results can be checked by the behavior of the R_S as function of V_{gs}. Since the corresponding $R_S(V_{gs})$ values all have the nearly the same V_{gs} dependence and absolute values, it can be concluded that the corresponding $L_{eff}(V_{gs})$ are accurate because $R_S(V_{gs})$ is extracted using L_{eff}.

4.5 Error analysis

Fig.14 shows the ΔL and R_S of experimental devices obtained by a conventional L-array method and the m-S&R algorithm. It can be clearly seen that the total spread (roughly equal to 4σ) is smaller for the m-S&R method. The σ is 30 nm for the conventional algorithm while only 5-10 nm for the m-S&R method.

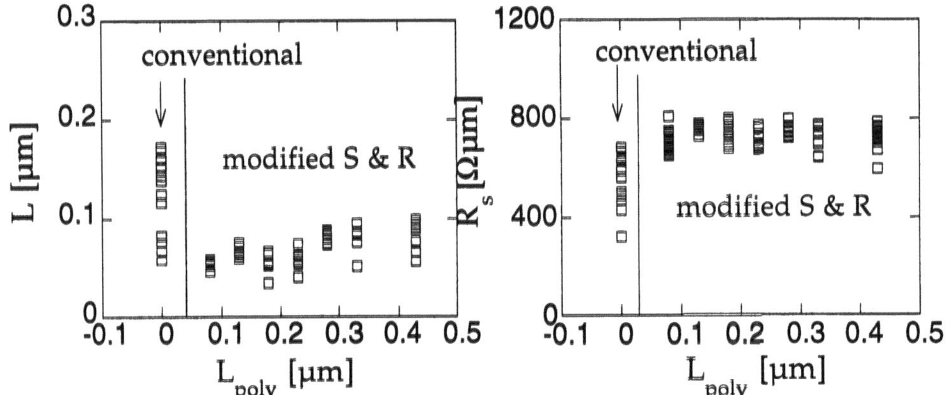

Fig.14 Comparison of the extracted ΔL and R_S for the conventional and m-S&R method

It is interesting to note that the statistical fluctuations of ΔL seem to decrease for smaller L_g. This can be understood from

$$\sigma_{\Delta L} \cong \frac{L_o}{r^2}\sigma_r = L_o\frac{\Delta r}{r^3} \tag{31}$$

As $L_{eff,i}$ decreases with respect to L_o, 'r' increases and consequently $\sigma_{\Delta L}$ decreases to a power of 3 for a constant Δr.

It is important to understand at this point what the total σ as shown in Fig.20 means. It contains at least two contributions: the real spread due to process variations and an extra introduced spread due to the extraction method itself. This latter term is smaller for the m-S&R as shown in the error analysis above. It is most probable that the remaining spread in the case of the m-S&R is mainly due to process variations.

5.CONCLUSIONS

Based on the results for a 0.1 μm NMOS devices it has been shown that classical methods are no longer sufficient for an accurate determination of Leff and Rs. Therefore a new "modified" shift and ratio method has been developed providing both process and device engineers a new tool for accurate process analysis and circuit prediction.

References:

[1] Y. Taur, D. Zicherman, D. Lombardi, P. Restle, C. Hsu, H. Hanafi, M. Wordeman, B. Davari, G. Shahidi, "A new shift & Ratio method for MOSFET channel length extraction", IEEE Electron Device Letters, Vol. 13, No. 5, pp. 267-269, 1992

[2] S. Kubicek, S. Biesemans, Q. Wang, K. Maex, K. De Meyer, "Sub 0.1 μm nMOSFETs fabricated using experimental design techniques to optimise performance and minimise process sensitivity," proceedings Symposium on VLSI Technology, pp. 105-106, 1995

[3] S. Biesemans, M. Hendriks, S. Kubicek, K. de Meyer, "Accurate determination of the gate voltage dependent effective channel length, series resistance and source drain doping profile for optimisation," proceedings Symposium on VLSI Technology, pp. 165-166, 1996

[4] S. E. Laux, "Accuracy of an effective channel length/external resistance extraction algorithm for MOSFETs," IEEE Trans. on Electron Devices, Vol. 31, No. 9, pp. 1245-1251, 1984

[5] P. Suciu, R. Johnston, "Experimental derivation of the source and drain resistance of MOS transistors," IEEE Trans. on Electron Devices, pp. 1846-1847, 1980

[6] K. Terada, H. Muta, "A new method to determine effective MOSFET channel length," Jap. J. Appl. Phys, Vol. 18, p. 953, 1979

[7] J. Chern, P. Chang, R. Motta, N. Godinho, "A new method to determine MOSFET channel length," IEEE Electron Device Lett., Vol. 1, No. 9, pp.170-173, 1980

[8] G. Hu, C. Chang, Y. Chia, "Gate-voltage-dependent effective channel length and series resistance of LDD MOSFET," IEEE Trans. on Electron Devices, Vol. 34, No.12, pp. 2469-2475, 1987

[9] S. Biesemans, K. De Meyer,"Comparison of an L-array and a single transistor method to extract Leff and Rs in deep submicron MOSFETs," proceeding ESSDERC, pp. 660-663, 1997

[10] J. Whitfield, "A modification on an improved method to determine MOSFET channel length", IEEE Electron Device Lett., pp.109-110, 1985

[11] A. Sabnis, J. Clemens., “Characterisation of the electron mobility in the inverted <100> Si Surface,“ proceedings IEDM, pp. 219-222, 1979

Recent Developments in BSIM for CMOS RF ac and Noise Modeling

Jeffrey J. Ou, Xiaodong Jin, Paul R. Gray, and Chenming Hu

Department of Electrical Engineering and Computer Sciences
University of California, Berkeley, USA

ABSTRACT

A major barrier to the successful realization of commercial CMOS RF communication systems is the relatively slow progress in the development of CMOS RF compact models. This paper reviews the recent activities of the BSIM development team in addressing this need. Both RF ac and noise models based on BSIM3v3 were proposed and evaluated with 2-D simulations and experimental data. Good agreement has been achieved in the GHz range. These models will enable the accurate simulations of new emerging CMOS RF LSI communication systems using SPICE.

I. Introduction

Commercial high-integration CMOS RF systems which attempt to integrate the radio-frequency (RF) building blocks together with baseband circuits and, potentially, the back-end DSP, will soon play an important role in new emerging wireless personal communications market [1][2]. The increasing popularity of digital wireless communication electronics has created a large demand for lower cost, lower power, and smaller form-factor solutions. Moreover, main-stream CMOS technologies are now capable of operation at radio frequencies. The current 0.18μm CMOS technology has achieved an f_T of 30GHz with reasonable power consumption, which may be sufficient for the commercial RF applications up to 2.5GHz. With the continuous scaling of deep-submicron CMOS technologies, projected 0.1μm in year 2006 [3], soon it will be possible to implement commercial systems operating around 5GHz or above.

To date, research efforts worldwide, particularly from universities, have responded to this trend very well. Examples of high-integration RF systems for cordless as well as cellular phones, and GPS standards have been presented in the literature [4]-[9]. However, the semiconductor industry has continued to focus on more expensive, traditional RF technologies, such as Si or SiGe BiCMOS, for high integration. An exception was reported this year, which is the first commercial-ready, single-chip, CMOS product for moderate-performance cordless phone applications [10]. One of the major barriers which has slowed the transition from experimental CMOS prototypes to high-performance, commercial implementations is the inadequate CMOS RF compact models [11]. Given the fact that the time-to-market pressure for the commercial product is larger than ever, such accurate modeling has become the key to the success of first-run working silicon.

This paper is intended to summarize the efforts been made to date by the BSIM development team at the University of California, Berkeley, to close this gap between technology and model development. Section II briefly describes the history of BSIM and the latest model BSIM3v3.2. Section III focuses on the RF ac model development based on BSIM3v3, followed by the noise model development in Section IV. Summary and future work will be highlighted in Section V.

II. BSIM3 Model

BSIM3 is the third-generation BSIM (Berkeley Short-channel IGFET Model), developed and maintained by the Device Research Group in the Department of Electrical Engineering and Computer Sciences at the University of California, Berkeley, for MOSFET device modeling in circuit simulations. BSIM3 is based on Poisson's equation using gradual channel approximation and coherent quasi-2D analysis, taking into account the effects of device geometry and process parameters [12]. Because it is a physics-based model, BSIM3 is robust and scalable, only one set of model parameters required to obtain the desired accuracy across a wide range of device geometries. Its third iteration, BSIM3 version 3 (BSIM3v3) was first introduced in October 1995 and later became the first industry standard of its kind for CMOS device modeling and integrated circuit design. BSIM3v3 greatly improves the accuracy of simulating the small-signal output resistance by considering three physical mechanisms (channel length modulation, drain-induced barrier lowering, and substrate current induced body effect). This improvement is not only important for analog IC design, but also essential for accurate RF ac and noise modeling. Although the proposed models introduced in Section III and IV can potentially be applied to other MOSFET models, an accurate low-frequency dc/ac model such as BSIM3v3 provides a solid foundation for RF device modeling. The latest version of BSIM3 is BSIM3v3.2, released in June 1998, in which a new intrinsic capacitance model (charge thickness model) is introduced to improve the accuracy of gate capacitance at the weak-to-inversion transition, and for the advanced thin gate-oxide CMOS technologies.

III. BSIM3v3 RF ac Model

A. Evaluation of Device RF ac Behavior

Conventionally, the small-signal ac behavior of a MOSFET in the GHz range is characterized by *scattering parameters* (s-parameters) in a two-port network [13], as shown in Fig. 1, where the gate is connected to port 1 and drain is port 2, with source and bulk tied to ground. Since s-parameters are measured under a matched condition (typically to 50Ω), it is relatively easier to control at high frequencies, compared to a *true* open or short (to obtain z-parameters and y-parameters, respectively) in the measurement. In addition, a great series of microwave techniques have been developed based on s-parameters, such as matching network design and stability analysis, which makes s-parameter method the favorable representation in RF measurement and microwave circuit design.

However, evaluating the accuracy of an RF model based solely on s-parameters may not be sufficient. For example, Fig. 2(a) shows a typical S_{11} of a NMOS device alone, referred to as an *untuned* case, and a model by fitting s-parameters. Fig. 2(b) is the S_{11} of the same device and the model under the *tuned* condition, in which an inductor is placed in series

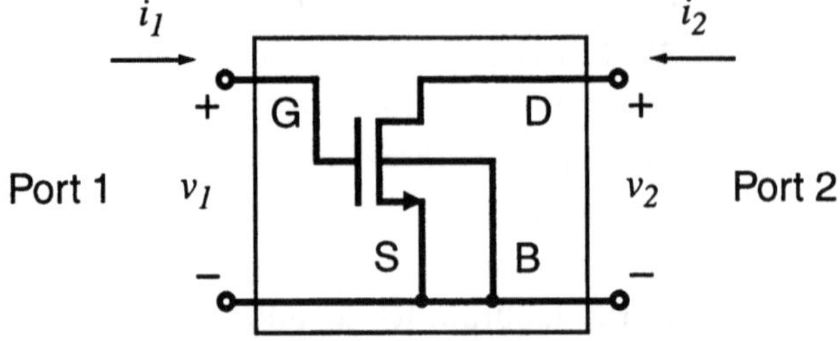

Figure 1. A 2-port representation of a NMOS device.

with C_{gs}, as shown in Fig. 2(c), a common circuit exercise for matching purpose. Clearly the model does not predict this tuned circuit correctly, even though the S_{11} of the device alone exhibits a fairly good agreement with data. Another method of evaluation is to plot the y-parameters, which can be converted from s-parameters. Fig. 2(d) and Fig. 2(e) shows the real and imaginary parts of Y_{11}. The difference between the model and data can now be observed in the real part of Y_{11}. However, the meaning of Y_{11} is hard to appreciate and therefore difficult to be used as a criterion for model evaluation. A better way to evaluate the model and to obtain more physical insight of a device is to write Y_{11} as

$$R_{in} = \mathrm{real}\left(\frac{1}{Y_{11}}\right), \text{ and } C_{in} = \frac{-1}{\mathrm{imag}\left(\frac{1}{Y_{11}}\right)\omega}, \tag{1}$$

where ω is the frequency in rad/sec. R_{in} can be interpreted as input resistance and C_{in} as input capacitance, which have clear meanings to both circuit designers and device engineers. Fig. 2(f) and Fig. 2(g) show the R_{in} and C_{in} of the same device, respectively. Now it is obvious that the model predicts the wrong value of R_{in}, which causes the inaccurate simulation.

The same philosophy can be applied to all y-parameters, as summarized in Table 1. Those parameters will be referred to as *circuit parameters* (c-parameters) in the discussion later.

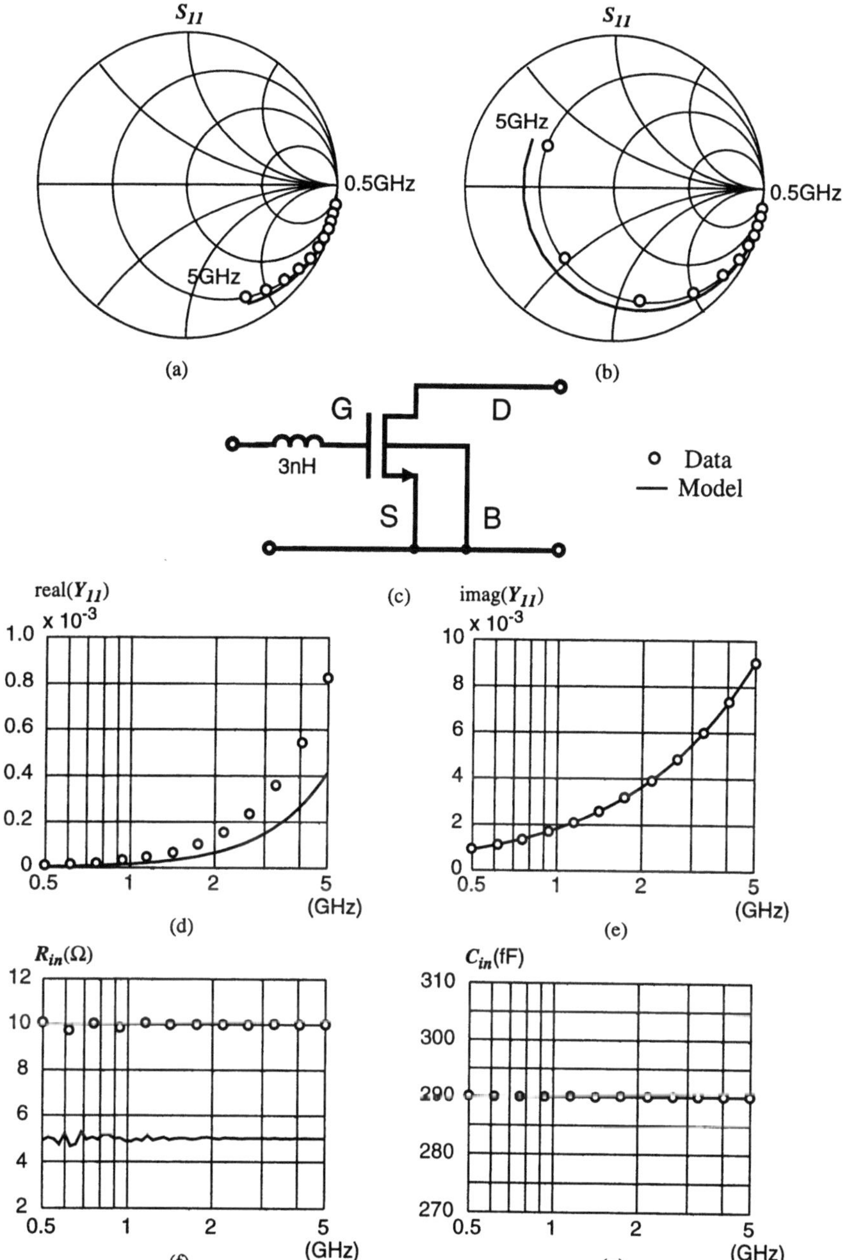

Figure 2. (a) S_{11} of an untuned device. (b) S_{11} of the same device under a tuned condition. (c) Illustration of the tuned device. (d)-(g) Real part of Y_{11}, imaginary part of Y_{11}, R_{in}, and C_{in} of the untuned device, respectively.

Table 1. Definitions of c-parameters.

Input Resistance $$R_{in} = \mathrm{real}\left(\frac{1}{Y_{11}}\right)$$	Input Capacitance $$C_{in} = \frac{-1}{\mathrm{imag}\left(\frac{1}{Y_{11}}\right)\omega}$$
Output Resistance $$R_{out} = \frac{1}{\mathrm{real}(Y_{22})}$$	Output Capacitance $$C_{out} = \frac{\mathrm{imag}(Y_{22})}{\omega}$$
Magnitude of Transconductance $$G_{mm} = \lvert Y_{21} + Y_{12}\rvert$$	Phase of Transconductance $$G_{mp} = \angle(Y_{21} + Y_{12})$$
Feedback Resistance $$R_{fb} = \mathrm{real}\left(\frac{-1}{Y_{12}}\right)$$	Feedback Capacitance $$C_{fb} = \frac{1}{\mathrm{imag}\left(\frac{1}{Y_{12}}\right)\omega}$$

B. Proposed RF ac model

To overcome the shortcomings of the low-frequency models, several CMOS RF ac models based on BSIM3v3 have been recently proposed [14]-[16]. All of them add a gate resistance R_g and a substrate resistance network to core BSIM3v3. Fig. 3 shows the one proposed in [16], in which three resistors R_g, R_{subd}, and R_{subs} are added to the existing BSIM3v3

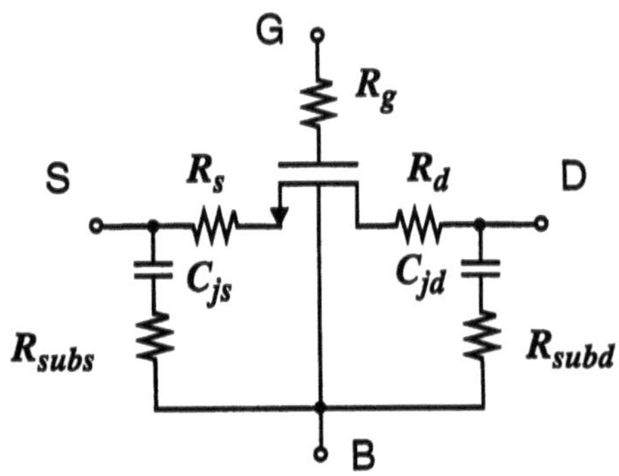

Figure 3. Proposed RF ac model with a simple substrate network.

model. The source/drain resistance (R_s, R_d) and the source/drain junction capacitance (C_{js}, C_{jd}) are part of the low-frequency model. R_g represents both the physical gate resistance and the *non-quasi-static* (NQS) effect of channel resistance. R_{subd} and R_{subs} are the lumped substrate resistors between source/drain junctions and substrate contacts. The values of R_{subd} and R_{subs} are not necessarily equal as they are functions of transistor layout. The addition of three resistors is the minimum requirement to obtain sufficient accuracy for the device RF ac behavior. As an illustration, Fig. 4 shows the c-parameters of a 240/0.35 NMOS device with 24 fingers. The proposed simple model provides fairly accurate simulation results compared to measured data. R_g is responsible for the non-zero R_{in}; while the roll-off in R_{out} is due to R_{subd}. Since the source is shorted to bulk, R_{subs} in this case has no effect on the overall device ac behavior.

C. Physics-Based Effective Gate Resistance

One observation from Fig. 4 is that constant R_g models R_{in} very well across a wide range of frequencies. However, the value of R_g has a strong bias dependency, which is related to channel resistance R_{ch} [17]. This section will describe the bias and geometry relations of R_g.

R_g consists of two parts: the distributed gate electrode resistance, R_{geltd}, and the distributed channel resistance seen from the gate, R_{gch}, as illustrated in Fig. 5,

$$R_g = R_{geltd} + R_{gch}. \tag{2}$$

The value of R_{geltd} is insensitive to the bias and frequency and can be obtained from the gate electrode sheet resistance (R_{eltd}),

$$R_{geltd} = R_{eltd}\left(\alpha\frac{W}{L} + \beta\right), \tag{3}$$

where the first term models the gate electrode resistance on top of gate oxide, and β models the external gate resistance in the field oxide region. R_{geltd} together with gate capacitance (C_{gs} and C_{gd}) forms a RC network that to the first order models the distributed effect along the gate. A simple

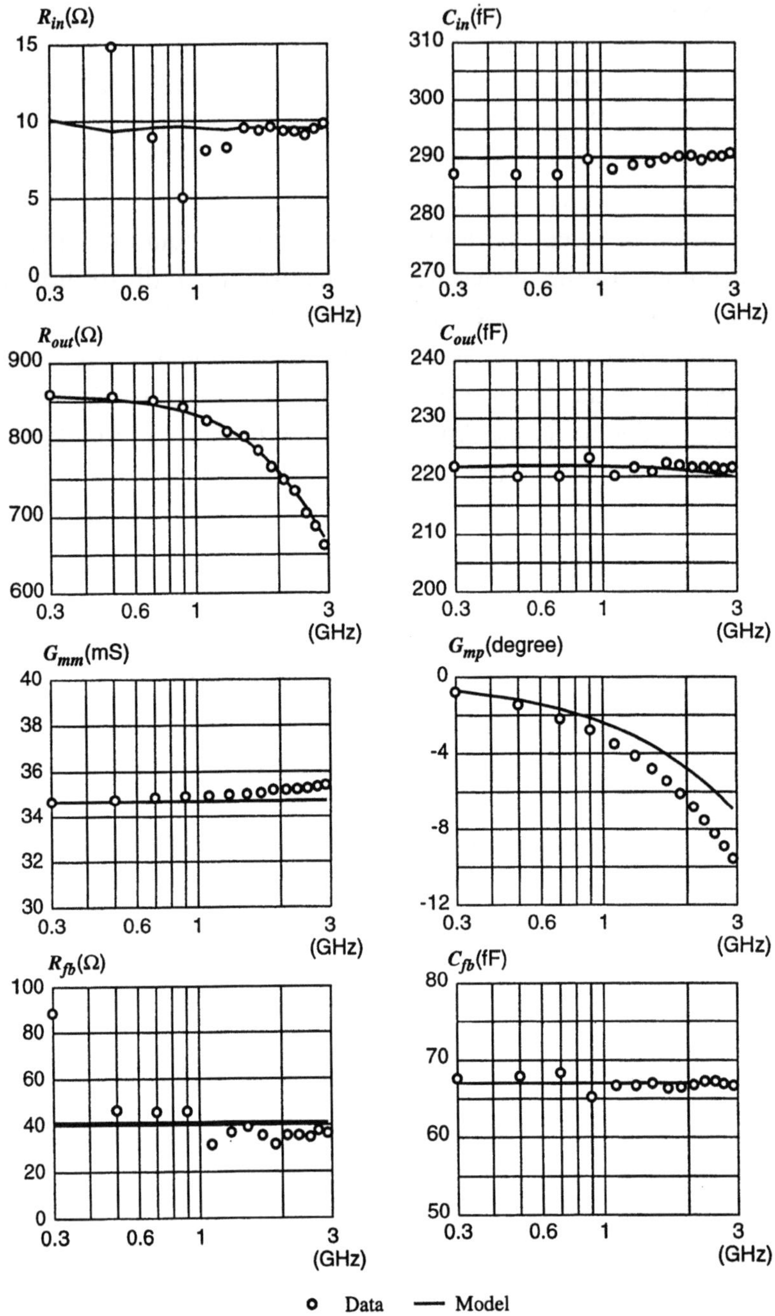

Figure 4. C-parameters of a 24-finger 240/0.35 NMOS biased at $V_{gs} = 0.9V$ and $V_{ds} = 2V$.

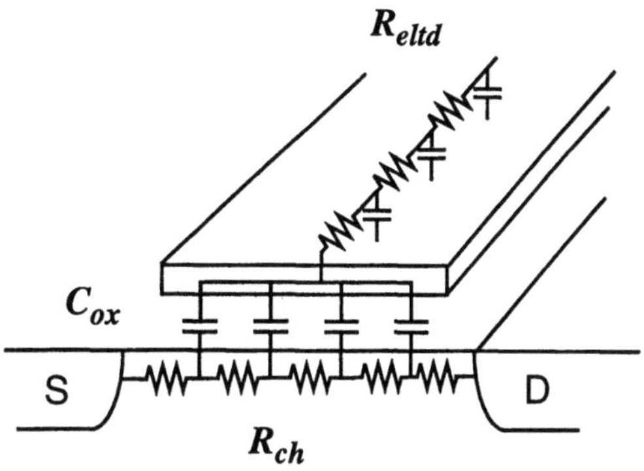

Figure 5. Illustration of the distributed gate resistance.

analysis leads to the value of α equal to 1/3 when the gate terminal is brought out from one side [18], and 1/12 when connected on both sides. A detailed derivation based on transmission-line theory resulting in the same first-order conclusion can be found in [19].

Furthermore, R_{gch} with the gate capacitance forms another RC network that to the first order models the NQS effect,

$$R_{gch} = \frac{R_{ch}}{\delta}, \tag{4}$$

where δ is the parameter accounting for NQS effect, equal to 12 if the resistance is uniformly distributed along the channel. Since the device in saturation violates this assumption, δ is left as a fitting parameter, with a typical value between 12 and 15.

R_{ch} can be derived from the channel conductance G_{ch},

$$R_{ch} = \frac{1}{G_{ch}}, \tag{5}$$

which involves two mechanisms. The first is the *static channel conductance* G_{st}, which accounts for the dc channel conductance. G_{st} can be obtained by integrating the dc resistance along the channel using a quasi-static assumption,

$$G_{st} = \frac{1}{\int dR} = \frac{I_d}{\int dV} = \begin{cases} I_d/V_{dsat} & \text{in saturation} \\ I_d/V_{ds} & \text{in triode} \end{cases}, \tag{6}$$

where V_{dsat} is the saturation drain voltage. Note that G_{st} describes only the channel conductance under dc condition. When an ac excitation occurs, an additional *excess-diffusion conductance* G_{ed} is presented in the channel,

$$G_{ed} = \eta\frac{kT}{q}\mu_{eff}C_{ox}\frac{W}{L}, \tag{7}$$

where η is a technology-dependent parameter with a typical value of 1. The overall conductance is then the sum of the two,

$$G_{ch} = G_{st} + G_{ed}. \tag{8}$$

The strong bias dependency of R_{gch}, particularly at low V_{gs}, can be easily seen in Fig. 6(a), where R_{gch} of a 10/0.5 NMOS is plotted against V_{gs}. Good agreement between 2-D simulations and the proposed R_g model is achieved with $\eta = 1$ and $\delta = 14$. Note the model will deviate from data at low V_{gs} if the excess-diffusion component is not considered. Fig. 6(b) shows the proposed model agrees with 2-D simulations fairly well both in saturation and triode regions.

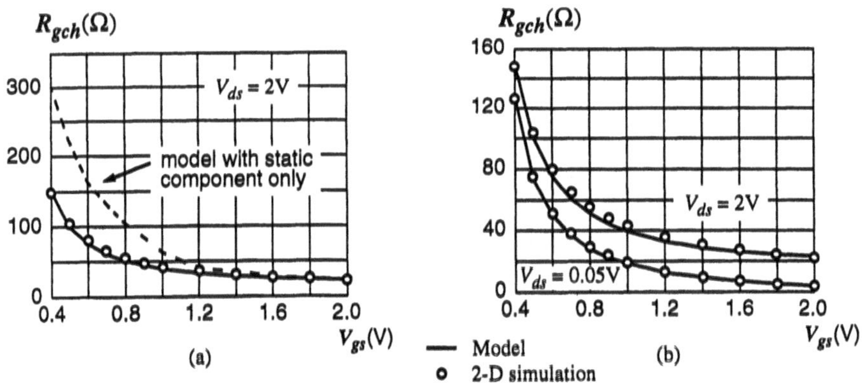

Figure 6. (a) Bias dependency of R_{gch} of a 10/0.5 NMOS.
(b) R_{gch} in saturation and triode regions.

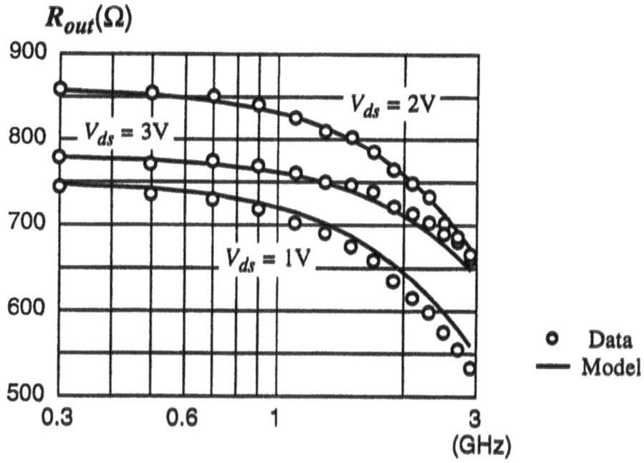

Figure 7. R_{out} of a 240/0.35 NMOS with V_{gs} = 0.9V and V_{ds} = 1V, 2V, and 3V. A fixed value of R_{subd} = 50Ω is used for all three cases.

In general the presence of R_g is undesired because it reduces the overall transconductance, degrades the quality-factor (Q) of device input impedance, and also contributes additional noise to devices (see Section IV). In the case that the total width of a transistor is fixed by other design constraints, R_{geltd} can be reduced by using shorter unit finger length, which results in less resistance per finger and more fingers in parallel. Using salicided poly-gate or future metal-gate technologies will drastically lower R_{geltd} as well. However, R_{gch} remains unchanged, which places a fundamental limit to the speed and the noise of transistors.

D. Substrate Resistance Network

As stated earlier, the roll-off in the output resistance of a device at RF is due to the substrate resistance R_{subd}. The value of R_{subd} is not a function of frequencies, nor the bias. Fig. 7 shows the measured R_{out} of a 24-finger 240/0.35 NMOS under different V_{ds}. The proposed model can fit data fairly well with one *fixed* value of R_{subd}, which is 50Ω in this case. The effect of this roll-off is directly responsible for the reduction of device gain and the non-optimal output matching, both of which may cause unexpected loss of the overall gain in circuits.

One difficulty in modeling substrate resistance is its geometry dependency. The transistor layout (W, L, number of fingers, number and the location of substrate contacts, *etc.*) and the profile of technologies (epi *vs.* non-epi, trench *vs.* LOCOS) both affect the substrate network. One solution to obtain an accurate substrate representation is to generate a resistive substrate mesh by dividing the entire substrate into grids, and combine this with the device model for SPICE simulations [20]. However, this approach is very time consuming and memory intensive. As a result it is almost impossible to simulate VLSI implementations. To trade the accuracy for efficiency, one possible compromise may be the substrate network with a 5-resistor implementation [21], as shown in Fig. 8.

IV. BSIM3v3 RF Noise Model

A. Noisy Two-Port Network

Similar to the RF ac performance, the RF noise behavior of a device can also be represented by a *noisy* 2-port network, as shown in Fig. 9(a). One convenient way to equivalently describe this network is to replace it by an input-referred noise voltage u, a noise current i, followed by a *noiseless* network, as shown in Fig. 9(b) [22]. Then the equivalent input noise resistance R_n and the equivalent input noise conductance G_n can be defined such that,

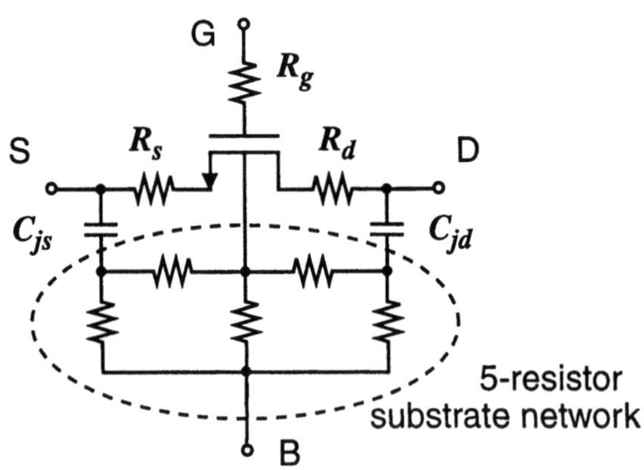

Figure 8. An RF ac model with a 5-resistor substrate network.

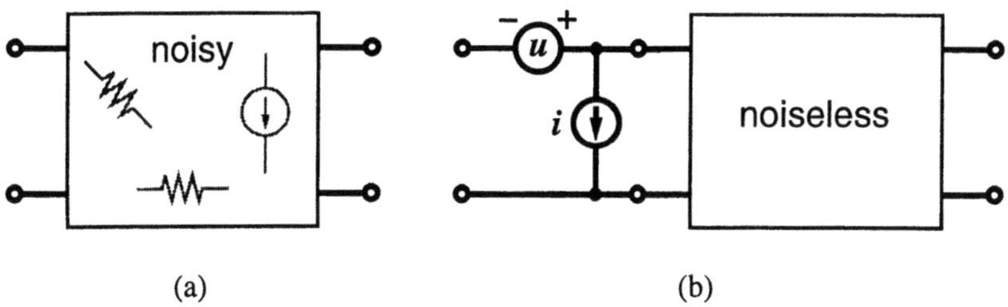

Figure 9. (a) A noisy 2-port network. (b) One equivalent representation.

$$\overline{u^2} = 4kTBR_n, \text{ and } \overline{i^2} = 4kTBG_n, \tag{9}$$

where k is the Boltzmann's constant, T is the electron temperature, and B is the bandwidth of interest. Since u and i represent all the noise sources in the 2-port network, they are correlated. To describe this correlation, the noise voltage u can be separated into two parts,

$$u = u_u + u_c, \tag{10}$$

where u_u is uncorrelated to the noise current i, and u_c is correlated to i. u_u can be further defined by an uncorrelated equivalent noise resistance R_u,

$$\overline{u_u^2} = 4kTBR_u; \tag{11}$$

and u_c can be related to i by a correlation impedance Z_{cor},

$$u_c = Z_{cor}i = (R_{cor} + jX_{cor})i. \tag{12}$$

Since the device noise figure (NF) is a strong function of the impedance looking back to the generator, for a given device with a fixed bias, an optimum generator impedance Z_{opt} exists such that a minimum noise figure (NF_{min}) can be found. If the device noise parameters R_u, G_n, and Z_{cor} are known, Z_{opt}, NF_{min}, and R_n can be calculated by

$$Z_{opt} = R_{opt} + jX_{opt}, \text{ where} \tag{13}$$

$$R_{opt} = \sqrt{\frac{R_u}{G_n} + R_{cor}^2}, \text{ and } X_{opt} = -X_{cor};$$

$$NF_{min} = 1 + 2G_n(R_{cor} + R_{opt}); \text{ and} \tag{14}$$

$$R_n = R_u + |Z_{cor}|^2 Gn. \tag{15}$$

On the other hand, if Z_{opt}, NF_{min}, and R_n are known, noise parameters R_u, G_n, Z_{cor} can also be obtained. Therefore, Z_{opt}, NF_{min}, and R_n is sufficient and necessary to describe the device noise behavior at a fixed bias.

Conventionally, optimum generator reflection coefficient Γ_{opt} is more commonly used than Z_{opt},

$$\Gamma_{opt} = \frac{Z_{opt} - Z_0}{Z_{opt} + Z_0}, \tag{16}$$

where Z_0 is typically 50Ω. Then *NF* for arbitrary generator reflection coefficient Γ_G is

$$NF = NF_{min} + \frac{4r_n|\Gamma_G - \Gamma_{opt}|^2}{(1 - |\Gamma_G|^2)|1 + \Gamma_{opt}|^2}, \tag{17}$$

where r_n is the normalized noise resistance, $r_n = R_n/Z_0$. It is clear that from Eqt. 17, NF_{min} is found when Γ_G is equal to Γ_{opt}.

B. Proposed RF Noise Model

Fig. 10 shows the noise sources contribute to the overall RF thermal noise of a MOS device based on the proposed RF ac model with the lumped substrate network. The noise generated from resistors can be represented by noise current sources shunted to the corresponding resistors; and the

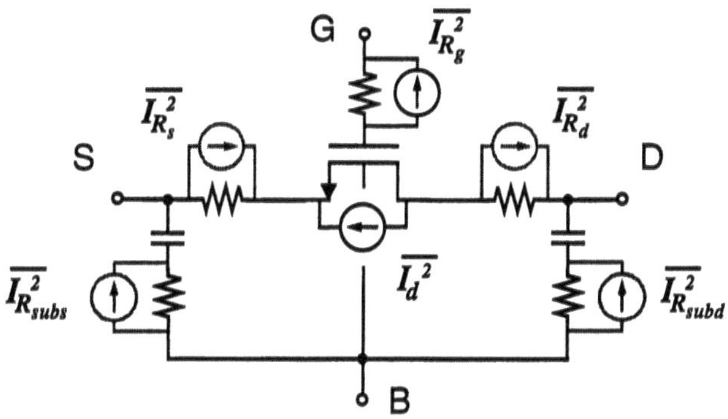

Figure 10. Noise sources in a MOS device.

thermal noise generated from the channel appearing at drain is commonly modeled by a noise current source $\overline{i_d^2}$. Since the channel thermal noise is caused by the random thermal movement of the carriers in the channel, $\overline{i_d^2}$ is directly related to the channel conductance G_{ch},

$$\overline{i_d^2} = 4kTB\gamma G_{ch}, \tag{18}$$

where γ is the noise factor. Note that this equation is valid only when the electron temperature is constant, which is a fair approximation along the channel up to the point where pinch-off occurs.

For long-channel devices, G_{ch} is approximately equal to g_m. As a result the conventional equation for channel noise current is

$$\overline{i_d^2} = 4kTB\gamma g_m, \tag{19}$$

with γ equal to 2/3 in saturation region [23]. However for short-channel devices, γ has been shown to be larger than 2/3, and more importantly, the value of γ varies with bias conditions [24].

Fig. 11 shows γ extracted from measured RF noise data of a 24-finger 240/0.35 NMOS at different bias conditions. With the proposed channel thermal noise model (Eqt. 18), γ is found to be insensitive to V_{ds} and has a weak dependency on V_{gs}.

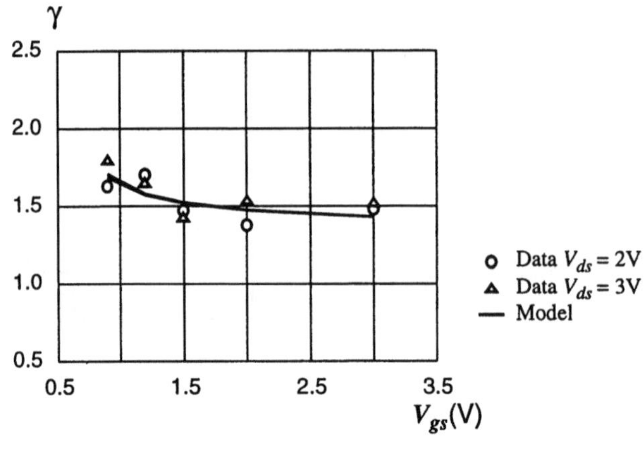

Figure 11. Noise factor γ of a 24-finger 240/0.35 NMOS under different bias conditions.

As described in the previous section, the gate resistance R_g has two components: R_{geltd} and R_{gch}, both of which contribute thermal noise. It has been shown that the noise generated from the distributed gate electrode has an equivalent thermal resistance equal to R_{geltd} [18]. Therefore the noise is

$$\overline{i^2_{R_{geltd}}} = 4kTB\frac{1}{R_{geltd}}. \tag{20}$$

The noise associated with R_{gch} is so called *induced gate noise*, first revealed by van der Ziel [25]. A gate noise current generator shunted to C_{gs} was used by van der Ziel to model this phenomenon, which is shown to be equivalent to the noise generated by a resistor, R_{gch} in our case, in series with C_{gs} [26]. Since the induced gate noise comes from the same sources as the channel thermal noise, they are correlated. This can be modeled by assigning a correlation factor c to separate the noise into two parts,

$$\begin{cases} \overline{i^2_{R_{gch}u}} = 4kTB\dfrac{1}{R_{gch}}(1-|c|^2) & \text{uncorrelated} \\ \overline{i^2_{R_{gch}c}} = 4kTB\dfrac{1}{R_{gch}}|c|^2 & \text{correlated} \end{cases}, \tag{21}$$

where c is generally a complex number.

Since R_s, R_d, R_{subd}, and R_{subs} are all physical resistors, their noise contributions are simply

$$\overline{i_{R_i}^2} = 4kTB\frac{1}{R_i}, \tag{22}$$

where $R_i = R_d$, R_s, R_{subd}, or R_{subs}.

Fig. 12 shows the comparison of the proposed RF noise model with measured data of a 240/0.35 NMOS, biased at $V_{gs} = 0.9\text{V}$ and $V_{ds} = 2\text{V}$. The model fits data well with the correlation factor c set to $0.4\angle{-120°}$.

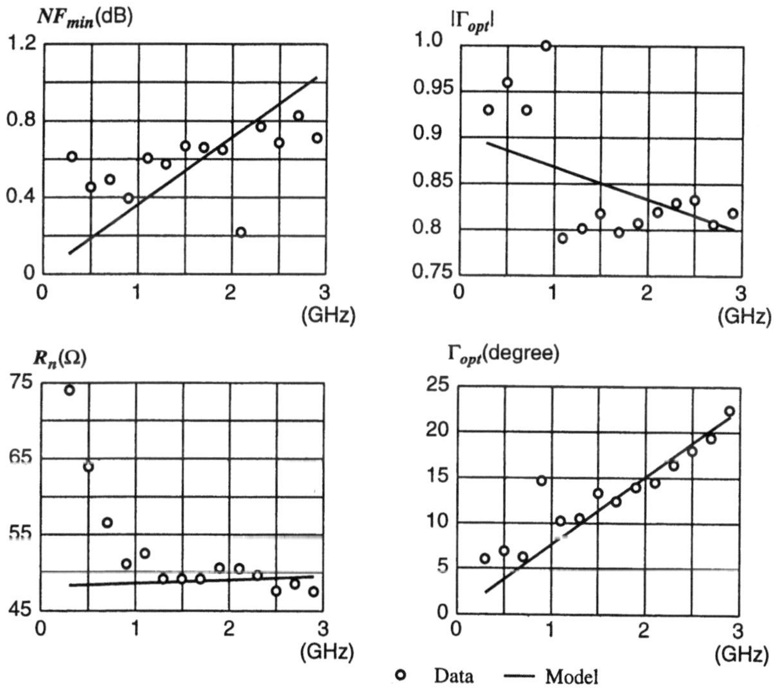

Figure 12. Noise parameters of a 24-finger 240/0.35 NMOS at $V_{gs} = 0.9\text{V}$ and $V_{ds} = 2\text{V}$.

V. Summary and Future Work

The proposed physics-based CMOS RF ac and noise models using BSIM3v3 are reviewed and summarized. The core of these models is the newly developed channel conductance expression. Based on this an effective gate resistance was developed to model its strong bias dependency; the bias and geometry dependencies of the channel noise are also modeled correctly. The accuracy of the models was evaluated with 2-D simulations and experimental data. Good agreement has been obtained in the GHz range. The BSIM3v3 CMOS RF ac and noise models will enable the simulations of the emerging CMOS RF LSI communication systems using SPICE.

The future work includes, but is not limited to: the study of the geometry dependency in the substrate resistance network, further investigation of the correlation in the induced gate noise, and the evaluation of RF large-signal behavior of CMOS devices, such as distortion, intermodulation, and power-added efficiency (PAE).

VI. Acknowledgment

This work is supported by Semiconductor Research Corporation, California MICRO, STMicroelectronics, Conexant Systems, National Semiconductor, Texas Instruments, and Berkeley Technology Associates. The authors would like to thank Weidong Liu at UC Berkeley, Chih-Hung Chen at Simon Fraser University, Canada, Gang Zhang at BTA, and Ingrid Ma at Intel for their contributions to this project.

VII. References

[1] P. Gray and R. Meyer, "Future directions of silicon IC's for RF personal communications," in *CICC*, pp. 83-90, May 1995.

[2] F. Brianti *et al.*, "High integration CMOS RF transceivers," in *AACD*, Lausanne-Ouchy, Switzerland, April 1996.

[3] *The National Technology Roadmap for Semiconductors*, SIA, 1997.

[4] J. Rudell *et al.*, "Recent developments in high integration multi-standard CMOS transceivers for personal communication systems," in *Symp. Low Power Electronics*, pp. 149-154, August 1998.

[5] J. Rudell *et al.*, "A 1.9GHz wide-band IF double conversion CMOS integrated receiver for cordless telephone applications," in *ISSCC*, pp. 304-305, February 1997.

[6] A. Abidi *et al.*, "The future of CMOS wireless transceivers," in *ISSCC*, pp. 118-119, February 1997.

[7] D. Shaeffer *et al.*, "A 115mW CMOS GPS receiver," in *ISSCC*, pp. 122-123, February 1998.

[8] M. Steyaert *et al.*, "A single-chip CMOS transceiver for DCS-1800 wireless communications," in *ISSCC*, pp. 48-49, February 1998.

[9] P. Orsatti *et al.*, "A 20mA-receive 55mA-transmit GSM transceiver in 0.25μm CMOS," in *ISSCC*, pp. 232-233, February 1999.

[10] T. Cho *et al.*, "A single-chip CMOS direct-conversion transceiver for 900 MHz spread-spectrum digital cordless phones," in *ISSCC*, pp. 228-229, February 1999.

[11] T. Manku, "Microwave CMOS - devices and circuits," in *CICC*, pp. 59-66, May 1998.

[12] W. Liu *et al.*, *BSIM3v3.2 MOSFET Model Users' Manual*, University of California, Berkeley, Memo. no. UCB/ERL M98/51, August 1998.

[13] *S-parameter Techniques for Faster, More Accurate Network Design*, Hewlett Packard, Test and Measurement, Application Note 95-1.

[14] W. Liu *et al.*, "RF MOSFET modeling accounting for distributed substrate and channel resistance with emphasis on the BSIM3v3 SPICE model," in *IEDM*, pp. 309-312, December 1997.

[15] D. Pehlke *et al.*, "High frequency application of MOS compact model and their development for scalable RF model libraries," in *CICC*, pp. 219-222, May 1998.

[16] J. Ou *et al.*, "CMOS RF modeling for GHz communication IC's," in *VLSI Symp. Tech.*, pp. 94-95, June 1998.

[17] X. Jin *et al.*, "An effective gate resistance model for CMOS RF and noise modeling," in *IEDM*, pp. 961-964, December 1998.

[18] B. Razavi, R. Yan, and K Lee, "Impact of distributed gate resistance on the performance of MOS devices," *IEEE Trans. Circuits and Systems I*, vol. 41, no. 11, pp. 750-754, November 1994.

[19] E. Abou-Allam and T. Manku, "A small-signal MOSFET model for radio frequency IC applications," *IEEE Trans. Computer-Aided Design*, vol. 16, no. 5, pp. 437-447, May 1997.

[20] R. Gharpurey and R. Meyer, "Modeling and analysis of substrate coupling in integrated circuits," *IEEE J. Solid-State Circuits*, vol. 31, no. 3, pp. 344-353, March 1996.

[21] L. Tiemeijer and D. Klaassen, "Geometry scaling of the substrate loss of RF MOSFETs," in *ESSDERC*, pp. 480-483, September 1998.

[22] G. Vendelin, A. Pavio, and U. Rohde, *Microwave Circuit Design Using Linear and Nonlinear Techniques*, New York: Wiley, 1990.

[23] P. Gray and R. Meyer, *Analysis and Design of Analog Integrated Circuits*, 3rd edition, New York: Wiley, 1993.

[24] A. Abidi, "High-frequency noise measurements on FET's with small dimensions," *IEEE Trans. Electron Devices*, vol. 33, no. 11, pp. 1801-1805, November 1986.

[25] A. van der Ziel, *Noise in Solid State Devices and Circuits*, New York: Wiley, 1986.

[26] D. Shaeffer and T. Lee, "A 1.5-V, 1.5-GHz CMOS low noise amplifier," *IEEE J. Solid-State Circuits*, vol. 32, no. 5, May 1997.

MOS Transistor Modeling Issues for RF Circuit Design

Christian Enz and Yuhua Cheng

Conexant Systems Inc., Newport Beach, U.S.A.

Abstract

This paper presents the basis of the modeling of the MOS transistor for circuit simulation at radio-frequency (RF). A physical equivalent circuit that can easily be implemented as a Spice subcircuit is first derived. In addition to the intrinsic device and the source and drain series resistance, the model also has a gate resistance, that is fundamental to correctly model the frequency and noise behavior. The subcircuit also includes a substrate network that accounts for the signal coupling occurring at HF from the drain to the source and the bulk. It is shown that the latter mainly affects the output admittance y_{22}. The bias and geometry dependence of all the subcircuit components, leading to a scalable model, are then discussed with emphasis on the substrate resistances. Analytical expressions of the Y-parameters, the transit frequency and the maximum oscillation frequency are established and compared to measurements made on a 0.25 μm CMOS process. The Y-parameters and transit frequency simulated with this scalable model versus frequency, geometry and bias are in good agreement with measured data. The non-quasi-static effects and their practical implementation in the Spice subcircuit are then briefly discussed. Finally, a new thermal noise model is introduced and compared to noise models available in BSIM3v3. The parameters used to characterize the noise at HF are then presented and the scalable model is favorably compared to measurements made on the same devices.

I. Introduction

The recent explosive communication market puts a strong demand for low-cost, small-form factor and low-power transceivers. Although until recently most of the RF circuits used in products were integrated in bipolar or BiCMOS technologies, todays deep-submicron CMOS processes typically reach the 50 GHz f_t region and offer low noise figures, making them serious alternative for RF circuits integration [1]-[5]. In addition, CMOS offers VLSI capabilities allowing for a high level of integration. Due to the different properties of MOS and bipolar technologies and devices, new architectures suitable for the integration of CMOS transceivers have to be used [6][9]. The feasibility of RF CMOS circuits integration has been demonstrated with prototypes that used even older CMOS processes [10]. There is no doubt that in the coming years new wireless products will appear with CMOS as the technology used for the integration of the RF portion [6]-[12].

Nevertheless, the design of RF circuits (MOS or bipolar) for real products remains a challenge due to the strong constraints on power consumption and noise that leave very little margins for the RF IC designer. Therefore, it is crucial to be able to accurately predict the performance of the CMOS RF circuits in order to reduce design cycles and have first time success. Although good results can be obtained for lower frequency circuits (typically below 100 MHz), the simulation of RF circuits in the GHz frequency range with the available MOS compact models such as BSIM3v3 [13][14], MOS Model 9 [15][16] or EKV [17]-[20] without consideration of the parasitic components, gives inaccurate or even wrong results. These models have therefore to be enhanced in order to be used for RF CMOS IC design. It can be shown that the strict minimum to do to improve the models used for RF circuit simulation, is to add terminal resistances, mainly a gate resistance [21][22]. But this is not sufficient. At high-frequency (HF), the signal coupling through the substrate has to be accounted for by adding adequate substrate resistances [21][23]-[31]. Also non-quasi-static (NQS) effects have to be accounted for at least for the slower p-channel device and/or for non-minimum channel length devices used for example in bias circuits† [21][34]-[49].

As is the case for the existing models used at low-frequency, the RF MOS model should be scalable (at least over the geometry range used for RF devices) and should predict the HF behavior correctly for any bias point [21][50]. RF MOS transistor modeling and parameter extraction therefore require a special effort, but it remains a difficult multidimensional problem since frequency is added to the already existing geometries (W, L) and bias dimensions.

Section II presents the modeling of the MOS transistor at RF with emphasis on the small-signal operation in strong inversion and in saturation. Simple expressions for the Y-parameters are derived that are useful for direct parameter extraction and for circuit analysis and optimization. Section III discusses the noise properties of the MOS at HF and gives simple expressions for the four noise parameters.

---

† One may argue that bias circuits do not need to be simulated at RF, which is partially true. However, sometimes dc control loops have to be checked for stability and therefore require careful simulation at RF.

II. MOS TRANSISTOR MODELING AT RF

A. MOST Equivalent Circuit at RF

Although it is always possible to have a detailed equivalent circuit that accounts for all the physical elements that are part of the RF MOS transistor, it is often too complex to be implemented as a compact model or a subcircuit for circuit simulation purposes. Moreover, many of the component values would be difficult or even impossible to extract and the subcircuit would contain too many internal nodes which would significantly reduce the simulation efficiency. Like it is often the case in modeling, a trade-off has to be found between accuracy and efficiency. A good compromise is obtained when simplifying the complete detailed equivalent circuit to the one presented in Fig. 1 [21][25]-[31], which from experience has shown to be sufficient for most RF circuit simulation. In order to implement this equivalent circuit in a Spice simulator as a subcircuit (SUBCKT), all the extrinsic components have been pulled out of the MOS transistor compact model, so that the MOS transistor symbol only represents the intrinsic part of the device. This allows to have access to internal nodes and model extrinsic components such as series resistances and overlap capacitances in a different way than what is available in the compact model.

The source and drain series resistors have been added outside the MOS model since the internal series resistances are only "soft" resistances embedded in the expression used to calculate the drain current to account for the dc voltage drop across the source and drain resistances, but they do not add any poles and are therefore invisible for ac simulation. The gate resistance is usually not part of the MOS compact model, but plays a fundamental role in RF circuits and has therefore to be added to. The substrate resistors R_{dsb}, R_{sb} and R_{db} have been added to account for the signal coupling through the substrate. This effect will be discussed in more details in Section F. The source-to-bulk and drain-to-bulk diodes are also part of the compact model but their anodes are connected to the same substrate node. The diodes internal to the compact model have therefore also been turned off and two external diodes D_{sb} and D_{db} have been added in order to account for the substrate resistance R_{dsb} existing between the source and the drain diffusions. Note that the intrinsic substrate node should be connected at some point along the R_{dsb} resistor, but simulations have shown that connecting the intrinsic substrate to the source or the drain side of R_{dsb} have little influence on the Y-parameters. Therefore the intrinsic substrate has been connected to the source side of R_{dsb} in order to save one node. In Fig. 1 b) two bias dependent overlap capacitances C_{gso} and C_{gdo} have been added. This is not always required, depending on the compact model used. For example, BSIM3v3 accounts for bias dependent overlap capacitances that if extracted correctly have shown a sufficient accuracy. Pulling them out allows also to correct for the inac-

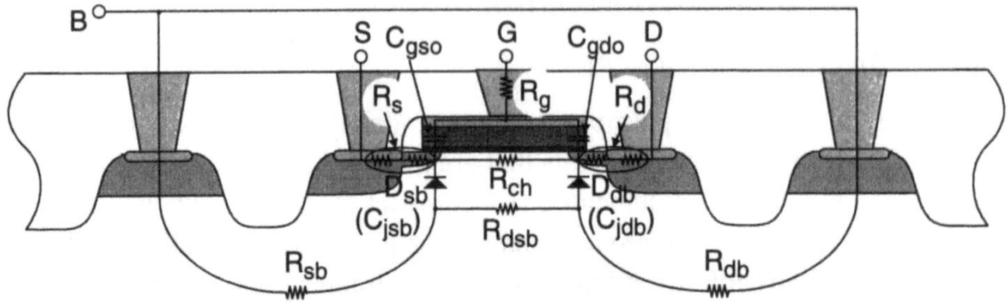

a) MOS Transistor Cross-Section.

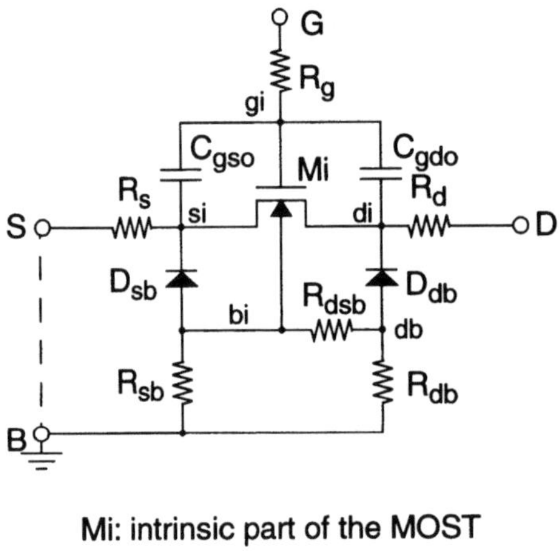

b) Equivalent subcircuit.

```
.SUBCKT NFET_RF D G S B WF=5E-6 LF=0.36E-6
NF=4
MI DI GI SI BI MIMODEL W=WF L=LF M=NF
RG G GI {RG}
RS S SI {RS}
RD D DI {RD}
CGSO GI SI {CGSO}
CGDO GI DI {CGDO}
RDSB DB BI {RDSB}
RSB BI B {RSB}
RDB DB B {RDB}
DSB BI SI DSBMODEL
DDB DB DI DDBMODEL
.MODEL DSBMODEL D
+...
.MODEL DDBMODEL D
+...
.MODEL MIMODEL NMOS
+...
.ENDS
```

c) Spice implementation of the subcircuit.

Fig. 1: MOS Transistor Equivalent Circuit.

curacies of the intrinsic capacitances appearing for short-channel devices. The circuit of Fig. 1 b) can then easily be implemented in Spice as a subcircuit as shown in Fig. 1 c).

B. Small-Signal Circuit in Saturation

The small-signal equivalent circuit of Fig. 1 b) is shown in Fig. 2 where it has been assumed that the MOS transistor operates in saturation and in quasi-static regime [21]. The voltage-controlled current sources (VCCS) are defined in Fig. 2 for both substrate referenced models (such as EKV) or source referenced models (such as BSIM3v3 or MOS model 9). The slope factor† n used in the transadmittances Y_m, Y_{ms} and Y_{mb} defined in Fig. 2 is equal to [18][21]

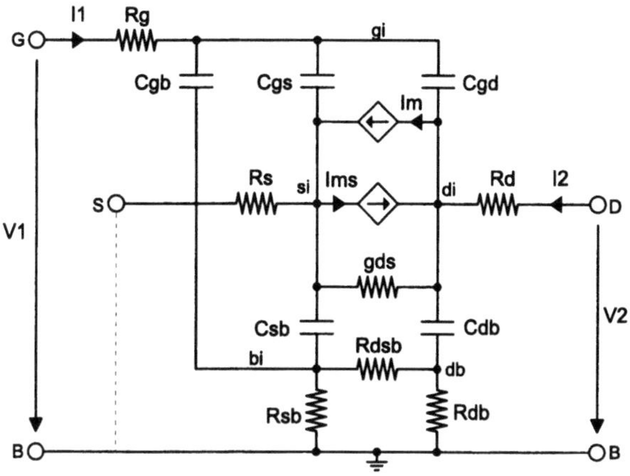

For bulk referenced model:

$$I_m = Y_m \cdot (V(gi) - V(bi))$$

$$I_{ms} = Y_{ms} \cdot (V(si) - V(bi))$$

For source referenced model:

$$I_m = Y_m \cdot (V(gi) - V(si))$$

$$I_{ms} = -Y_{mb} \cdot (V(bi) - V(si))$$

with:

$$Y_{ms} = nY_m$$

$$Y_{mb} = (n-1)Y_m$$

Fig. 2: Quasi-static small-signal circuit of circuit shown in Fig. 1 b) valid in saturation.

$$n = g_{ms}/g_m = g_{mb}/g_m + 1 \tag{1}$$

which depends on the gate-to-bulk voltage and typically ranges from 1.6 in weak inversion to 1.3 in strong inversion for an n-channel transistor (1.4 to 1.2 for p-channel) [18][21].

C. Component Bias Dependence

The capacitances shown in Fig. 2 include both intrinsic and extrinsic parts

$$\begin{gathered} C_{gs} = C_{gsi} + C_{gso} \qquad C_{gd} = C_{gdi} + C_{gdo} \qquad C_{gb} = C_{gbi} + C_{gbo} \\ C_{sb} = C_{bsi} + C_{jsb} \qquad C_{db} = C_{bdi} + C_{jdb} \end{gathered} \tag{2}$$

where C_{gsi}, C_{gdi}, C_{gbi}, C_{bsi} and C_{bdi} are the intrinsic capacitances, C_{gso}, C_{gdo} and C_{gbo} the overlap capacitances, C_{jsb} and C_{jdb} the junction capacitances. Note that all these capacitances (including the overlap capacitances) are bias dependent. The long-channel intrinsic capacitances normalized to the total gate oxide capacitance $C_{ox} = N_f \cdot W_{eff} \cdot L_{eff} \cdot C'_{ox}$ are plotted versus the gate overdrive voltage $V_G - V_{TO}$ in Fig. 3 a). N_f is the number of fingers, W_{eff} and L_{eff} the effective width and length of a single finger and C'_{ox} the oxide capacitance per unit area. Capacitance C_{ggi} corresponds to the total intrinsic gate capacitance $C_{ggi} = C_{gsi} + C_{gbi} + C_{gdi}$. The shape of the intrinsic capacitances versus bias are slightly different for short-channel devices due mainly to polydepletion [21][51] and short-channel effects such as velocity saturation [21][52]. As an example, capacitance C_{gsi} may typically look as shown by the dashed line

† The slope factor is related to the slope of the $\log(I_D) - V_G$ characteristic in weak inversion which is equal to $1/(nU_T)$ where $U_T = kT/q$ [18]

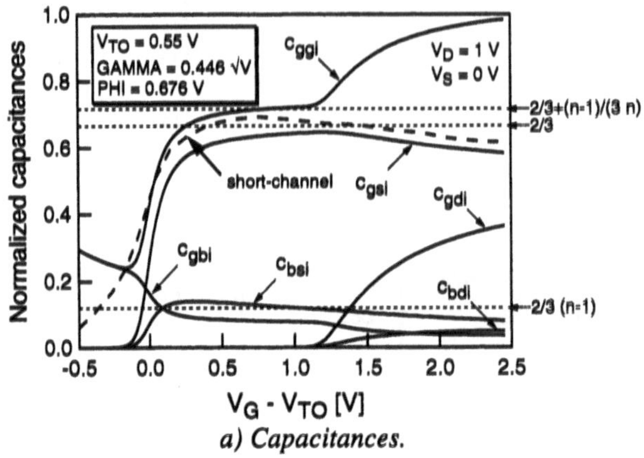

a) Capacitances.

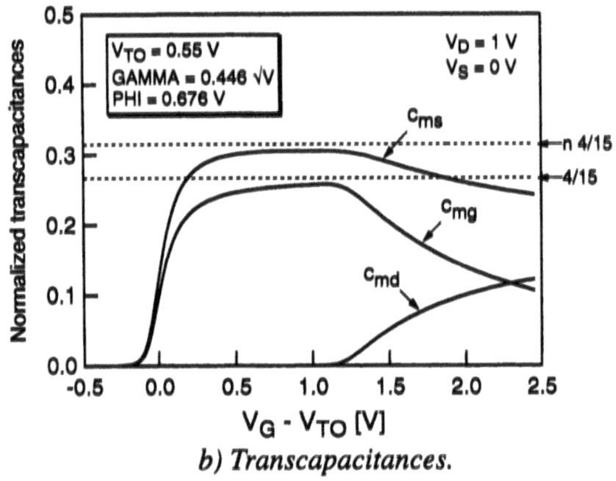

b) Transcapacitances.

Fig. 3: Normalized intrinsic capacitances and transcapacitances.

in Fig. 3 a). The long-channel intrinsic capacitances in strong inversion and in saturation can be approximated by

$$C_{gsi} \cong \frac{2}{3} \cdot C_{ox} \qquad C_{gdi} \cong 0 \qquad C_{gbi} \cong \frac{n-1}{3n} \cdot C_{ox}$$
$$C_{bsi} = (n-1) \cdot C_{gsi} \qquad C_{bdi} = (n-1) \cdot C_{gdi} \cong 0 \tag{3}$$

The gate and source transadmittances Y_m and Y_{ms} are given by

$$Y_m = g_m - j\omega \cdot C_m \qquad Y_{ms} = g_{ms} - j\omega \cdot C_{ms} \tag{4}$$

where C_m and C_{ms} are respectively the gate and source transcapacitances [21]. According to (4), the transcapacitances add a certain phase shift to the drain current variation, that partly accounts for the propagation delay along the channel. They can be interpreted as a 1st-order approximation of the NQS effects [21]. The long-channel transcapacitances are plotted versus the overdrive voltage

$V_G - V_{TO}$ in Fig. 3 b). It is interesting to note that to 1st-order, the ratios between the transcapacitances and the related transconductances are equal

$$\tau = C_m/g_m = C_{ms}/g_{ms} \tag{5}$$

so the transadmittances can be rewritten as

$$Y_m = g_m \cdot (1 - j\omega \cdot \tau) \qquad Y_{ms} = g_{ms} \cdot (1 - j\omega \cdot \tau). \tag{6}$$

In strong inversion and saturation, the transcapacitances can be approximated by their long-channel values $C_m \cong 4/15 \cdot C_{ox}$, $C_{ms} \cong n \cdot C_m$ leading to $Y_{ms} \cong n \cdot Y_m$. Since the gate transconductance is given by $g_m \cong n\beta(V_G - V_{TO} - nV_S)$, the time constant τ becomes

$$\tau \cong \tau_0 \cdot \frac{4}{15} \cdot \frac{2nU_T}{V_G - V_{TO} - nV_S} \qquad \text{where} \qquad \tau_0 = L_{eff}^2/(2\mu_{eff}U_T) \tag{7}$$

is proportional to the square of the effective gate length and only slightly bias dependent through the effective mobility μ_{eff}.

It is worth mentioning that any charge-based compact model, such as BSIM3v3, MOS Model 9 or EKV, inherently have transcapacitances and therefore transadmittances given by (6). Although the phase of the transadmittances given by (6) are equal to that of a 1st-order NQS model, their magnitude increases with frequency instead of decreasing as it would in the 1st-order (or higher-order) NQS model. This difference has to be accounted for when implementing an NQS model on top of a charge-based model to avoid wrong magnitude and phase characteristics.

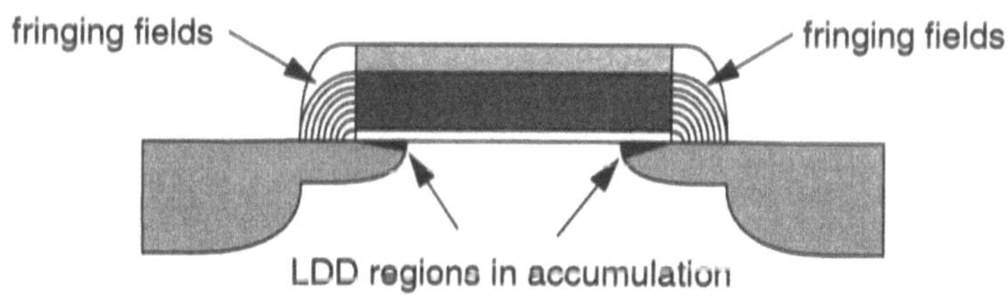

Fig. 4: Overlap capacitances.

The overlap capacitances are also bias dependent due to the LDD regions [53]. The overlap portion of the LDD regions behave similarly to a MOS capacitor of the opposite type than the channel. Therefore, when the transistor is biased in inversion, the overlap LDD regions may be in accumulation. Eventhough this bias dependence is somewhat accounted for in the BSIM3v3 compact model, it may be advantageous to include it into the separate overlap capacitances appearing in the subcircuit. By doing this way, the capacitances C_{gso} and C_{gdo} can not

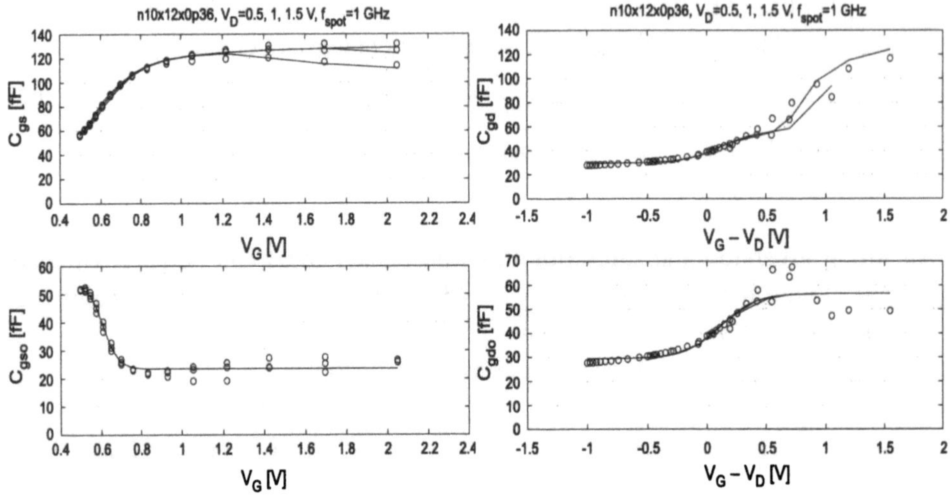

$C_{gso\text{-}min} = 53$ fF $\quad \Delta C_{gso} = 29.8$ fF

$V_{gso\text{-}norm} = 69$ mV $\quad \Delta V_{gso} = 609$ mV

a) Gate-to-source capacitance.

$C_{gdo\text{-}min} = 29.5$ fF $\quad \Delta C_{gdo} = 27$ fF

$V_{gdo\text{-}norm} = 293$ mV $\quad \Delta V_{gdo} = 117$ mV

b) Gate-to-drain capacitance.

Fig. 5: Bias dependence of the gate-to-source and gate-to-drain capacitances.

only account for the bias dependent overlap capacitances but can also correct the intrinsic capacitance bias dependence for short-channel effects in the case they are not correctly accounted for.

Good results for the bias dependence of the gate-to-source and gate-to-drain overlap capacitances have been obtained by using the following simple empirical relation [32][33]

$$C_{gso} = C_{gso\text{-}min} - \frac{\Delta C_{gso}}{2} \cdot \left(1 + \tanh\left(\frac{V_G - V_S - \Delta V_{gso}}{V_{gso\text{-}norm}}\right)\right)$$

$$C_{gdo} = C_{gdo\text{-}min} + \frac{\Delta C_{gdo}}{2} \cdot \left(1 + \tanh\left(\frac{V_G - V_D - \Delta V_{gdo}}{V_{gdo\text{-}norm}}\right)\right) \tag{8}$$

where $C_{gso\text{-}min}$, ΔC_{gso}, ΔV_{gso}, $V_{gso\text{-}norm}$ $C_{gdo\text{-}min}$, ΔC_{gdo}, ΔV_{gdo} and $V_{gdo\text{-}norm}$ are parameters. Fig. 5 shows the gate-to-source and gate-to-drain capacitances extracted from an S-parameter measurement made at 1 GHz (symbols of the upper plots). The overlap capacitances are then obtained by subtracting the intrinsic capacitance obtained from simulations using a compact model (EKV v2.6 in this case) (symbols of lower plots). The result is then fitted with equations (8). The fitted capacitances are shown in Fig. 5 by the lines. In the scalable model, parameters $C_{gso\text{-}min}$, ΔC_{gso}, $C_{gdo\text{-}min}$ and ΔC_{gdo} are scaled with geometry.

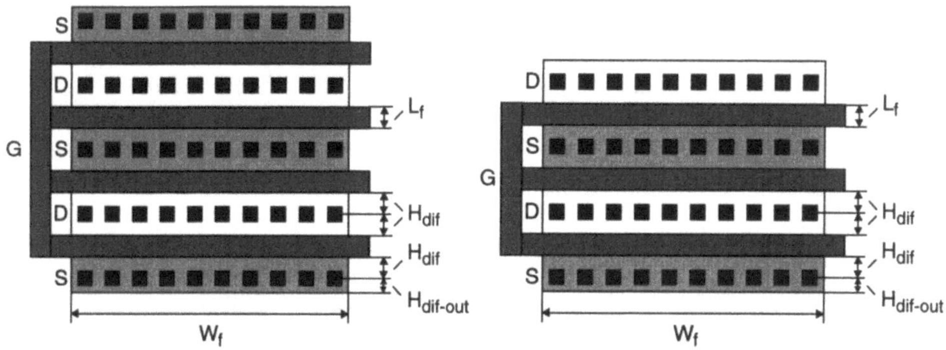

a) Even (4) number of fingers. *b) Odd (3) number of fingers.*

Fig. 6: RF MOS transistor layout.

Note that all the terminal resistances as well as the substrate resistances are also bias dependent. The source and drain resistances bias dependence is mostly due to the LDD regions. The gate resistance bias dependence results primarily from the NQS effects [21][31]. The bias dependence of the substrate resistances comes from the variations of the depletion regions below the gate and surrounding the source and drain diffusions. Although they could be implemented in the subcircuit of Fig. 1, good results have been obtained without accounting for the bias dependencies of these resistances.

D. Component Scaling

The layout of typical RF MOS transistors are sketched in Fig. 6 for an even and an odd number of fingers. The capacitances are given by

$$\begin{aligned} C_{gs} &= C_{gsi} + C_{gso} \cong N_f \cdot W_f \cdot (L_f \cdot c_{gsi} + L_{ov} \cdot c_{gso}) \cdot C'_{ox} \\ C_{gd} &= C_{gdi} + C_{gdo} \cong N_f \cdot W_f \cdot (L_f \cdot c_{gdi} + L_{ov} \cdot c_{gdo}) \cdot C'_{ox} \\ C_{gb} &= C_{gbi} + C_{gbo} \cong C_{gbi} = N_f \cdot W_f \cdot L_f \cdot c_{gbi} \cdot C'_{ox} \\ C_{sb} &= C_{sbi} + C_{jsb} \qquad C_{db} = C_{dbi} + C_{jdb} \end{aligned} \tag{9}$$

where L_{ov} is the overlap region length. c_{gsi}, c_{gdi}, c_{gbi}, c_{bsi} and c_{bdi} are the normalized intrinsic capacitances and c_{gso}, c_{gdo} are the normalized overlap capacitances. The junction capacitances C_{jsb}, C_{jdb} have to be split into components from area, gate perimeter and isolation (LOCOS or STI) perimeter. Let $N_{s\text{-}in}$ and $N_{s\text{-}out}$ ($N_{d\text{-}in}$ and $N_{d\text{-}out}$) be the number of inner and respectively outer source (drain) diffusions with $N_s = N_{s\text{-}in} + N_{s\text{-}out}$ ($N_d = N_{d\text{-}in} + N_{d\text{-}out}$). Assuming that $H_{dif\text{-}out} = H_{dif}$, the perimeters and area are given by

$$P_{s\text{-}gate} = (2N_{s\text{-}in} + N_{s\text{-}out}) \cdot W_f \qquad P_{s\text{-}isol} = 2N_s \cdot 2H_{dif} + N_{s\text{-}out} \cdot W_f$$
$$P_{d\text{-}gate} = (2N_{d\text{-}in} + N_{d\text{-}out}) \cdot W_f \qquad P_{d\text{-}isol} = 2N_d \cdot 2H_{dif} + N_{d\text{-}out} \cdot W_f \quad (10)$$
$$A_s = N_s \cdot 2H_{dif} \cdot W_f \qquad A_d = N_d \cdot 2H_{dif} \cdot W_f$$

and the junction capacitances are then calculated as

$$C_{jsb} = A_s \cdot C_J + P_{s\text{-}isol} \cdot C_{JSW} + P_{s\text{-}gate} \cdot C_{JSWG}$$
$$C_{jdb} = A_d \cdot C_J + P_{d\text{-}isol} \cdot C_{JSW} + P_{d\text{-}gate} \cdot C_{JSWG} \quad (11)$$

where C_J is the bias dependent bottom junction capacitance per unit area, C_{JSW} the bias dependent side-wall junction capacitance per unit length on the field oxide edge and C_{JSWG} the bias dependent side-wall junction capacitance per unit length on the gate oxide edge.

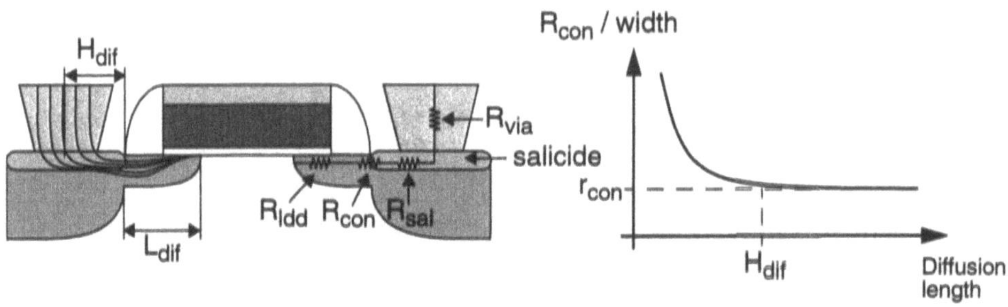

Fig. 7: Source and drain resistances.

As shown in Fig. 7, the source and drain resistances are made of different parts including the via resistance R_{via}, the salicide resistance R_{sal}, the salicide-to-silicon contact resistance R_{con} and the LDD resistance R_{ldd}. The total resistance is usually dominated by the contact and LDD resistances

$$R_s = R_{ldd} + R_{con} + R_{sal} + R_{via} \cong R_{ldd} + R_{con} \quad (12)$$

with $$R_{ldd} = \frac{L_{dif}}{N_f W_f} \cdot R_{ldd\text{-}sh} \qquad R_{sal} = \frac{H_{dif}}{N_f W_f} \cdot R_{sal\text{-}sh} \quad (13)$$

where the typical values of the sheet resistances are $0.9\ldots1.1k\Omega/\square$ for and for The total via resistance depends on the number of via per finger with typically resistance per via. As can be seen from the above numbers, the total resistance is usually dominated by the contact and the LDD resistances. Note that the contact resistance does not scale with the length of the salicide above a certain minimum value (typically). This is due to the fact that most of the current flows within the salicide instead of going to the diffusion because of the latter higher resistivity. Therefore, increasing the salicide width above the minimum H_{dif} does not reduce the contact resistance eventhough the bottom contact area between salicide and silicon is increased. Since the LDD region length L_{dif} is almost constant, the total source and drain resistances R_s and R_d only scales with the finger width and the number of fingers according to

$$R_s = R_d \cong R_{ldd} + R_{con} \cong \frac{0.5 R_{dsw}}{N_f W_f} \tag{14}$$

where R_{dsw} is the drain-to-source resistance which is typically ranging from $0.8\ldots1 k\Omega\mu m$ for a 0.25µm process and may become much less for more advanced deep-submicron processes.

The gate resistance is made of the poly-silicon resistance and a bias dependent resistance due to NQS effects (often named R_i in MESFETs). The latter will be discussed more in details below. The poly-silicon resistance is given by

$$R_g = \kappa \cdot \frac{1}{3} \cdot \frac{W_f}{N_f \cdot L_f} \cdot R_{g\text{-}sh} \tag{15}$$

RF MOS transistor are usually very wide and factor $1/3$ accounts for the distributed nature of the RC line across the channel (in the width direction) [21][55]. As shown in Fig. 8 a) and b), if the poly gate resistance is connected on both side, the gate resistance is four times smaller than that obtained for a gate connected only on one-side [21]. Factor κ accounts for this and is equal to unity for a gate contacted on one side and to $1/4$ for a two-side contacted gate.

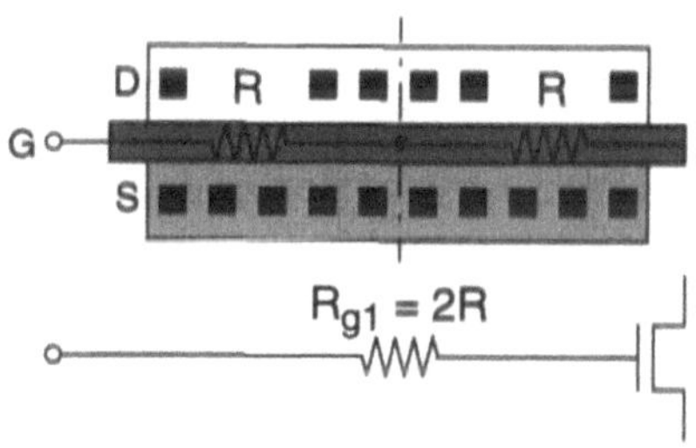

a) One side-contacted gate resistance.

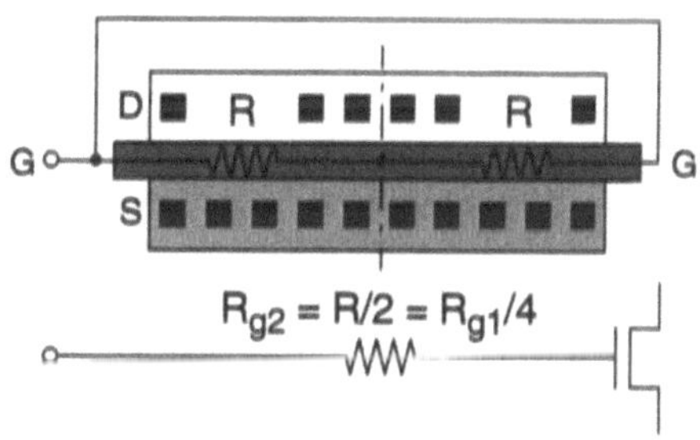

b) Double side-contacted gate resistance.

Fig. 8: Gate resistance due to poly-Si.

E. Y-Parameters Analysis

Since the capacitances are all approximately proportional to the total gate width $N_f \cdot W_f$ and since the terminal resistances are inversely proportional to $N_f \cdot W_f$, the time constant due to the terminal resistances depend only on the gate length L_f, the overlap length L_{ov} or the diffusion width H_{dif}. The latter dimensions are usually taken as minimum to achieve the highest cutoff frequency.

Therefore the poles due to the terminal resistances are at a much higher frequency than typically the transit frequency, so that they basically can be neglected when calculating the Y-parameters and the related quantities. Neglecting also the substrate resistances in the small-signal circuit of Fig. 2 (i.e. assuming that they are zero), allows to derive the following analytical expressions for the Y-parameters

$$
\begin{aligned}
y_{11} &\cong \frac{j\omega C_{gg}}{1+j\omega R_g C_{gg}} \cong \omega^2 (R_g C_{gg}^2) + j\omega C_{gg} \\
y_{12} &\cong \frac{-j\omega C_{gd}}{1+j\omega R_g C_{gg}} \cong -\omega^2 R_g C_{gg} C_{gd} - j\omega C_{gd} \\
y_{21} &\cong \frac{g_m - j\omega(C_m + C_{gd})}{1+j\omega R_g C_{gg}} \cong g_m - \omega^2 R_g C_{gg}(C_m + C_{gd}) - j\omega(C_m + C_{gd} + g_m R_g C_{gg}) \\
y_{22} &\cong \frac{g_{ds} + j\omega(C_{bd} + C_{gd})}{1+j\omega R_g C_{gg}} \cong g_{ds} + \omega^2 R_g C_{gg}(C_{bd} + C_{gd}) + j\omega(C_{bd} + C_{gd} - g_{ds} R_g C_{gg})
\end{aligned}
\tag{16}
$$

where it has also been assumed that $\omega R_g C_{gg} << 1$. Equations (16) have been verified experimentally in Fig. 9, which shows a comparison between the measured (after a two-step de-embedding process), the simulated and the analytical Y-parameters for a n-channel transistor with $N_f = 10$, $W_f = 12\mu m$, $L_f = 0.36\mu m$ (n10x12x0p36). The simulation have been performed with the complete QS model of Fig. 1 b), with the EKV v2.6 compact model. As can be seen from Fig. 9, the analytical expressions match the measurements very well even up to 10 GHz except for $Re\{y_{22}\}$. This discrepancy is due to the substrate coupling effect discussed in the next Section. Note that Fig. 9 also shows that there may be a big discrepancy in $Im\{y_{21}\}$ if the transcapacitance C_m in (16) is neglected.

Note that the simplified Y-parameters given by (16) can be used for a direct extraction of the RF model parameters from measurements as presented in [30]. For example C_{gd} can be extracted from $|Im\{y_{12}\}/\omega|$ and C_{gg} from $|Im\{y_{11}\}/\omega|$. In the linear region, $C_{gs} \cong C_{gd}$ and $C_{gb} \cong C_{gg} - C_{gs} - C_{gd}$, whereas in saturation, $C_{gb} << C_{gs}, C_{gd}$ and therefore $C_{gs} \cong C_{gg} - C_{gd}$. The gate resistance can be extracted as $Re\{y_{11}\}/(Im\{y_{11}\})^2$. The extraction of the substrate resistances requires a more complicated procedure described in [30].

F. Substrate Coupling

At high frequency, the signal at the drain couples to the source and the bulk terminals through the junction capacitance and the substrate resistances (c.f. Fig. 1). As mentioned above and observed in Fig. 9, this coupling mainly affects the output admittance y_{22}. Fig. 10 shows the cross-sections of multi-fingers RF MOS transistors where only the most important substrate resistances have been

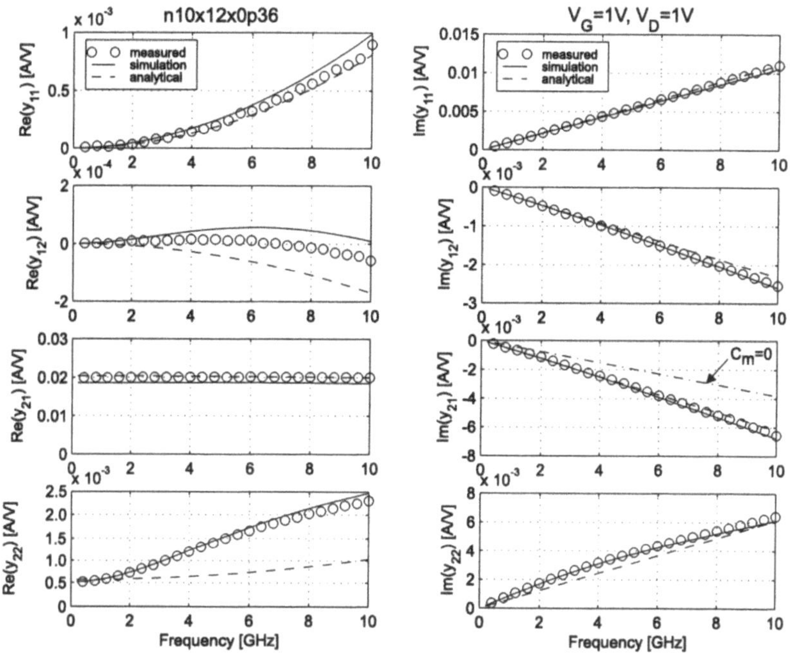

Fig. 9: Comparison between the measured, simulated and analytical Y-parameters (n-channel, $N_f = 10$, $W_f = 12\mu m$, $L_f = 0.36\mu m$ and $V_G = V_D = 1V$).

drawn. In order to match the equivalent circuit shown in Fig. 1 b), equivalent capacitance and resistances have to be calculated from the individual capacitances and resistances shown in Fig. 10. Since all the source (drain) diffusions are connected together via metal layers (assumed to have negligible resistances compared to the substrate resistances), the junction capacitances C_{jsb} and C_{jdb} as well as the substrate resistances R_{dsb}, R_{sb} and R_{db} are given by

$$C_{jsb} = \sum_{k=1}^{N_s} C_{js\text{-}k} \qquad C_{jdb} = \sum_{k=1}^{N_d} C_{jd\text{-}k}$$
$$\frac{1}{R_{sb}} = \sum_{k=1}^{N_s} \frac{1}{R_{sb\text{-}k}} \qquad \frac{1}{R_{db}} = \sum_{k=1}^{N_d} \frac{1}{R_{db\text{-}k}} \qquad \frac{1}{R_{dsb}} = \sum_{k=1}^{N_f} \frac{1}{R_{dsb\text{-}k}} \tag{17}$$

By symmetry, all the individual source (drain) junction capacitances are equal to the one of a single source (drain) diffusion $C_{js\text{-}k} = C_{js\text{-}f}$ ($C_{jd\text{-}k} = C_{jd\text{-}f}$)†. The same is valid for the individual drain-to-source substrate resistances $R_{dsb\text{-}k}$, leading to

† Note that in general $C_{js\text{-}f} \neq C_{jd\text{-}f}$ due to different source-to-bulk and drain-to-bulk voltages.

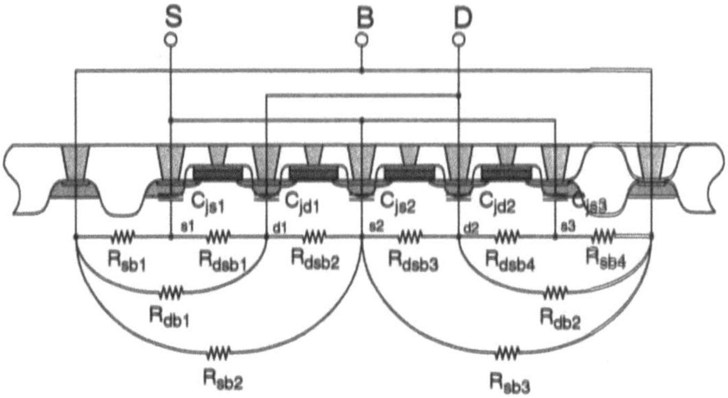

a) Even number of fingers.

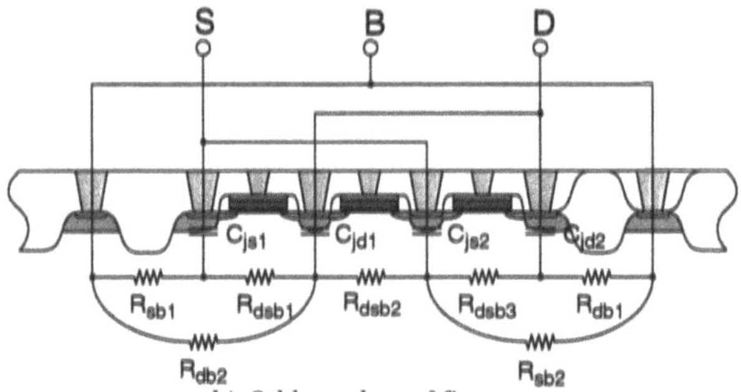

b) Odd number of fingers.

Fig. 10:Substrate resistances.

$$C_{jsb} \cong N_s \cdot C_{js\text{-}f} \qquad C_{jdb} \cong N_d \cdot C_{jd\text{-}f} \qquad R_{dsb} \cong \frac{L_f}{N_f \cdot W_f} \cdot R_{dsb\text{-}sh} \tag{18}$$

where $C_{js\text{-}f}$ and $C_{jd\text{-}f}$ are the junction capacitances of a single source and drain diffusion and $R_{dsb\text{-}sh}$ is the drain-to-source substrate sheet resistance. The calculation of the source-to-bulk and drain-to-bulk substrate resistances needs to distinguish between even and odd-number of fingers as shown in Fig. 10 a) and b). For the even number of fingers transistor, $R_{sb1} << R_{sb2}$ and $R_{sb4} << R_{sb3}$ since the outer source diffusions are closer to the bulk than the inner source diffusions and by symmetry $R_{sb1} \cong R_{sb4}$ and $R_{db1} \cong R_{db2}$ resulting in

$$\begin{aligned} \frac{1}{R_{sb}} &\cong N_f \cdot \left(\frac{1}{R_{sb1}} + \frac{1}{R_{sb4}}\right) \cong \frac{2}{R_{sb1}} = \left(\frac{r''_{sb}}{2 \cdot W_f}\right)^{-1} \\ \frac{1}{R_{db}} &\cong N_f \cdot \left(\frac{1}{R_{db1}} + \frac{1}{R_{db2}}\right) \cong \frac{2}{R_{db1}} = \left(\frac{r''_{db}}{2 \cdot W_f}\right)^{-1} \end{aligned} \qquad \textit{for } N_f \textit{ even} \tag{19}$$

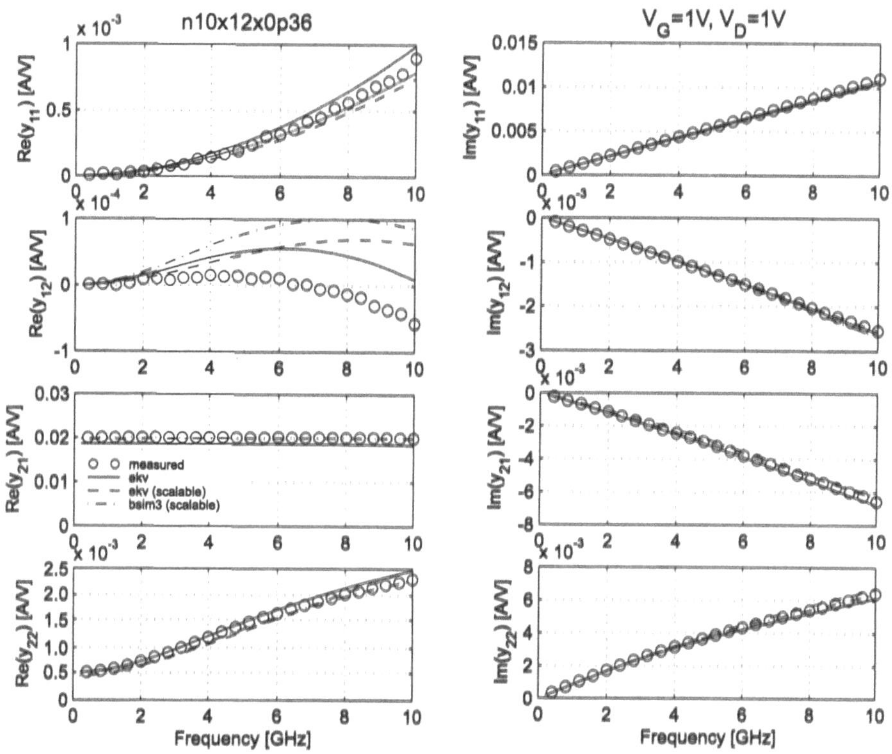

Fig. 11:Comparison between the measured and simulated Y-parameters versus frequency (n-channel, $N_f = 10$, $W_f = 12\mu m$, $L_f = 0.36\mu m$ and $V_G = V_D = 1V$).

where r''_{sb} and r''_{db} are the source-to-bulk and drain-to-bulk substrate resistances per unit length (transistor width). r''_{sb} and r''_{db} typically range from $100\ldots300\Omega m$ (for a bulk technology). For an odd number of fingers, $R_{sb1} << R_{sb2}$, $R_{db1} << R_{db2}$ and $R_{sb1} \cong R_{db1}$ which results in

$$\frac{1}{R_{sb}} \cong \frac{1}{R_{sb1}} = \left(\frac{r''_{sb}}{W_f}\right)^{-1} \cong \frac{1}{R_{db}} \qquad \textit{for } N_f \textit{ odd}. \tag{20}$$

As mentioned earlier, r''_{sb} and r''_{db} are also bias dependent, but this discussion is beyond the scope of this paper.

The subcircuit of Fig. 1 has been used to build a scalable RF MOS model using the above scaling rules. Both BSIM3v3 and EKV v2.6 have been used for the intrinsic compact model (M_i in the schematic of Fig. 1). The Y-parameters have been obtained from the measured S-parameters which have been de-embedded using a two-step procedure [56]. DC parameters, including the series resistance parameters R_{dsw}, have been extracted from dc measurements [57].

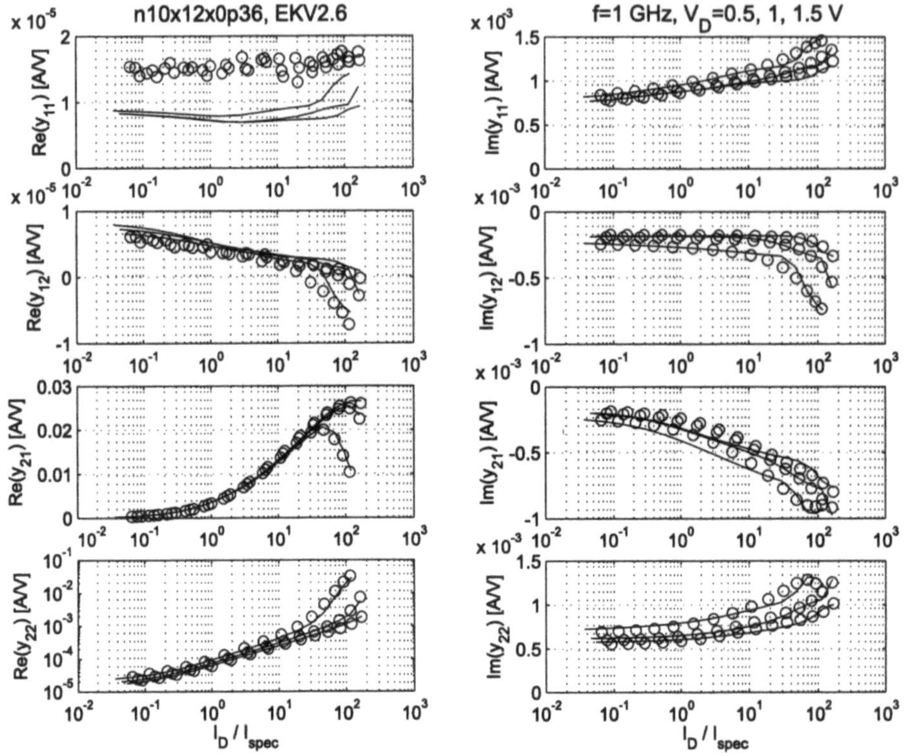

Fig. 12:Comparison between the measured and simulated Y-parameters versus inversion factor I_D/I_{spec} (n-channel, $N_f = 10$, $W_f = 12\mu m$, $L_f = 0.36\mu m$ and $V_G = V_D = 1V$).

Most parameters specific to the RF part have been extracted using the methodology presented in [30]. A comparison between measured and simulated Y-parameters versus frequency is presented in Fig. 11. The simulation have been performed with three different models: a) EKV v2.6 with parameters extracted for that particular device (line), b) scalable RF subcircuit with EKV v2.6 (dashed line) and c) scalable RF subcircuit with BSIM3v3. Both BSIM3v3 and EKV v2.6 show a very good match with measurements including the output admittance y_{22}. There are small differences between the scalable models and the model optimized for the particular device mainly due to the very simple scaling rules used for the substrate resistances. Note that the discrepancies in $Re\{y_{12}\}$ are not critical since y_{12} is dominated by its imaginary part corresponding approximately to ωC_{gd}. Similar results have been obtained for other operating points and other device geometries using the *same scalable model*. The bias dependence has been checked by measuring the S-parameters at 1 GHz and sweeping the gate voltage. The de-embedded Y-parameters are plotted versus the inversion factor I_D/I_{spec} in Fig. 12, where $I_{spec} = 2n\beta U_T^2$ is the specific current delimiting the regions of weak and strong inversion with $U_T = kT/q$ [18]. The inversion factor is a useful way to characterize the state of inversion of the channel of a MOST in saturation: it is much larger than one in strong inversion, close to unity in the moderate inversion and much smaller than one in weak inversion [18]. Note that I_{spec} depends on the transistor geometry according to $\beta = \mu_{eff} C'_{ox} W_{eff}/L_{eff}$ [18]. A good match is obtained in Fig. 12 for EKV v2.6 except for $Re\{y_{11}\}$. The reason for this discrepancy is that the gate resistance used in the subcircuit of Fig. 1 assumes that R_g is only due to the poly-silicon gate resistance, which is incorrect. It can be shown that the gate resistance is also affected by NQS effects which are the main cause of its bias dependence. In the model of Fig. 1, R_g had to be overestimated in order to fit $Re\{y_{11}\}$ at higher frequency. This resulted in a value of $Re\{y_{11}\}$ that is smaller than the measured one as can be observed in Fig. 12. The only way to correct this is to account for NQS effects as discussed below or in [31].

G. Transit Frequency

The performance of RF devices are often evaluated by looking at their transit frequency f_t and maximum frequency of oscillation f_{max}. The transit frequency can be defined as the frequency for which the extrapolated small-signal current gain h_{21} drops to unity. h_{21} can be evaluated directly from the Y-parameters given in (16) resulting in

$$h_{21} \equiv \left.\frac{I_2}{I_1}\right|_{V_2 = 0} = \frac{y_{21}}{y_{11}} = \frac{g_m - j\omega(C_m + C_{gd})}{j\omega C_{gg}} \cong \frac{g_m}{j\omega C_{gg}} \equiv \frac{\omega_t}{j\omega} \tag{21}$$

and

$$f_t \cong \frac{g_m}{2\pi \cdot C_{gg}} \tag{22}$$

where $C_{gg} = C_{gs} + C_{gd} + C_{gb}$ is the total gate capacitance. The transit frequency f_t can be extrapolated from the measured gain h_{21} according to

$$f_t = -Im\{h_{21}\} \cdot f_{spot} \tag{23}$$

where f_{spot} is a sufficiently low frequency (typically 1 GHz) where the slope of h_{21} is -20 dB/dec. The f_t measured for two n-channel transistors having different gate length are plotted in Fig. 13. The measurements are compared to simulations using the subcircuit of Fig. 1 with EKV v2.6 for the compact model. The simulation agree well with the measured data over a wide bias range. The peaking of f_t is due to the transistor leaving saturation and entering into the triode region. At this point the transconductance starts to saturate or even decrease while the gate-to-drain intrinsic capacitance starts to increase making the f_t reach a maximum and then sharply decrease. In strong inversion and in saturation f_t can be approximated by

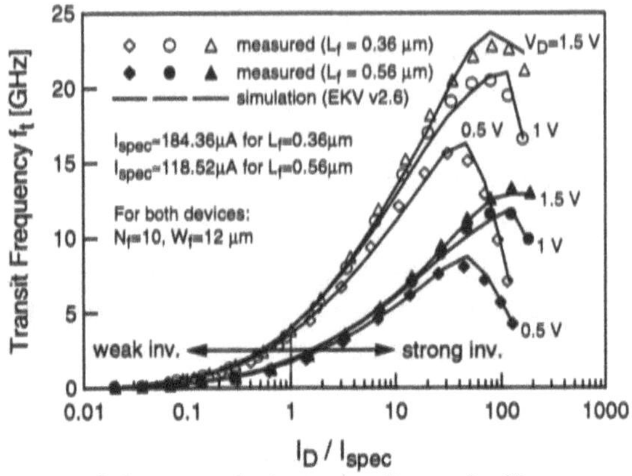

a) f_t *versus the inversion factor* I_D/I_{spec}.

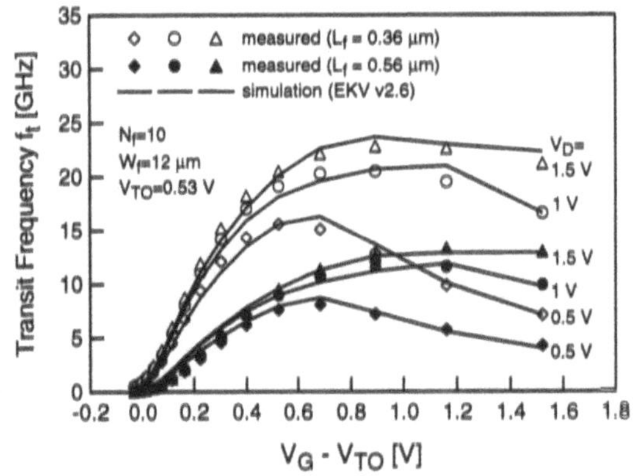

b) f_t *versus the overdrive voltage* $V_G - V_{TO}$.

Fig. 13:Transit frequency f_t versus bias for two different transistors.

$$f_t \cong \frac{1}{2\pi\tau_0} \cdot \frac{V_G - V_{TO} - nV_S}{2\eta n U_T} \qquad where\ \eta = \frac{2}{3} + \frac{n-1}{3n} + \frac{2L_{ov}}{L_{eff}} \cong 0.7\ldots0.9 \quad (24)$$

Note that eventhough (24) does not give very accurate results for f_t, it is useful for having a rough estimation and to evaluate how f_t scales. It is difficult to obtain a simple analytical expression relating f_t directly to the bias due to all the short-channel effects affecting g_m and C_{gg}. Nevertheless, accurate results are obtained when using (22) with the values of g_m and C_{gg} taken from each operating point information provided by a dc simulation using the complete compact model formulation (.OP command). This confirms the validity of (22).

According to (7) and (24), f_t should scale as $1/L_{eff}^2$, but this is only true at low field where there is no velocity saturation. If the carriers' velocity is saturated at v_{sat}, the peak f_t is

$$f_{t\text{-}peak} \cong \frac{v_{sat}}{2\pi \cdot L_{eff}} \tag{25}$$

where v_{sat} is the carrier saturated velocity (typically 10^{+5} m/s). According to (25), f_t only scales as $1/L_{eff}$ due to velocity saturation. This means that scaling is more effective for devices biased in the weak and moderate inversion region than in strong inversion. For future scaled down processes, the RF transistors could then take advantage of the better scaling achieved in moderate inversion to bias the devices in this region of operation. This is also good in terms of power consumption since the current efficiency (measured by the g_m/I_D ratio) improves when moving the bias toward weak inversion and also in terms of voltage swing. Of course this is not always possible for all RF applications.

It is also interesting to look at the transit time which is defined as $\tau_t \equiv 1/(2\pi f_t)$ and can be decomposed into $\tau_t = \tau_i + \tau_e$ where τ_i corresponds to the transit time of the intrinsic part

$$\tau_i = C_{ggi}/g_m . \tag{26}$$

The time constant τ_e corresponds to the additional delay introduced by the extrinsic part of the device due to the overlap capacitances and to the series resistances

$$\tau_e \cong \frac{C_{ggo}}{g_m} + R_d C_{gd} + nR_s\left(C_{gb} + C_{gd} + \frac{n-1}{n} C_{gs}\right) \tag{27}$$

where C_{ggo} is the total gate overlap capacitance

$$C_{ggo} = C_{gso} + C_{gdo} + C_{gbo} . \tag{28}$$

The transit times τ_i, τ_e and τ_t are plotted together with the transit frequency f_t in Fig. 14 a) versus the inversion factor I_D/I_{spec}. The ratio between the extrinsic and the intrinsic transit times is plotted in Fig. 14 b) which shows that extrinsic parasitics account for about 40% of the total transit time in strong inversion and about 50% in moderate inversion. This means that the increase of parasitic to intrinsic time constant ratio does not degrade dramatically when moving the operating point from strong to moderate inversion. This is another good reason for moderate inversion to be considered for RF operation with deep-submicron devices in order to meet the low-voltage and low-power requirements.

Note that although the peak f_t is an interesting figure of merit, RF MOS transistors are seldom biased close to the maximum f_t.

It is also important to mention that f_t does not depend on the gate resistance and is therefore only a incomplete figure of merit since the gate resistance strongly affects the frequency and noise performance. That is one reason why another figure of merit has to be defined and will be discussed below.

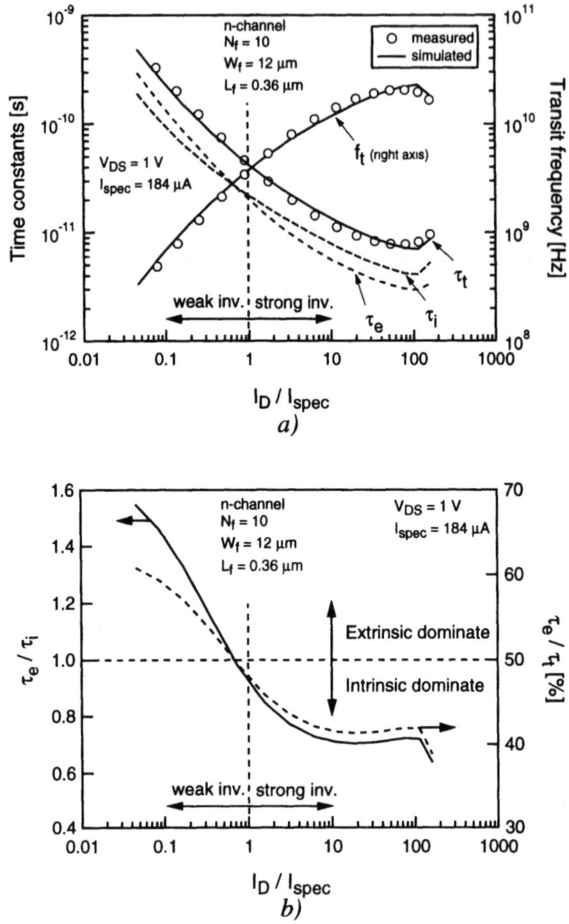

Fig. 14:Transit time versus inversion factor I_D/I_{spec}.

H. Maximum Oscillation Frequency

Another figure of merit that accounts for R_g can be defined from the unilateral power gain U corresponding to the maximum available gain of a twoport (corresponding to the transducer gain with matched source and load impedances i.e. $Y_G = y_{11}^*$ and $Y_L = y_L^*$) with its feedback transadmittance neutralized (i.e. $y_{12} = 0$). The advantage of using the unilateral power gain is that it can be defined even if the twoport is unstable in matched condition. The unilateral power gain U can be expressed from the Y-parameters as [58]

$$U = \frac{|y_{21} - y_{12}|^2}{4(g_{11}g_{22} - g_{12}g_{21})} \cong \frac{g_m^2}{4R_g C_{gg}(g_{ds}C_{gg} + g_m C_{gd})\omega^2} \cong \frac{g_m}{4R_g C_{gg} C_{gd}\omega^2} \equiv \left(\frac{\omega_{max}}{\omega}\right)^2 \quad (29)$$

where $g_{kl} = Re\{y_{kl}\}$.

The maximum frequency of oscillation f_{max} is defined as the frequency for which the extrapolated unilateral power gain becomes unity

$$f_{max} \cong \frac{g_m}{4\pi\sqrt{R_g C_{gg}(g_{ds}C_{gg} + g_m C_{gd})}} \cong \frac{1}{4\pi} \cdot \sqrt{\frac{g_m}{R_g C_{gg} C_{gd}}} \cong \sqrt{\frac{f_t}{8\pi R_g C_{gd}}} \tag{30}$$

Unlike f_t, it is not always obvious how to extrapolate the unilateral power gain for f_{max} because the asymptotic behavior is not always clearly defined. Nevertheless, (30) usually gives reasonable results and can be used as a figure of merit. Since most of the time $2\pi f_t \cdot R_g C_{gd} < 1/\sqrt{2}$, f_{max} is usually larger than f_t. Equation (30) shows that the smaller the $R_g C_{gd}$ product is, the larger f_{max}. Therefore $R_g C_{gd}$ can also be used as a figure of merit.

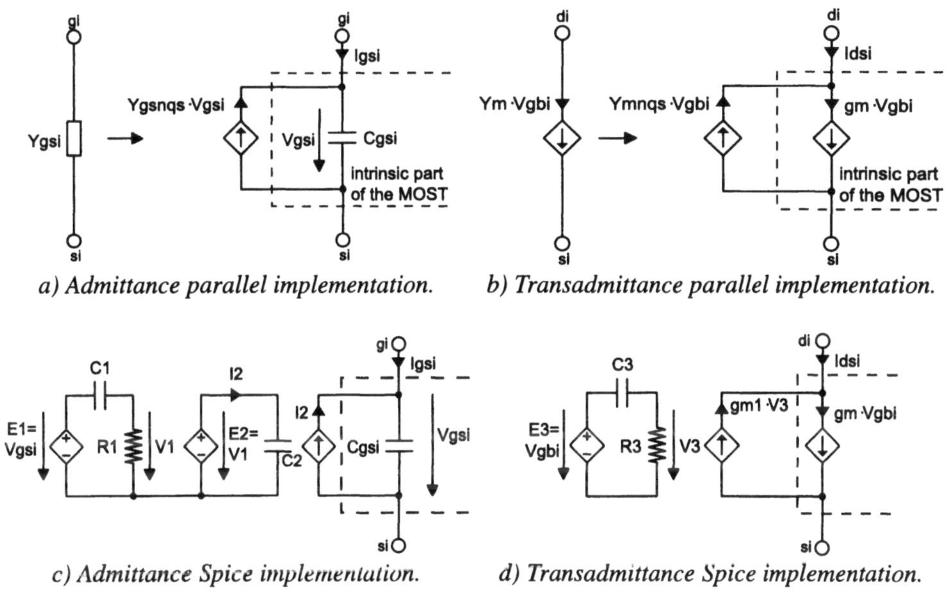

Fig. 15:NQS admittances and transadmittances implementation.

I. Non-Quasi-Static Effects

The quasi-static assumption used for deriving the small-signal circuit of Fig. 1 and the associated components does not hold anymore when the frequency gets close to the intrinsic cutoff frequency defined as $f_\tau = 1/(2\pi\tau)$ of the device where τ is given by (7) (or the length of the device is increased reducing its cut-off frequency). Non-Quasi-Static (NQS) effects may therefore appear for non-minimum channel length transistors, particularly for PFETs which have a smaller mobility and therefore a smaller f_τ than NFETs. Many NQS models have been proposed in the literature taking into account different effects and with different degrees of approximation [21][34]-[49]. These models all require to replace the

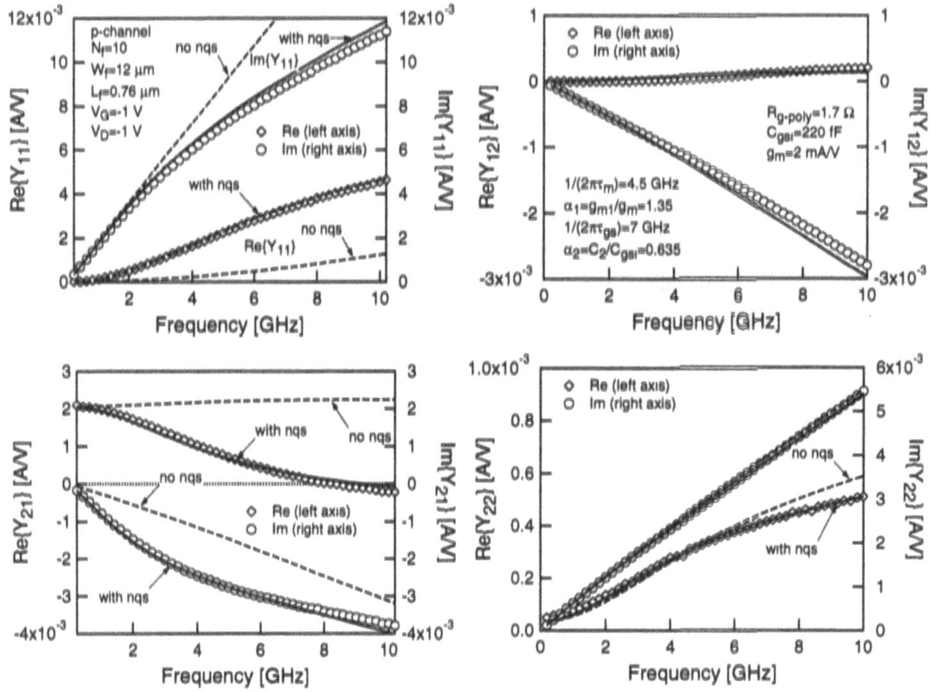

Fig. 16:Y-parameters for a NQS model using VCCS shown in Fig. 15.

intrinsic capacitances and transconductances with higher-order admittances and transadmittances. This implementation is not straight-forward in compact models since the poles and zeroes require additional nodes internal to the compact model that have to be solved either locally or globally depending on the simulator. NQS effects can be modeled at 1st-order by replacing each intrinsic capacitance by a RC series network. But this is impossible to do without having access to the compact model implementation. This problem can be circumvented by using VCCS connected in parallel to the compact model intrinsic capacitances and transconductances as shown in Fig. 15 a) and b). The VCCS can then be synthesized in many different ways in order to be implemented in Spice. An example is shown in Fig. 15 c) and d). The admittance and transadmittance corresponding to the circuits of Fig. 15 c) and d) are given by

$$
\begin{aligned}
Y_{gsi}(j\omega) &= j\omega C_{gsi} \cdot \frac{1 + j\omega\tau_{gs} \cdot (1 - \alpha_2)}{1 + j\omega\tau_{gs}} \\
Y_m(j\omega) &= g_m \cdot \frac{1 + j\omega\tau_m \cdot (1 - \alpha_1)}{1 + j\omega\tau_m}
\end{aligned}
\tag{31}
$$

with $\tau_{gs} = R_1 C_1$, $\alpha_2 = C_2/C_{gsi}$, $\tau_m = R_3 C_3$ and $\alpha_1 = g_{m1}/g_m$.

A 1st-order NQS model would require $C_2 = C_{gsi}$ ($\alpha_2 = 1$), $g_{m1} = g_m$ ($\alpha_1 = 1$) and $\tau_{gs} = \tau_m/2$ [21], but the domain of validity of this 1st-order NQS model can be extended beyond the limit of f_τ by using the zero of (31) by setting $C_2 < C_{gsi}$ and $g_{m1} > g_m$. The circuits of Fig. 15 c) and d) have been added to the subcircuit of Fig. 1 for the simulation of a p-channel MOS transistor with a non-minimum gate length $L_f = 0.76\mu m$. The simulation results are very close to the measured Y-parameters as can be seen from Fig. 16. It should be noticed that these results have been obtained with EKV v2.6 with the XQC switch [17] set to select the capacitive model instead of the charge-based model in order to avoid the left-hand side zero of the intrinsic admittance as already discussed earlier.

III. Noise Modeling at RF

A. Noise Sources in the MOS Transistor

The different noise sources in the MOS transistor are shown in Fig. 17 together with their power spectral densities (PSD). They include: the noise at the drain S_{ind}, constituted by the channel thermal noise $4kT \cdot G_{nch}$ and the flicker noise $K_f/(C'_{ox} W_{eff} L_{eff} f^{\alpha})$, the terminal resistances thermal noise S_{vnrg}, S_{inrs}, S_{inrd} and the substrate resistances thermal noise S_{inrsb}, S_{inrdb} and S_{inrdsb}. The flicker noise mainly affects the low-frequency performance of the device and can be ignored at high-frequency†. In addition, there is also a gate noise current source that represents the effects of the shot noise due to the gate leakage current and the induced gate noise accounting for the capacitive coupling of the local thermal noise sources within the channel to the gate terminal [21][59]-[64]. Usually the gate leakage current is small enough and the corresponding shot noise can be neglected. Hence, S_{ing} is dominated by the induced gate noise.

B. Channel Thermal Noise

Although all the noise sources contribute to the total noise at high-frequency, the dominant contribution still comes from the channel thermal noise having a PSD given by [21][18]

$$S_{ind} = 4kT \cdot G_{nch} \qquad with \qquad G_{nch} = \gamma \cdot g_{do} \tag{32}$$

where $k = 1.38\times10^{-23} J/K$ is the Boltzman constant and T the absolute temperature in K. G_{nch} is the channel thermal noise conductance and g_{do} is equal to the source transconductance g_{ms}. Note that g_{ms} becomes equal to the channel conductance when the transistor is biased in the linear region.

† Flicker noise is nevertheless important for some RF circuits such as mixers or oscillators that up-convert the low-frequency noise around the carrier and deteriorate the phase noise or the signal-to-noise ratio.

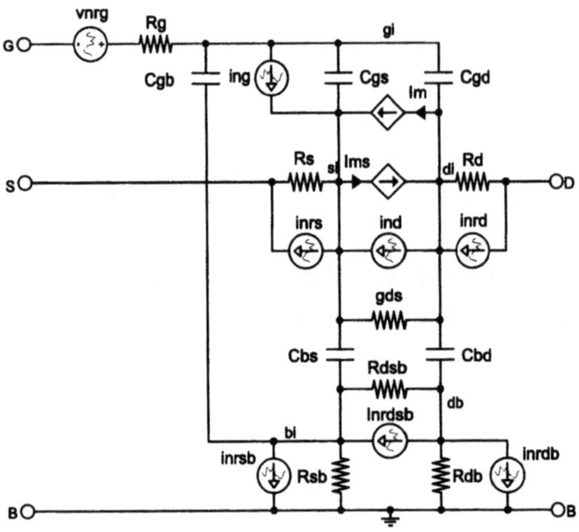

$$S_{ind} = 4kT \cdot G_{nch} + \frac{K_f}{C'_{ox} W_{eff} L_{eff} f^{\alpha}}$$

$$S_{ing} = 4kT \cdot G_{ng} + 2qI_g \cong 4kT \cdot G_{ng}$$

$$S_{vnrg} = 4kT \cdot R_g$$

$$S_{inrs} = 4kT/R_s$$

$$S_{inrd} = 4kT/R_d$$

$$S_{inrsb} = 4kT/R_{sb}$$

$$S_{inrdb} = 4kT/R_{db}$$

$$S_{inrdsb} = 4kT/R_{dsb}$$

Fig. 17:Noise sources in the MOS transistor.

γ is a bias dependent factor, which for long-channel devices is equal to unity in the linear region and to $2/3$ in saturation [59]. γ_{sat} is related to another factor defined as [59]

$$\alpha_{sat} \equiv g_m \cdot R_{Nin} = \frac{G_{nch}}{g_m} = \frac{g_{ms}}{g_m} \cdot \gamma_{sat} = n \cdot \gamma_{sat}. \tag{33}$$

α_{sat} is a good figure of merit to compare the thermal noise performance of different transconductors. It shows how much noise is generated by the device at the input for a given transconductance. As stated by (33) α_{sat} is proportional to γ_{sat} and is typically equal to unity in strong inversion. For short-channel devices in saturation γ_{sat} (α_{sat}) might become larger than the long-channel value $2/3$ ($n \cdot 2/3$) due to both velocity saturation and hot electrons [59][66][69]. This effect has already been analyzed by van der Ziel twenty years ago [59] and has been verified experimentally by Abidi more than ten years ago [66]. Some models have been proposed to account for the velocity saturation effect [67] and hot carrier effects [68], but they have not been implemented in any compact model yet. More recently Klein proposed a simple model for the thermal noise that accounts for both velocity saturation and hot carriers and can be easily implemented in a compact model [69]. Klein's noise model was originally developed for a transistor biased in saturation and in strong inversion, but it can be extended to cover weak to strong inversion by rewriting the γ_{sat} noise factor as

$$\gamma_{sat}(i_f) \cong \gamma_{sat\text{-}long} \cdot \left(1 + \frac{1}{G(i_f)} \cdot \frac{v_{sat} \cdot \tau_r}{L_{eff}}\right) \tag{34}$$

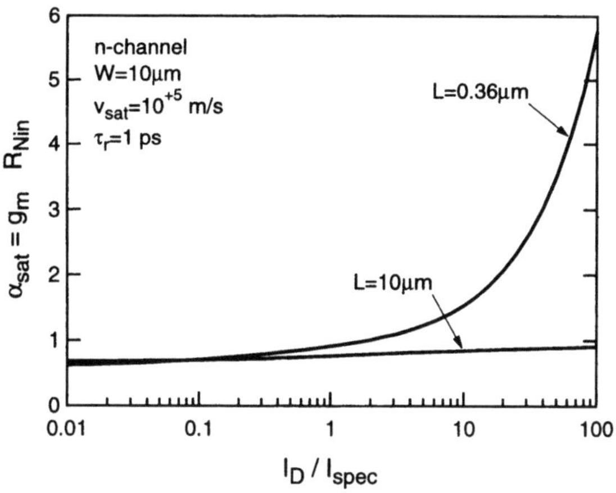

Fig. 18: α_{sat} noise factor computed from (34) for a long- ($L_f = 10\mu m$) and a short channel ($L_f = 0.36\mu m$) device with $v_{sat} = 10^5\ m/s$ and $\tau_r = 1ps$.

where τ_r is a relaxation time (of the order of 1ps) used as a fitting parameter. Note that $(v_{sat} \cdot \tau_r)/L_{eff}$ is approximately equal to unity for $L_{eff} \cong 0.1\mu m$. Factor $G(i_f)$ is the normalized g_m/I_D ratio defined as [20]

$$G(i_f) \equiv \frac{g_m \cdot nU_T}{I_D} = \frac{2}{1 + \sqrt{4i_f + 1}} \tag{35}$$

with $i_f = I_D/I_{spec}$ being the inversion factor in saturation [20]. $\gamma_{sat\text{-}long}$ also depends on the bias according to

$$\gamma_{sat\text{-}long} = \frac{2}{3} \cdot \left(1 + \frac{1 - \sqrt{4i_f + 1}}{8 \cdot i_f}\right) = \begin{cases} 2/3 & for\ i_f >> 1\ (strong\ inv.) \\ 1/2 & for\ i_f << 1\ (weak\ inv.) \end{cases} \tag{36}$$

This simple model assumes that the carriers velocity is saturated and that the lateral field is equal to the critical field along all the channel from source to drain. Although these assumptions are questionable, the resulting model can fit the measured data over bias and geometry [69].

The noise factor α_{sat} calculated from (34) is plotted versus the inversion factor in Fig. 18 for a long- and a short-channel device. For long-channel, the second term in (34) becomes negligible and therefore α_{sat} reduces to $n \cdot \gamma_{sat\text{-}long}$. For short-channel, α_{sat} increases in strong inversion to about 6 times the long-channel value. This is in good agreement with the data measured by Abidi in [66].

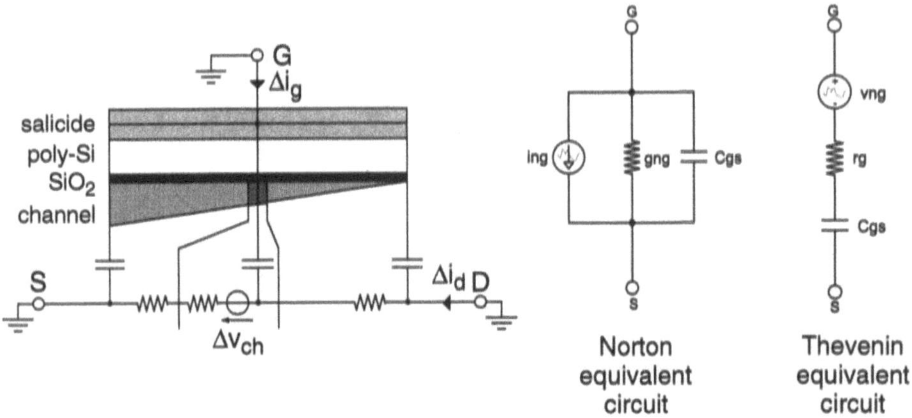

Fig. 19:Induced gate noise and equivalent model.

C. Induced Gate Noise

At high-frequency, the local channel voltage fluctuations due to thermal noise couple to the gate through the oxide capacitance and cause an induced gate noise current to flow (c.f. Fig. 19) [21][59]-[64]. This noise current can be modeled by a noisy current source i_{ng} connected in parallel to C_{gsi} and having a PSD given by [62][63]

$$S_{ing} = 4kT \cdot \delta G_{ng}(\omega) \qquad with \qquad G_{ng}(\omega) = \omega^2 C_{gs}/(5g_m) \tag{37}$$

and where δ is a bias dependent factor that is equal to $4/3$ for a long-channel device [62][63]. Note that PSD S_{ing} is frequency dependent. The parallel equivalent circuit shown in Fig. 19 can also be transformed into a series equivalent circuit with

$$S_{rng} = 4kT \cdot \delta R_{ng} \qquad with \qquad R_{ng} = 1/(5g_m) \tag{38}$$

where now S_{rng} corresponds to a white noise PSD [63]. This last result shows that the induced gate noise can be calculated directly from a 1st-order NQS model (without performing any integration as it is done in [59]) and by assuming that resistance R_{ng} implementing the NQS effect is noisy [21].

Since the physical origin of the induced gate noise is the same as for the channel thermal noise at the drain, the two noise sources i_{nd} and i_{ng} are partially correlated with a correlation factor [59][62][63]

$$c \equiv \frac{S_{ing,ind}}{\sqrt{S_{ind} S_{ing}}} = jc_g \qquad with\ c_g = \sqrt{5/32} = 0.395\ (long\text{-}channel) \tag{39}$$

where $S_{ing,ind}$ is the cross-power spectral density [21]. The induced gate noise and moreover its correlation to the thermal noise at the drain are not implemented

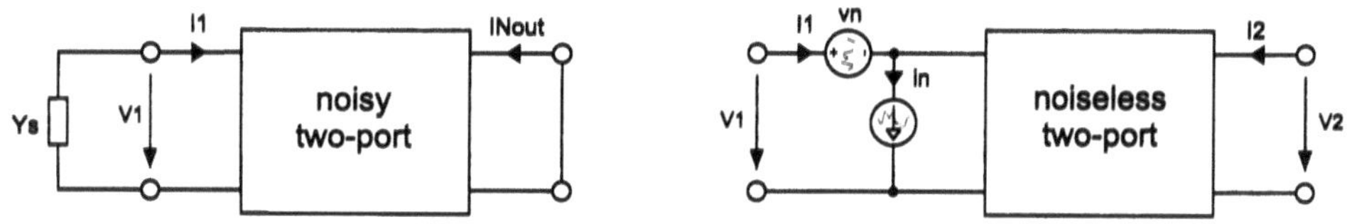

Fig. 20:Noisy twoport and its ABCD-parameter representation.

in compact models yet†. This is probably not to severe for frequencies much smaller than the device f_t, since besides the thermal noise at the drain, the other most important contributor to the total noise are the substrate and the gate resistances.

D. HF Noise Parameters

The noise at HF is characterized by four parameters: the minimum noise factor F_{min} (or minimum noise figure $NF_{min} = 10 \cdot \log(F_{min})$), the input referred noise resistance R_v and the optimum source admittance $Y_{opt} = G_{opt} + jB_{opt}$ for which the minimum noise figure is obtained [64]. The latter are usually the parameters obtained from measurement. They are not very appealing from a circuit designer's point of view. Designers prefer to use the parameters related to the noisy twoport equivalent circuit shown in Fig. 20, namely, the PSD of the input noise voltage source $S_v = 4kTR_v$, the PSD of the input noise current source $S_i = 4kTG_i$ and the correlation admittance $Y_c = G_c + jB_c$ accounting for the correlation existing between i_n and v_n. Noise source i_n can also be split into a noise source i_{nu} which is uncorrelated to v_n and a noise source i_{nc} that is fully correlated with v_n

$$S_{iu} = S_i - |Y_c|^2 \cdot S_v = (1 - |c|^2) \cdot S_i \qquad S_{ic} = |Y_c|^2 \cdot S_v = |c|^2 \cdot S_i \quad (40)$$

were c is the correlation factor‡ defined as $c = Y_c \cdot \sqrt{R_v / G_i}$. The rules to transform R_v, F_{min}, G_{opt} and B_{opt} into R_v, G_i, G_c and B_c are given below††

$$G_i = |Y_{opt}|^2 R_v = (G_{opt}^2 + B_{opt}^2) R_v$$

$$G_c = \frac{F_{min} - 2R_v G_{opt} - 1}{2R_v} \quad (41)$$

$$B_c = -B_{opt}$$

† Except MOS model 9 which includes gate induced noise but does not account for the correlation between the gate induced noise and the drain noise.

‡ The correlation factor c denoting the correlation existing between noise sources i_n and v_n should not be confused with the correlation factor c_g accounting for the correlation existing between the induced gate noise and the drain thermal noise.

†† Note that R_v, G_i, G_{iu} and G_{ic} are usually frequency dependent.

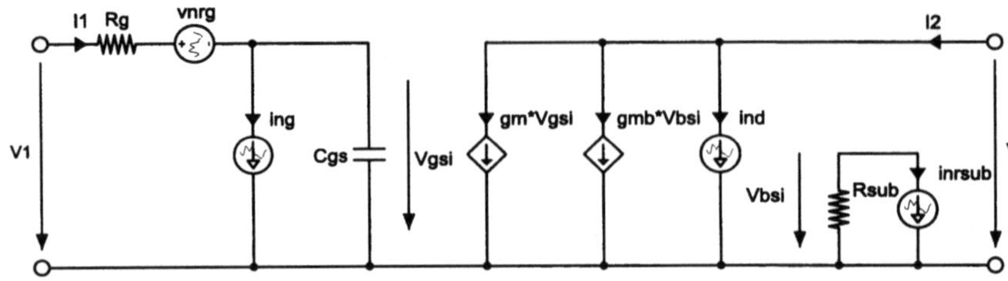

Fig. 21:Simplified small-signal schematic for noise calculation.

and inversely from R_v, G_i, G_c and B_c into R_v, F_{min}, G_{opt} and B_{opt}

$$\begin{aligned} F_{min} &= 1 + 2R_v(G_{opt} + G_c) = 1 + 2R_v \cdot \left(\sqrt{G_{iu}/R_v + G_c^2} + G_c\right) \\ G_{opt} &= \sqrt{G_{iu}/R_v + G_c^2} = \sqrt{G_i/R_v - B_c^2} \\ B_{opt} &= -B_c \end{aligned} \tag{42}$$

According to (42), the smaller the products $R_v G_{opt}$ and $R_v G_c$ are, the smaller F_{min}.

The noise parameters can be calculated analytically using the simplified small-signal circuit shown in Fig. 21, where it has been assumed that $R_{dsb} << R_{sb}, R_{db}$ leading to the definition $R_{sub} = R_{sb} // R_{db}$. The capacitances C_{bs} and C_{bd} have also been neglected. The following noise parameters are obtained

$$\begin{aligned} R_v &= \frac{\alpha_{sat}}{g_m} \cdot D_c \qquad G_i = \alpha_{sat} g_m \Omega^2 \cdot \psi \\ G_c &= \frac{R_g \cdot (g_m \Omega)^2 \cdot \psi}{D_c} \qquad B_c = \frac{g_m \Omega \cdot \chi}{D_c} \end{aligned} \tag{43}$$

$$D_c = 1 + \alpha_g + \alpha_{sub} + (g_m R_g \Omega)^2 \cdot \psi \cong 1 + \alpha_g + \alpha_{sub} \qquad \Omega = \omega/\omega_t$$

where α_g is the ratio of the noise PSD of the gate resistance $4kTR_g$ to the input referred channel noise $4kT\alpha_{sat}/g_m$ and α_{sub} the ratio of the output referred substrate resistance noise PSD to the output referred channel noise

$$\alpha_g = \frac{g_m R_g}{\alpha_{sat}} \qquad \alpha_{sub} = \frac{g_{mb}^2 R_{sub}}{\alpha_{sat} g_m} = \frac{(n-1) g_{mb} R_{sub}}{\alpha_{sat}} \tag{44}$$

where $g_{mb} = (n-1)g_m$ has been used. Parameters ψ and χ account for the induced gate noise and its correlation to the drain noise

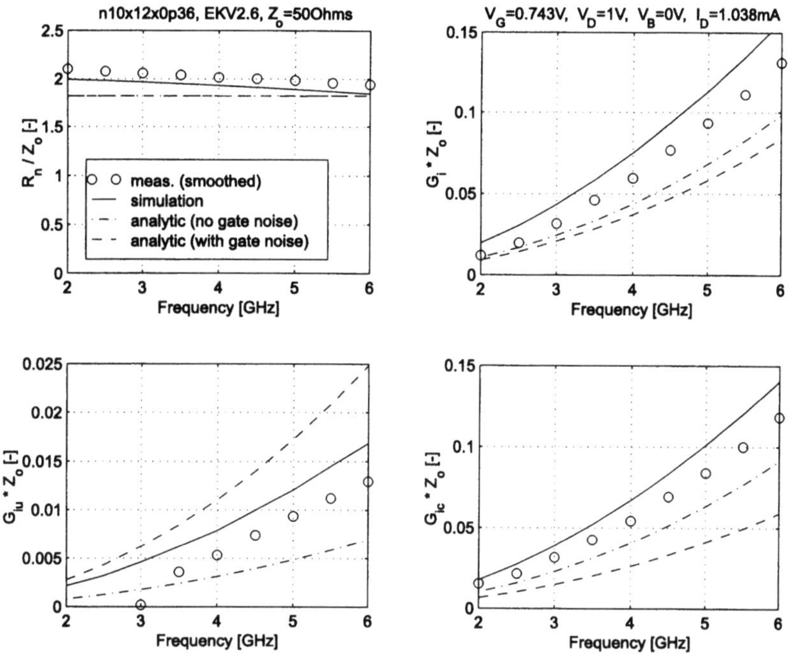

Fig. 22:Comparison between measured and simulated noise parameters.

$$\psi = 1 + \alpha_{sub} + \frac{\delta}{5n\alpha_{sat}} - 2c_g\sqrt{\frac{\delta}{5n\alpha_{sat}}} \qquad \chi = 1 + \alpha_{sub} - c_g\sqrt{\frac{\delta}{5n\alpha_{sat}}} \tag{45}$$

ψ and χ both reduce to $1 + \alpha_{sub}$ when the induced gate noise is ignored ($\delta = 0$). As expected by looking at the schematic of Fig. 21, the induced gate noise contributes mainly to G_i through the factor ψ, the gate resistance contributes to R_v and the channel noise and substrate noise contribute to both G_i and R_v. Substrate noise may typically contribute to 20% of R_v whereas R_g typically contributes to about 5%. It is therefore important to account for the substrate resistance when doing noise calculation and noise optimization.

The noise parameters of an n-channel device have been measured and carefully de-embedded using the methodology presented in [64][70][65]. They are presented in Fig. 22 and compared to the results obtained from simulation using the complete subcircuit of Fig. 1. Also plotted are the results obtained from (43) including the correlation between induced gate noise and channel drain noise.

IV. CONCLUSION

The design of RF CMOS integrated circuits requires MOS transistor models that are accurate up to the GHz frequency range. The compact models currently available do not account for some important effects caused by components such as the gate resistance and the substrate coupling resistances, that strongly affect the behaviour of the MOS transistor at GHz frequencies. All these effects can be accounted for by adding in a subcircuit all the extrinsic components to the existing compact model which is then used only as the intrinsic part of the device (meaning all the extrinsic components within the compact model are turned off, by selecting the adequate parameters). The RF MOS transistor subcircuit can be made scalable by carefully describing the geometry dependence of all the external extrinsic components, assuming that the intrinsic compact model is also scalable. Although all these components show a bias dependence, it is generally sufficient to account for the bias dependence of the junction and overlap capacitances, leaving the resistances bias independent. A simple model for the bias dependence of the overlap capacitance is proposed. In some case it might be required to also account for the bias dependence of the gate resistance which is mainly due to non-quasi static effects.

Simple expressions for the Y-parameters are derived that give a better insight into the device frequency behaviour at high frequency and also allow for a direct extraction (meaning no optimization) of almost all the RF model parameters. The only effect that this analytical model does not capture is the coupling of the signal from the drain to the source diffusions and to the substrate contacts. This effect mainly affects the output admittance y_{22} and can be modeled by a simple Π resistive network. A scalable model for these substrate resistances is proposed that gives excellent results for both n- and p-channels up to 10 GHz. The substrate resistances can be extracted from measurement according to the procedure described in [30]. The subcircuit presented in this paper has been successfully validated over frequency up to 10 GHz, over geometry and over bias on a 0.25 μm CMOS process for both n- and p-channel devices.

The transit frequency f_t and the maximum oscillation frequency f_{max} are derived and successfully checked against measurements. A closer look at the transit time $\tau_t \equiv 1/f_t$ shows that it can be decomposed into the sum of an intrinsic and an extrinsic component. It is also shown that in strong inversion, the extrinsic transit time accounts for about 40% of the total transit time, whereas in moderate inversion it increases to about 50%. This reveals that there is no strong penalty in transit time degradation due to extrinsic components when moving the operating point from strong inversion to moderate inversion.

Non-quasi-static (NQS) operation may be important to model for devices with non minimum channel length and particularly for p-channel transistors. The main problem with NQS is the implementation of the poles and zeroes required for obtaining the higher order admittances and transadmittances. A simple Spice subcircuit NQS implementation example, using voltage controlled current sources, is described. It shows excellent agreement up to 10 GHz with the Y-parameters measured on a p-channel transistor with 0.76 μm gate length.

Finally the different noise sources in the transistor are described, including the induced gate noise due to the capacitive coupling of the local noise source within the channel to the gate terminal. A new model for the channel thermal noise of short-channel transistors accounting for velocity saturation and hot electrons derived from [69] is described. The noise at high frequency is characterized by four parameters, namely the input referred noise resistance R_v, the minimum noise factor F_{min}, and the real and imaginary part G_{opt} and B_{opt} of the optimum source admittance for which the noise factor is minimum. These parameters can be transformed into the power spectral densities of the voltage and current noise sources and the correlation admittance of the corresponding noisy twoport ABCD-parameters representation. The noise parameters of a simplified MOS transistor model including the induced gate noise, the correlation between the induced gate noise and the drain thermal noise, the substrate resistance noise and the gate resistance noise are derived. It is shown that the induced gate noise can usually be neglected for frequencies much lower than the device transit frequency. The analytical results as well as the simulation results obtained from the complete subcircuit model are favorably compared to measurements. This analysis also shows that the contribution of the substrate resistance may be significant and should therefore not be neglected as it is often the case. An example shows that the substrate and gate resistances can typically contribute to about 20% and 5% respectively to the total input referred noise resistance.

For some RF applications in the GHz range using future deep-submicron processes, it may be advantageous to move the operating point of the RF devices from strong inversion to moderate inversion to take advantage of the higher g_m/I_D ratio and therefore of the higher current efficiency and benefit from the lower electrical fields within the transistor. This avoids velocity saturation and hot carriers effects, resulting in smaller excess noise factor α_{sat}. It also allows to take full advantage of the $1/L_{eff}^2$ scaling of the transit frequency compared to the $1/L_{eff}$ scaling when velocity saturation is present. Finally, moderate inversion also better accommodates the more and more stringent low-voltage constraint.

ACKNOWLEDGMENTS

The authors would like to thank their colleagues Michael Schroter, Mishel Matloubian and Valentina Delatorre for fruitful discussions and Dave Pehlke, James Chen, Len Tocci for doing excellent measurements.

REFERENCES

[1] S. P. Voinigescu, S. W. Tarasewicz, T. Mac Elwee and J. Ilowski, "An Assessment of the State-of-the-Art 0.5μm Bulk CMOS Technology for RF Applications," *Proc. of the Int. Electron Devices Meeting*, pp. 29.5.1-29.5.4, Dec. 1995.

[2] H. S. Momose, E. Morifuji, T. Yoshitomi, T. Ohguro, M. Saito, T. Morimoto, Y. Katsumata and H. Iwai, "High-Frequency AC Characteristics of 1.5nm Gate Oxide MOSFETs," *Proc. of the Int. Electron Devices Meeting*, pp. 105-108, Dec. 1996.

[3] H. S. Momose, R. Fujimoto, S. Otaka, E. Morifuji, T. Ohguro, T. Yoshitomi, H. Kimijima, S. I. Nakamura, T. Morimoto, Y. Katsumata, H. Tanimoto and H. Iwai, "RF Noise in 1.5 nm Gate Oxide MOSFETs and Evaluation of the NMOS LNA Circuit Integrated on a Chip," *Proc. of the VLSI Symposium on Technology*, June 1998.

[4] L. E. Larson, "Integrated Circuit Technology Options for RFIC's - Present Status and Future Directions," *IEEE Journal of Solid-State Circuits*, vol. 33, pp. 387-399, March 1998.

[5] H. S. Momose, R. Fujimoto, S. Otaka, E. Morifuji, T. Ohguro, T. Yoshitomi, H. Kimijima, S. I. Nakamura, T. Morimoto, Y. Katsumata, H. Tanimoto and H. Iwai, "RF Noise in 1.5 nm Gate Oxide MOSFETs and Evaluation of the NMOS LNA Circuit Integrated on a Chip," *Proc. of the VLSI Symposium on Technology*, June 1998.

[6] A. Abidi, "Direct-Conversion Radio Transceivers for Digital Communications," *IEEE Journal of Solid-State Circuits*, vol. 30, pp. 1399-1410, Dec. 1995.

[7] A. Abidi, A. Rofougaran, G. Chang, J. Rael, "The future of CMOS wireless transceivers," *Int. Solid-State Circ. Conf. Dig. of Tech. Papers*, pp. 118-119, Feb. 1997.

[8] J. Rudell, J. Ou, T. Cho, G. Chien, F. Brianti, J. Weldon, P. Gray, "A 1.9 GHz wide-band IF double conversion CMOS integrated receiver for cordless telephone applications," *Int. Solid-State Circ. Conf. Dig. of Tech. Papers*, pp. 304-305, Feb. 1997.

[9] B. Razavi, "Architectures and circuits for RF CMOS Receivers," *Proc. IEEE Custom Integrated Circuits Conf.*, pp. 393-400, May 1998.

[10] A. Rofougaran, J. Y.-C. Chang, M. Rofougaran and A. Abidi, "A 1GHz CMOS RF Front-End IC for Direct-Conversion Wireless Receiver," *IEEE Journal of Solid-State Circuits*, vol. 31, pp. 880-889, July 1996.

[11] P. Orsatti, F. Piazza, Q. Huang and T. Morimoto, "A 20mA-Receive, 55mA-Transmit, Single-Chip GSM Transceiver in 0.25μm CMOS," *Int. Solid-State Circ. Conf. Dig. of Tech. Papers*, Feb. 1999.

[12] T. Cho, E. Dukatz, M. Mack, D. Mac Nally, M. Marringa, S. Mehta, C. Nilson, L. Plouvier and S. Rabii, "A Single-Chip CMOS Direct-Conversion Transceiver for 900MHz Spread-Spectrum Digital Cordless Phones, *Int. Solid-State Circ. Conf. Dig. of Tech. Papers*, Feb. 1999.

[13] Y. Cheng et al., "BSIM3, Version 3.0 Manual," Department of Electrical Engineering and Computer Science, University of California, Berkeley, CA 94720, 1996. (http://www-device.EECS.Berkeley.EDU/~bsim3/)

[14] Y. Cheng et al., "A Physical and Scalable I-V Model in BSIM3v3 for Analog/Digital Circuit Simulation," *IEEE Transactions on Electron Devices*, Vol. 44, No. 2, pp. 277-287, February 1997.

[15] MOS MODEL 9 documentation. (information about MOS model 9 is available at http://www-eu.semiconductors.philips.com/Philips_Models/publications_mm9.stm).

[16] D. B. M. Klaassen, "Compact Modeling of Submicron CMOS'," *Proc. of the European Solid-State Circ. Conf.*, pp. 40-46, Edition Frontieres, 1996.

[17] M. Bucher, C. Lallement, C. Enz, F. Theodoloz and F. Krummenacher, "The EPFL-EKV MOSFET Model, Version 2.6, Technical Report, LEG-EPFL, 1997. (documentation and more information on EKV is available at http://legwww.epfl.ch/ekv/).

[18] C. C. Enz, F. Krummenacher and E. A. Vittoz, "An Analytical MOS Transistor Model Valid in All Regions of Operation and Dedicated to Low-Voltage and Low-Current Applications," *special issue of the Analog Integrated Circuits and Signal Processing journal on Low-Voltage and Low-Power Design*, vol. 8, pp. 83-114, July 1995.

[19] C. C. Enz, "MOS Transistor Modeling Dedicated to Low-Current and Low-Voltage Analog Circuit Design and Simulation," in *Low-power HF Microelectronics: Integrating Process, Device and Design for Manufacturability,* Edited by G. Machado, IEE Book Publishing, 1996, Chapter 7, pp. 247-299, ISBN 0 85296 874 4.

[20] M. Bucher, C. Lallement, C. Enz, F. Theodoloz and F. Krummenacher, "Scalable G_m/I Based MOSFET Model," *Proc. of the Int. Semiconductor Device Research Symp.*, Dec. 1997.

[21] Y. Tsividis, *Operation and Modeling of the MOS Transistor*, 2nd-Edition, Mc-Graw-Hill, 1999.

[22] M. C. Ho, K. Green, R. Culbertson, J. Y. Yang, D. Ladwig and P. Ehnis, "A Physical Large-Signal Si MOSFET Model for RF Circuit Design," *IEEE MTT-S Digest*, pp. 391-394, June 1997.

[23] R. R. J. Vanoppen, J. A. M. Geelen and D. B. M. Klaassen, "The high-frequency analogue performance of MOSFETs," *Proc. of the Int. Electron Devices Meeting*, pp. 173-176, 1994.

[24] D. B. M. Klaassen, B. Nauta and R. R. J Vanoppen, *RF modeling of MOSFETs in Analog Circuit Design: MOST RF Circuits, Sigma-Delta Converters and Translinear Circuits*, eds. W. Sansen, R. J. van de Plassche and J. H. Huijsing, pp. 3-24, Kluwer Academic Publishers, 1996.

[25] W. Liu, R. Gharpurey, M. C. Chang, U. Erdogan, R. Aggarwal and J. P. Mattia, "R.F.MOSFET Modeling Accounting for Distributed Substrate and Channel Resistances with Emphasis on the BSIM3v3 SPICE Model," *Proc. of the Int. Electron Devices Meeting*, pp. 309-312, Dec. 1997.

[26] D. R. Pehlke, M. Schroter, A. Burstein, M. Matloubian and M. F. Chang, "High-Frequency Application of MOS Compact Models and their Development for Scalable RF MOS Libraries," *Proc. IEEE Custom Integrated Circuits Conf.*, pp. 219-222, May 1998.

[27] J.-J. Ou, X. Jin, I. Ma, C. Hu and P. Gray, "CMOS RF Modeling for GHz Communication IC's," *Proc. of the VLSI Symposium on Technology*, June 1998.

[28] Y. Cheng, M. Schroter, C. Enz, M. Matloubian and D. Pehlke, "RF Modeling Issues of Deep-submicron MOSFETs for Circuit Design," *Proc. of the IEEE Int. Conf. on Solid-State and Integrated Circ. Tech.*, pp. 416-419, Oct. 1998.

[29] L. F. Tiemeijer and D. B. M. Klassen, "Geometry Scaling of the Substrate Loss of RF MOSFETs," *Proc. of the European Solid-State Dev. Res. Conf.*, pp. 481 483, Sept. 1998.

[30] S. H. Jen, C. Enz, D. R. Pehlke, M. Schroter, B. J. Sheu, "Accurate MOS Transistor Modeling and Parameter Extraction Valid up to 10-GHz," *Proc. of the European Solid-State Dev. Res. Conf.*, Bordeaux, pp. 484-487, Sept. 1998.

[31] X. Jin, J. -J. Ou, C.-H. Chen, W. Liu, M. J. Deen, P. R. Green and C. Hu, "An Effective Gate Resistance Model for CMOS RF and Noise Modeling," *Proc. of the Int. Electron Devices Meeting*, Dec. 1998.

[32] C. E. Biber, M. L. Schmatz, T. Morf, U. Lott and W. Bachtold, "A Nonlinear Microwave MOSFET Model for Spice Simulators," *IEEE Trans. Microwave Theory Tech.*, vol. 46, No. 5, pp. 604-610, May 1998.

[33] C. E. Biber, "Microwave Modeling and Circuit Design with Sub-micron CMOS Technologies," PhD Dissertation No. 12505, Swiss Federal Institute of Technology, Zurich (ETHZ), 1998.

[34] M. Bagheri and Y. Tsividis, "A Small-Signal dc-to-High Frequency Non-quasi-static

Model for the Four-Terminal MOSFET Valid in All Regions of Operations," *IEEE Transactions on Electron Devices*, vol. ED-32, No. 11, pp. 2383-2391, Nov. 1985.

[35] W. Sansen and P. Vandeloo, "Modeling the MOS Transistor at High Frequencies," *Electron. Lett.*, vol. 22, No. 15, pp. 810-812, July 1986.

[36] P. Mancini, C. Turchetti and G. Masetti, "A Non-Quasi-Static Analysis of the Transient Behavior of the Long-Channel MOST Valid in All Regions of Operation," *IEEE Transactions on Electron Devices*, vol. ED-34, No. 2, pp. 325-334, Feb. 1987.

[37] P. Vandeloo and W. Sansen, "Measuring and Fitting of the Small-Signal Model of the MOS Transistor for High Frequency Applications," *IEEE Instrum. Meas. Technology Conf., pp. 232-237, Apr. 1988.*

[38] H. J. Park, P. K. Ko and C. Hu, "A Non-Quasi-Static MOSFET Model for SPICE - Transient Analysis," *IEEE Transactions on Electron Devices*, vol. 36, pp. 561-576, March. 1989.

[39] P. J. V. Vandeloo and W. M. C. Sansen, "Modeling of the MOS Transistor for High Frequency Analog Design," *IEEE Transactions on Computer-Aided Design*, vol. 8, No. 7, pp. 713-723, July 1989.

[40] K. -W. Chai and J. Paulos, "Unified Nonquasi-Static Modeling of the Long-Channel Four-Terminal MOSFET for Large- and Small-Signal Analyses in All Operating Regimes," *IEEE Transactions on Electron Devices*, vol. 36, pp. 2513-2520, Nov. 1989.

[41] P. Roblin, S. C. Kang and H. Morkoc, "Analytical Solution of the Velocity-Saturated MOSFET/MODFET Wave Equation and Its Application to the Prediction of the Microwave Characteristics of MODFET's," *IEEE Transactions on Electron Devices*, vol. 37, No. 7, pp. 1608-1622, July 1990.

[42] L.-J. Pu and Y. Tsividis, "Small-Signal Parameters and Thermal Noise of the Four-Terminal MOSFET in Non-quasi-static Operation," *Solid-State Electronics*, Vol. 33, No. 5, pp. 513-521, 1990.

[43] H. J. Park, P. K. Ko and C. Hu, "A Charge Conserving Non-Quasi-Static (NQS) MOSFET Model for SPICE Transient Analysis," *IEEE Transactions on Computer-Aided Design*, vol. 10, pp. 629-642, May 1991.

[44] P. Roblin, S. C. Kang and W.-R. Liou, "Improved Small-Signal Equivalent Circuit Model and Large-Signal State Equations for the MOSFET/MODFET Wave Equation," *IEEE Transactions on Electron Devices*, vol. 38, No. 8, pp. 1706-1991, Aug. 1991.

[45] S. C. Kang and P. Roblin, "Optimal Second-Order Small-Signal Model for Long- and Short-Channel Three-Terminal MOSFET/MODFET Wave Equation," *IEEE Transactions on Electron Devices*, vol. 39, No. 8, pp. 1909-1915, Aug. 1992.

[46] H. J. Park, P. K. Ko and C. Hu, "A Non-Quasi-Static MOSFET Model for SPICE - AC Analysis," *IEEE Transactions on Computer-Aided Design*, vol. 11, pp. 1247-1257, Oct. 1992.

[47] R. Daniels, A. T. Yang and J. P. Harrang, "A Universal Large/Small Signal 3-Terminal FET Model Using a Nonquasi-Static Charge-Based Approach," *IEEE Transactions on Electron Devices*, vol. 40, pp. 1723-1729, Oct. 1993.

[48] T. Smedes and F. M. Klaassen, "An Analytical Model for the Non-Quasi-Static Small-Signal Behaviour of Submicron MOSFETs," *Solid-State Electronics*, vol. 38, No. 1, pp. 121-130, 1995.

[49] W. Liu, C. Bowen and M.-C. Chang, "A CAD-Compatible Non-Quasi-Static MOSFET Model," *Proc. of the Int. Electron Devices Meeting*, pp. 151-154, Dec. 1996.

[50] Y. Tsividis and K. Suyama, "MOSFET Modeling for analog circuit CAD: problems and prospects", *IEEE Journal of Solid-State Circuits*, vol. SC-29, pp. 210-216, March 1994.

[51] N. Arora, R. Rios and C.-L. Huang, "Modeling the Polysilicon Depletion Effect and its Impact on Submicrometer CMOS Circuit Performance," *IEEE Transactions on Electron De-*

vices, vol. 42, No. 5, pp. 935-943, May 1995.

[52] H. Iwai, M. R. Pinto, C. S. Rafferty, J. E. Oristian and R. W. Dutton, "Analysis of Velocity Saturation and Other Effects on the Short-Channel MOS Transistor Capacitances," *IEEE Transactions on Computer-Aided Design*, vol. 6, No. 2, pp. 173-184, March 1987.

[53] P. Klein, "A Compact-Charge LDD-MOSFET Model," *IEEE Transactions on Electron Devices*, vol. 44, No. 9, pp. 1483-1490, Sept. 1997.

[54] M. Pfost, H.-M. Rein and T. Holzwarth, "Modeling Substrate Effects in the Design of High-Speed Si-Bipolar IC's,",*IEEE Journal of Solid-State Circuits*, vol. 31, No. 10, pp. 1493-1501, Oct. 1996.

[55] E. Abou-Allam, T. Manku, "A Small-Signal MOSFET Model for Radio Frequency IC Applications," *IEEE Transactions on Computer-Aided Design*, vol. 16, pp. 437-447, May 1997.

[56] M. C. A. M. Koolen, J. A. M. Geelen and M. P. J. G. Versleijen, "An Improved De-Embedding Technique for On-Wafer High-Frequency Characterization," *IEEE Proc. Bipolar Circuit and Technology Meeting*, pp. 188-191, 1991.

[57] M. Bucher, C. Lallement and C. C. Enz, "An Efficient Parameter Extraction Methodology for the EKV MOST Model," *Proc. IEEE Int. Conf. on Test Structures*, pp. 145-150, March 1996.

[58] H. Beneking, *High-Speed Semiconductor Devices*, Chapman & Hall, 1994.

[59] A. van der Ziel, *Noise in Solid-State Devices and Circuits*, John Wiley, 1986.

[60] S. Tedja, J. Van der Spiegel and H. H. Williams, "Analytical and Experimental Studies of Thermal Noise in MOSFET's," *IEEE Transactions on Electron Devices*, vol. 41, No. 11, pp. 2069-2075, Nov. 1994.

[61] D. P. Triantis, A. N. Birbas and S. E. Plevridis, "Induced Gate Noise in MOSFETs Revisited: The Submicron Case," *Solid-State Electronics*, vol. 41, No. 12, pp. 1937-1942, 1997.

[62] D. K. Shaeffer and T. H. Lee, "A 1.5-V, 1.5-GHz CMOS Low-Noise Amplifier," *IEEE Journal of Solid-State Circuits*, vol. 32, No. 5, pp. 745-759, May 1997.

[63] T. H. Lee, *The Design of CMOS Radio-Frequency Integrated Circuits*, Cambridge University Press, 1998.

[64] C. Chen and M. J. Deen, "High Frequency Noise of MOSFETs I: Modeling," *Solid-State Electronics*, vol. 42, pp. 2069-2081, Nov. 1998.

[65] C. Chen, M. J. Deen, Z. X. Yan, M. Schroter and C. Enz, "High Frequency Noise of MOSFETs II: Experiments," *Solid-State Electronics*, vol. 42, pp. 2083-2092, Nov. 1998.

[66] A. Abidi, "High-Frequency Noise Measurements on FET's with Small Dimensions." *IEEE Transactions on Electron Devices*, vol. ED-33, No. 11, pp. 1801-1805, Nov. 1986.

[67] B. Wang, J. R. Hellums and C. G. Sodini, "MOSFET Thermal Noise Modeling for Analog Integrated Circuits," *IEEE Journal of Solid-State Circuits*, vol. 29, No. 7, pp. 833-835, July 1994.

[68] D. P. Triantis, A. Birbas and D. Kondis, "Thermal Noise Modeling for Short-Channel MOSFET's," *IEEE Transactions on Electron Devices*, vol. 43, No. 11, pp. 1950-1955, Nov. 1996.

[69] P. Klein, "An Analytical Thermal Noise Model of Deep-Submicron MOSFETs for Circuit Simulation with Emphasis on the BSIM3v3 SPICE Model," *Proc. of the European Solid-State Dev. Res. Conf.*, pp. 460-463, Sept. 1998.

[70] K. Aufinger and J. Boeck, "A Straightforward Noise De-Embedding Method and its Application to High-Speed Silicon Bipolar Transistors," *Proc. of the European Solid-State Dev. Res. Conf.*, pp. 957-960, Sept. 1996.

RF Modelling and characterisation of SOI and bulk MOSFET 's

R. Gillon[‡] (renaud.gillon@mie.alcatel.be)
D. Vanhoenacker[†] (vanhoenacker@emic.ucl.ac.be)
J.-P. Colinge[°] (colinge@ece.ucdavis.edu)

(‡) Alcatel Microelectronics, Oudenaarde, Belgium
(†) Laboratoire d'Hyperfrequences UCL, Louvain-la-Neuve, Belgium
(°) Dept of Electrical and Computer Engineering, UCD, Davis, California

Abstract

Shrinking device dimensions and ever higher working frequencies are creating a strong demand for sophisticated models and characterisation methods. New S-parameters de-embedding strategies allow a finer analysis of the measurement results and a more detailled device modelling.

1 Introduction

Submicrometer CMOS technologies are breaking through in the RF arena under the drive of the explosion in mobile communications. CMOS designers in the industry are working today on circuits operating up to 5 GHz, and plans are already being made to go beyond. This whole evolution has created a pressing need for new models and characterisation methods. As the devices are shrinking and the working frequency is increasing, parasitics are becoming critical. Accurate models are required and characterisation methods must evolve to meet the challenge. Models must also cover an ever-widening band, and new effects are coming into play, such as : channel propagation delays, substrate coupling, self-heating, etc.

It is the purpose of this paper to give an overview of the solutions which were elaborated at the UCL during the last five years, and which are now finding their way to the industry.

2 De-embedding techniques

To correct for the overwhelming influence of cables and probes on the S-parameters measurements, a calibration of the network analyser is performed using a commercially available impedance standards substrates (ISS). These calibration substrates are usually fabricated on an alumina substrate with a thick gold metallisation. To measure the MOS devices, the probes must then be moved onto the silicon wafer. Due to the different substrate materials and probe-pad geometries, additional corrections must be applied to obtain the characteristics of the DuT in conditions corresponding closely to normal use inside a circuit. The most commonly used approach for this second de-embedding step is the immitance correction method. This method suffers several serious drawbacks, which have been avoided in this work by introducing an alternative approach : *in-situ calibration*.

2.1 Immitance correction

This method was first proposed by van Wijnen for bipolar devices, [1]. It was later enhanced by Fraser, [2], and by Cho, [3]. The method is based on the adoption of the equivalent circuit depicted in figure 1 to describe the probing structure into which the DuT is embedded. Cho showed that all six elements of the equivalent

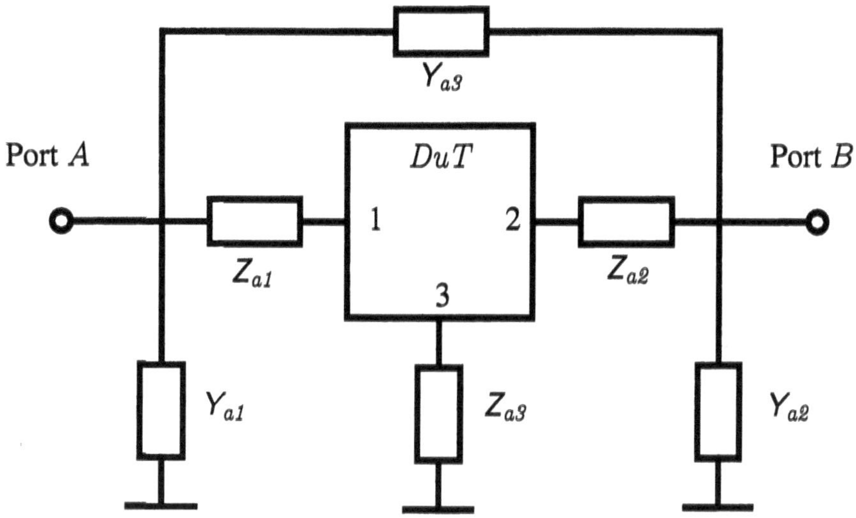

Figure 1: Equivalent circuit model for the immittance correction method.

circuit can be determined from the measured characteristics of three devices :

1. An open circuit, obtained by removing the DuT and leaving the lines corresponding to access nodes 1 and 2 open.
2. A short circuit, obtained by replacing the DuT by metal lines connecting all three access nodes together.
3. A through connection, obtained by replacing the DuT by a metal line connecting nodes 1 and 2 together, while node 3 is left dangling.

The method is unable to determine the true characteristics of the three devices used, so that ideal characteristics are assumed : There is no fringing capacitance for the open ends, the metal lines of the short and through have neither inductance nor resistance. These assumptions limit the applicability of the method to small device geometries where the parasitics of the short, the open and the through

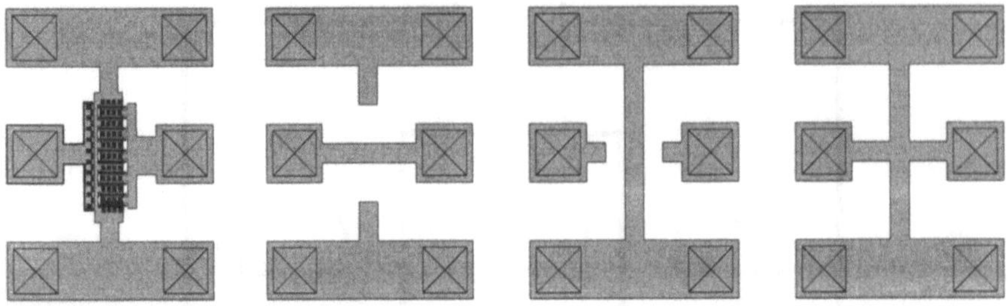

Figure 2: Calibration structures for the immittance correction method.

are negligible. These assumptions also result in a general *overestimation* of the corrections to apply which, in the case of transistors, may result in artificially high apparent values for several characteristics, such as the current-gain transit frequency and can lead to systematic offset between actual circuit performance and the design targets, as discovered by Kim, [4].

Further, the immitance correction approach relies on a model topology using only two complex parameters to characterise the transition from the probe-tips to the device itself. The two remaining parameters, Y_{a3} and Z_{a3}, account only for some parasitic coupling between input and output. The theory of the electro-magnetic reciprocity, [5], shows that *three* complex parameters are generally required to model the passive two-port transition corresponding to a single set of probe-pads, accommodating one probe. The model of figure 1 is thus an oversimplification which may not account properly for the characteristics of the probe-pads at high-frequencies.

Finally, the layout of the de-embedding structures is tightly linked to the geometry of the DuT, so that it is current practice to have several sets of such structures, one for each different device layout. This leads to a rather unefficient use of the test-chip area.

2.2 In-situ calibration

To avoid the inherent limitations of the immitance correction method, an alternative de-embedding strategy has been developped. It is based on scattering parameter calibration techniques belonging the TAN-family of algorithms described by Heuermann and Schiek, [6]. These techniques are renown for their accuracy, well documented and enjoy a sound theoretical foundation, [7]. The error-model used consists of a pair of two-port transfer matrices, $\mathbf{T}_{A(12)}$ and $\mathbf{T}_{B(12)}$, that can model any waveguide transisition where only a single waveguide-mode is propagating.

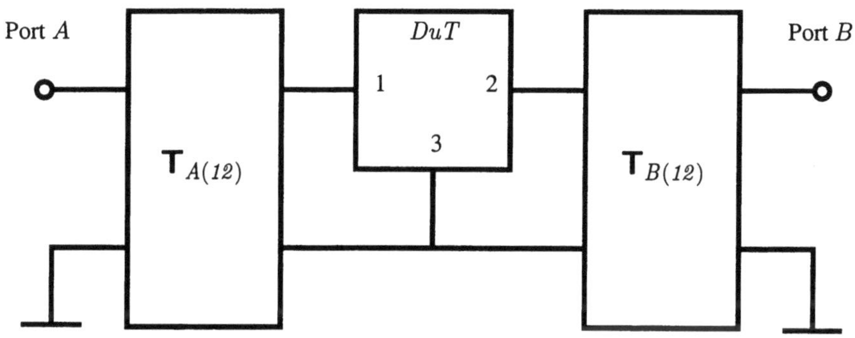

Figure 3: Equivalent circuit model for the in-situ TRL calibration.

The underlying assumptions are thus :

- that the on-wafer probes support only a single mode, which is the case if they are used within the bandwidth specified by the manufacturer.
- that the on-wafer coplanar waveguide (CPW) structures themselves operate with a single mode. In the case of the present work, the cut-off frequency of higher order CPW modes and surface waves were estimated to be higher than 60 GHz.

- that no direct coupling exists between the on-wafer probes. In the case of probes based on micro-coaxial waveguides, the direct coupling can be made negligibly small if the spacing between probes is kept sufficiently wide, ≈ 400 μm.

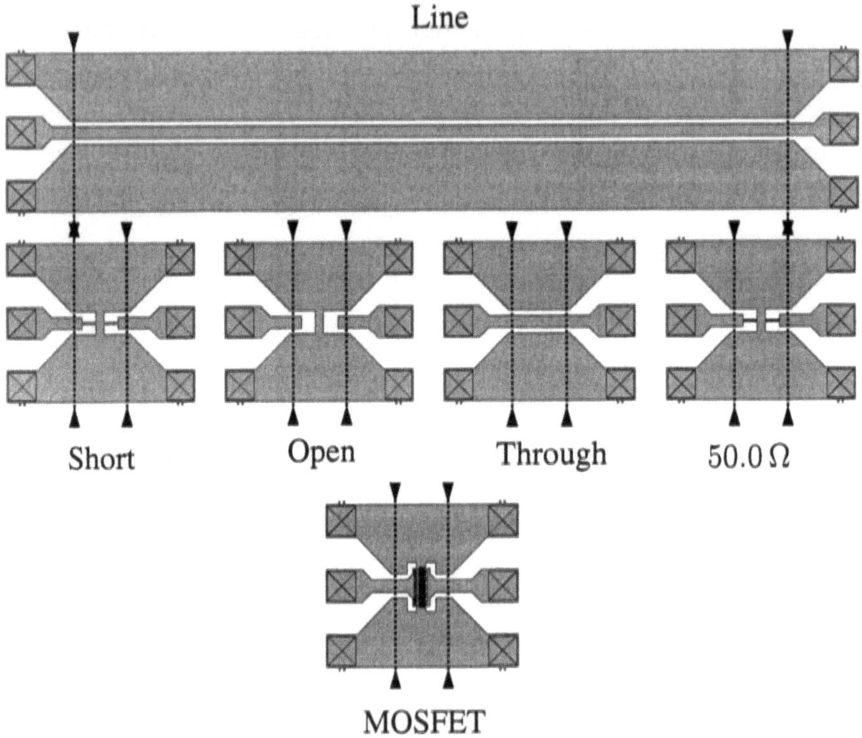

Figure 4: Calibration structures for the in-situ calibration. The reference planes of the calibration are materialised by the vertical lines.

Within the TAN-family, the Through-Reflect-Line (TRL) and the Through-Reflect-Match (TRM) algorithms are widely used for the calibration of on-wafer probes using commercial ISS. For the present de-embedding purposes, TRL was found to give the best results, because it has the least stringent requirements for the characteristics of the calibration structures. Figure 4 shows the layout of the calibration structures implemented on the wafers in the immediate vicinity of the DuT's, hence the name *in-situ*.

Aside from the comprehensive model for the transitions at both ports, the in-situ TRL calibration has the advantage to be able to locate the reference planes at *precise positions* along the CPW's feeding the test devices. This allows to minimise the input and output *adjacent* parasitics, while allowing to avoid the pitfall of overestimated corrections : The reference planes can always be located at some distance from the device themselves, so that probe placement errors do not result in an incursion of the reference plane inside the device region. This gives also some flexibility in the design of the RF probing structures, so that one single calibration set can be used for all devices, transistors, resistors, inductors, capacitors, ...

The practical implementation of the in-situ calibration for the de-embedding of devices was organised as follows :

1. Make a reference calibration on a commercial ISS.
2. Load the silicon wafer and measure the calibration structures shown in Fig. 4.
3. Apply the TRL algorithm to get the *error-boxes*, $\mathbf{T}_{A(12)}$ and $\mathbf{T}_{B(12)}$.
4. Measure the DuT and apply the correction, [8].

The result of this procedure are the de-embedded S-parameters of the DuT *referenced to the characteristic impedance of the CPW line standard,* Z_c, [7]. However Z_c is a-priori unknown, as it is a very complex function of susbtrate resistivity, metallisation and insulation layers thicknesses and actual layout of the CPW line. In order to use the de-embedded S-parameters it is necessary to know Z_c accurately to be able to apply transforms yielding admittance

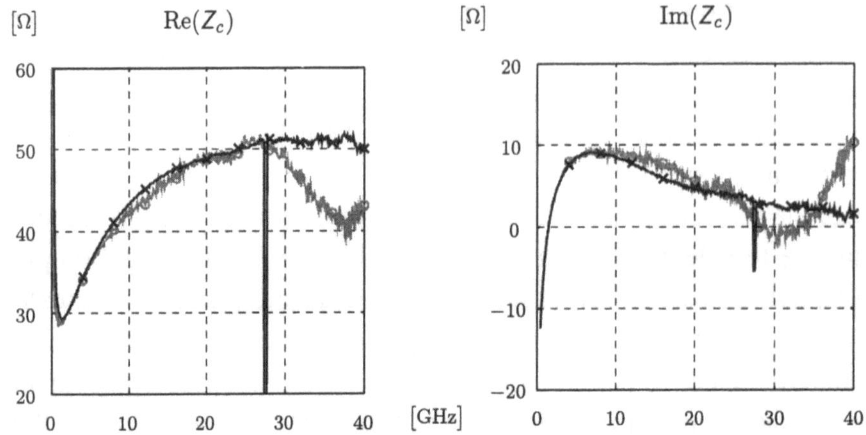

Figure 5: Comparing the *load measurement* (rings) and the *calibration comparison* (crosses) methods.

or impedance parameters or S-parameters reference to the classical 50 Ω value, [7].

In this work, two schemes have been tested for the determination of the characteristic impedance Z_c. The *load measurement* method developed by the author, [9], relies on the comparison between S-parameters and DC-resistance measurement of a resistor. The *calibration comparison method*, [10], extracts Z_c from the error-boxes relating the in-situ calibration to the preceding ISScalibration. This latter method requires a precise knowledge of the reference impedance of the ISS calibration. Both methods were found to agree well, and finally the *load measurement* method was preferred as it is more likely to ensure a coherent DC and *HF* measurements.

The *in-situ calibration* procedure together with the characteristic impedance determination have been used successfully for more than four years on several projects aiming at the development of microwave SOI technologies.

3 Extrinsic device parasitics

At high frequencies, extrinsic device parasitics have an important influence on the device performance. To retain the great scaling flexibility of MOS devices in RF design, it is essential to have accurate models for the layout dependencies of these parameters. Accurate models can be built more easily if the topology of the circuit model is chosen in close agreement with the scaling properties of the device structure.

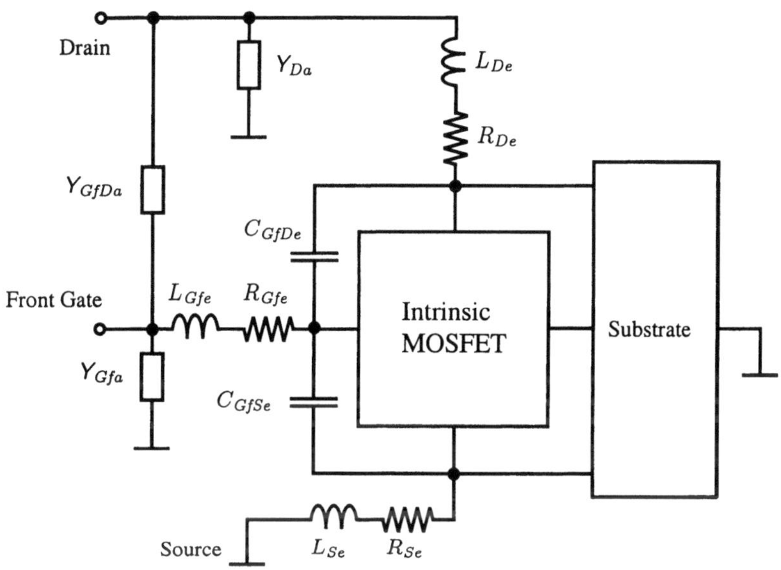

Figure 6: The complete equivalent circuit topology for the SOI MOSFET.

The equivalent circuit of figure 6 has been developped for the MOSFET shown in figure 4. The model accounts for the de-embedded device, which is enclosed between the reference planes materialised in 4. A precise physical interpretation was attributed to every element of the circuit and an extraction procedure was set-up accordingly. Very good correlation between parameter values extracted from S-parameters measurements and values independently using other technology characterisation procedures, [11].

A lot of attention was devoted to the elaboration of a broadband model for the substrate. On medium to high resistivity material, the behaviour of the substrate – oxide stack is quite complicated due to the presence of dielectric relaxation frequencies in the low microwave range. Analytical models for the substrate coupling between source, drain and back-gate, as well as for the substrate "capacitance" of metallisation zones, Y_{Gfa}, Y_{Da}, Y_{GfDa}, were obtained according to the methods described in [12].

4 Non-quasistatic intrinsic device models

The equations below form the basic framework describing charge repartition and the charging delays (NQS effects) in the channel of MOS devices, [13]. The drift-diffusion equation, 1a, is normally used in static operation to compute the charge density (Q''_{nf}) profile along the channel giving the well-known static current equation. The continuity equation 1b, combined with the assumption that the space-average of the channel current is equal to the static current, gives the quasi-static (QS) charge repartition scheme allowing to evaluate gate-source or gate-drain capacitances. The transcendental equation 1c is an empirical equation describing the relationship between the channel potential (V_C) and the inversion charge.

$$I_{Cf}(x) == -W\,\mu\,Q''_{nf}(V_C)\frac{dV_C}{dx} \tag{1a}$$

$$\frac{\partial I_{Cf}(x,t)}{\partial x} = W\left\{\frac{\partial Q''_{nf}(V_C(x,t))}{\partial t}\right\} \tag{1b}$$

$$V_{Gf} - V_{Thf}(V_C) = \kappa_f\,\phi_T\,\log(\frac{Q''_{nf}}{Q''_{nfTh}}) - \frac{Q''_{nf} - Q''_{nfTh}}{C''_{of}} \tag{1c}$$

To date, no general analytical solution of this set of equations has been published. In the small-signal case, an analytical solution of the channel potential *waves* — in the frequency domain —- exists, which lead to the development of a NQS small-signal model, [13]. There have also been attempts to expand the solution of the time-dependent potential profile in Fourrier-series, [14].

At zero drain-source bias, a simple analytical formulation of the admittance matrix can be obtained for small-signal excitations : the *distributed RC-channel* model borrows the equations from transmission line theory. Using this framework a four-terminal model has been built and successfully used for extraction of physical mobility parameters from S-parameters measurements, [15, 16].

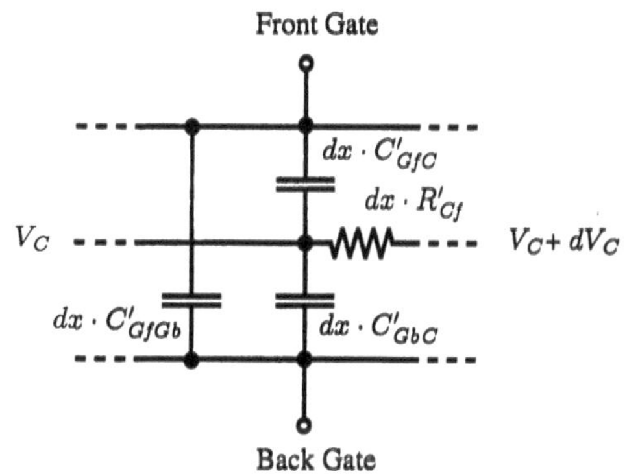

Figure 7: Distributed equivalent circuit for the channel at $V_{DS} = 0$.

Numerical solution schemes for this equation set have also been proposed. Dubois and Robilliart presented a model based on a fine discretisation of the channel, [17]. They use finite differences to evaluate the derivatives in equations 1a and 1b, together with an analytical charge-sheet model for the $Q''_{nf}(V_C)$ relationship at all points in the channel. They reported good results using ten channel nodes, but no comparison was made with *HF* device measurements.

In our work we use a similar approach, discretising the channel into smaller sections for which the QS charge repartition scheme is used, allowing to circumvent the numerical evaluation of derivatives. Q''_{nf} is evaluated analytically using the smooth approximation

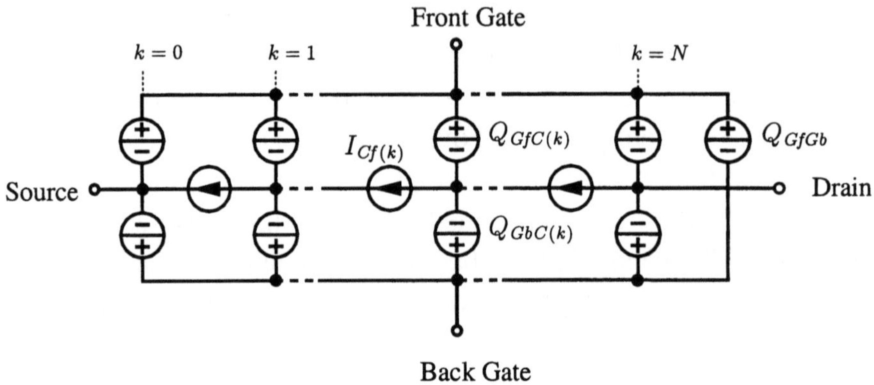

Figure 8: Multiple channel-sections model topology.

of equation 1c proposed by Iniguez,[18]. In order to have a valid model for submicrometer devices, DIBL corrections have been introduced and a channel-length modulation model was developed. Only the outer channel sections are capable of transition from linear to saturated operation and are equipped with the smooth interpolation function proposed in [19].

In order to choose the optimal number of channel sections the model was compared to the distributed channel model, at $V_{DS} = 0\,\text{V}$ which is the worst case situation concerning channel propagation delays. Three sections were found to give a satisfactory accuracy up to higher than the f_T of our devices, lying around 15 GHz.

Another innovative approach to NQS modelling has been recently released in the famous BSIM3v3 model, [20]. This model tries to evaluate the deficit occuring in the channel charge with respect to the prediction of the quasi-static model, which corresponds to the equilibrium level reached after all channel transients have settled. The idea is very appealing but the final implementation suffers a series of drawbacks :

1. The Elmore resistance concept assumes a uniform distribution

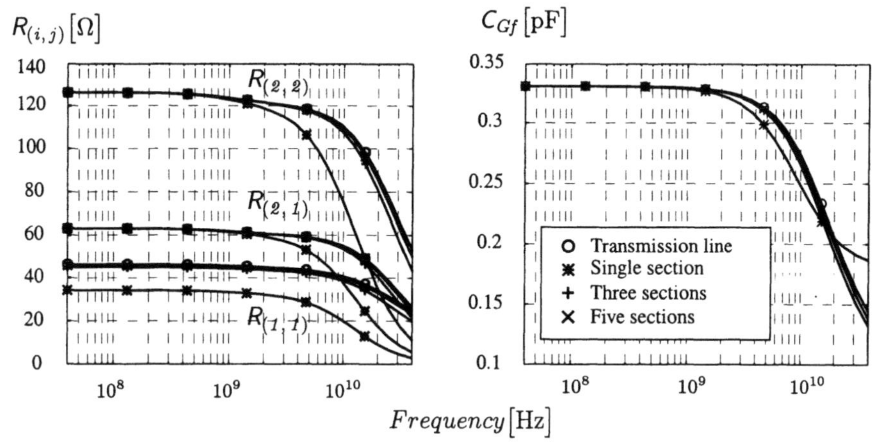

Figure 9: Comparing the analytical distributed channel model and the discrete model with $N = 1, 3, 5$, at $V_{Gf} = 1.0\,\text{V}$ and $V_{DS} = 0$.

of channel resistance across the channel. A situation that occurs only at $V_{DS} = 0$.

2. In [20] the Elmore constant ϵ which is normally 3, is set to 5, probably to compensate for the effects of non-uniform channel resistance. In doing so, accuracy on the prediction of the impedance at the gate is lost, probably to ensure better results in the forward transadmittance.

As a result, the NQS model included in BSIM3v3 seems better suited for long-channel devices, $L_{(eff)} > 3\,\mu\text{m}$, subjected to medium-speed transients with rise times around 1 ns.

5 Results

This section compares simulated S-parameter curves obtained from the model described in sections 3 and 4 with broadband on-wafer measurements performed according to the procedure outlined in section 2.

Figure 10 shows detailled result obtained at $V_{DS} = 0$ when NQS effects are the strongest, while figures 11 and 12 cover cases at $V_{DS} \neq 0$. At biasing voltages below 2 V, measured and simulated data agree well. Above 2 V, the fit starts to degrade with increasing biasing voltages. This can be attributed to the self-heating of the device which is not included in the model.

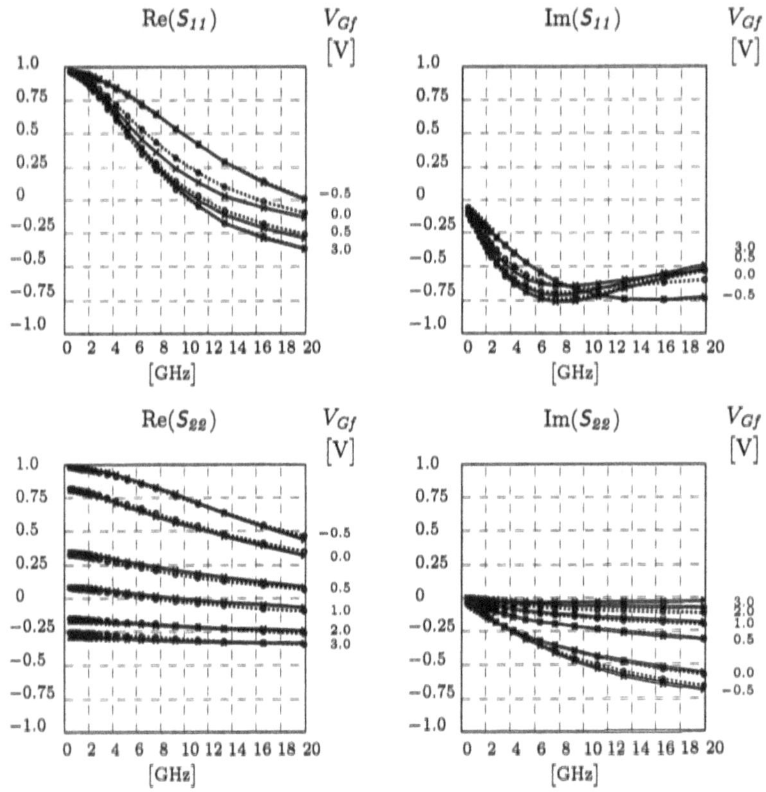

Figure 10: Comparing the measured (dotted lines with rings) and modelled (continuous lines with crosses) S-parameters after extraction of the unified model parameters for a $10 \times (24.0\mu m/1.0\mu m)$ SOI n-MOSFETat $V_{DS} = 0$.

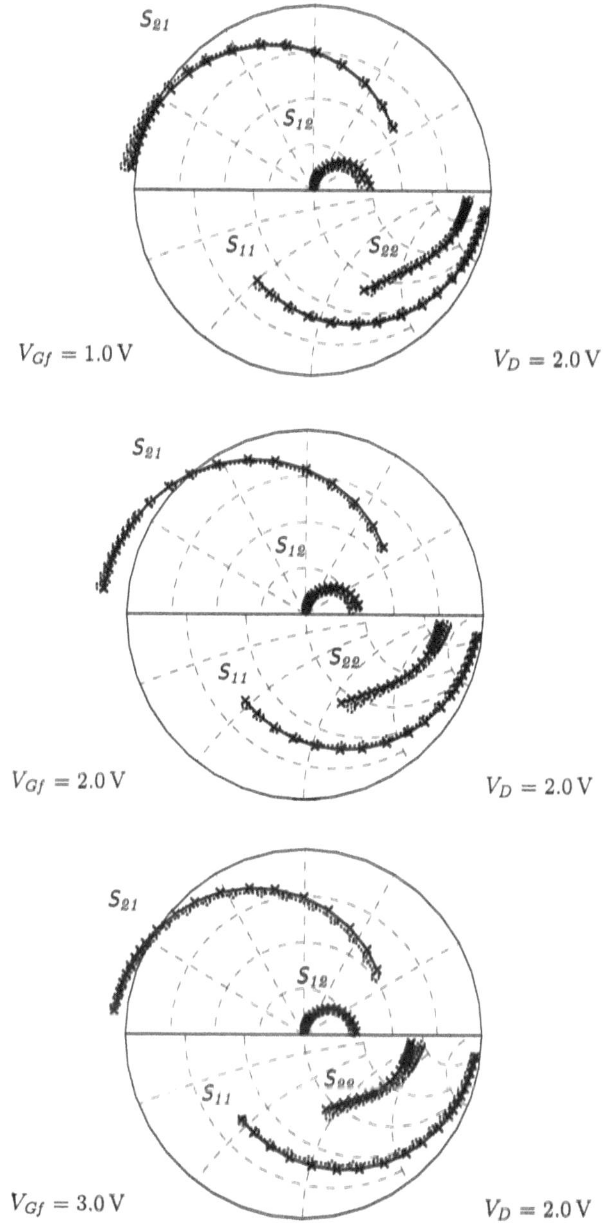

Figure 11: Comparing the measured (dotted lines with rings) and modelled (continuous lines with crosses) S-parameters after extraction of the unified model parameters for a $10 \times (24.0\mu m/1.0\mu m)$ SOI n-MOSFET.

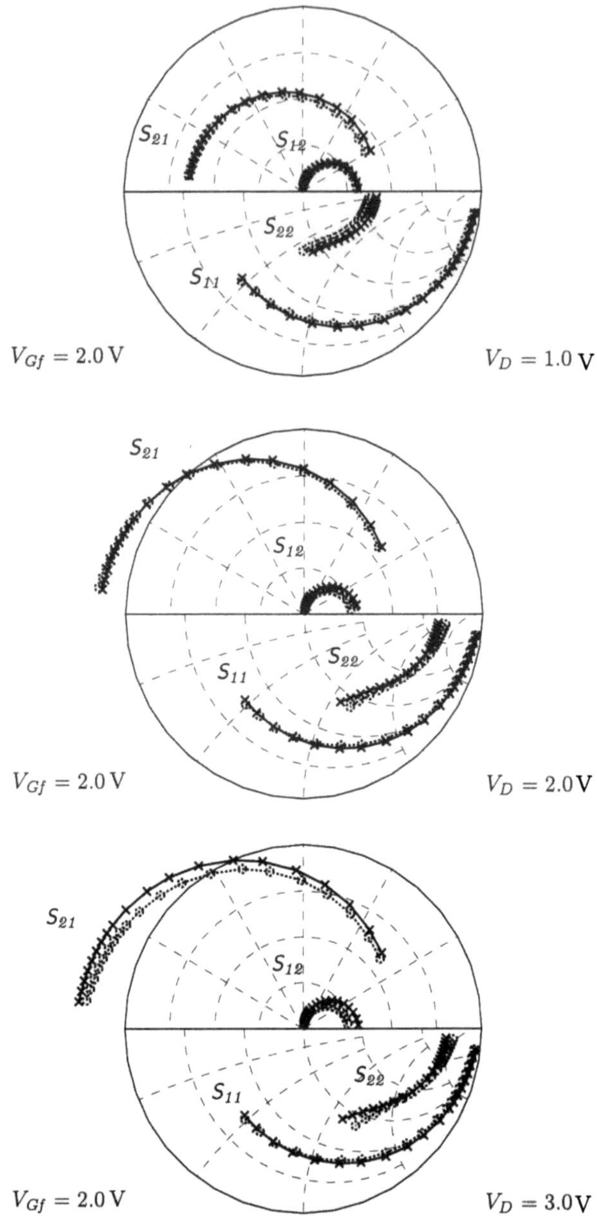

Figure 12: Comparing the measured (dotted lines with rings) and modelled (continuous lines with crosses) S-parameters after extraction of the unified model parameters for a $10 \times (24.0\mu m/1.0\mu m)$ SOI n-MOSFET.

6 Conclusion

This paper covered several important aspects of the RF modelling issue : de-embedding of on-wafer S-parameters measurements, scaleability of the devcie parasitics model, non-quasi-static modelling of submicrometer MOSFET's. The limitations of the immittance correction approach were discussed, and an alternative de-embedding method was presented. A short overview of the state-of-the-art in NQS modelling was given. The NQS model was extensively compared to high-frequency S-parameters data.

Current development include application of the channel discretisation scheme on top of BSIM3v3 for bulk MOSFET's. Modelling results for bulk devices will be shown at the workshop.

7 Acknowledgements

The authors wish to acknowledge fruitful discussions with J.-P. Raskin. Part of this work was funded by the "Communauté Française de Belgique" through the program "Actions de Recherche Concertées", and by the European Community through the SPACE (ESPRIT) program.

References

[1] P. J. van Wijnen, H. R. Claessen, and E. A. Wolsheimer, "A new straightforward calibration and correction procedure for on-wafer high frequency S-parameters measurements (45 MHz - 18 GHz)," in *IEEE 1987 Bipolar Circuits and Technology Meeting*, 1987.

[2] A. Fraser, R. Gleason, and E. W. Strid, "GHz on-silicon-wafer probing calibration methods," in *IEEE 1988 Bipolar Circuits and Technology Meeting*, 1988.

[3] H. Cho and D. Burk, "A three-step method for the de-embedding of high-frequency s-parameters measurements," *IEEE Trans. on Electron Devices*, vol. 38, pp. 1371–1375, June 1991.

[4] C.-H. Kim, C. S. Kim, H. K. Yu, and K. S. Nam, "An isolated-open pattern to de-embed pad parasitics," *IEEE Microwave and Guided Waves Letters*, vol. 8, pp. 96–98, Feb. 1998.

[5] D. F. Williams and R. B. Marks, "Reciprocity relations in waveguide junctions," *IEEE Microwave and Guided Waves Letters*, vol. 41, pp. 1105–1110, June 1993.

[6] H. Heuermann and B. Schiek, "Robust algorithms for txx network analyzer self-calibration procedures," *IEEE Trans. on Instrumentation and Measurement*, vol. 43, pp. 18–22, Feb. 1994.

[7] R. B. Marks and D. F. Williams, "A general waveguide circuit theory," *J. Res. of the Natl Inst. Stand. and Technol.*, vol. 97, pp. 533–562, Sep–Oct 1992.

[8] H. Heuermann and B. Schiek, "Procedures for the determination of the scattering parameters for network analyzer calibration," *IEEE Trans. on Instrumentation and Measurement*, vol. 42, pp. 528–531, Apr. 1993.

[9] R. Gillon, J.-P. Raskin, D. Vanhoenacker, and J.-P. Colinge, "Modelling and optimizing the soi mosfet in view of mmic applications," in *25th European Microwave Conference Digest*, (Bologna, Italy), pp. 543–547, Sep. 4–7 1995.

[10] D. F. Williams, R. B. Marks, and A. Davidson, "Comparison of on-wafer calibrations," in *38th ARFTG Conference Digest*, pp. 68–81, Dec. 1991.

[11] J.-P. Raskin, R. Gillon, J. Chen, D. Vanhoenacker, and J.-P. Colinge, "Accurate SOI MOSFET characterization at mi-

crowave frequencies for device performance optimisation and analogue modelling," *IEEE Trans. on Electron Devices*, May 1998.

[12] J.-P. Raskin, A. Viviani, D. Flandre, and J.-P. Colinge, "Substrate crosstalk reduction using SOI technology," *IEEE Trans. on Electron Devices*, vol. 44, pp. 2252–2261, Dec. 1997.

[13] M. Bagheri and Y. Tsividis, "A small-signal dc-to-high-frequency nonquasistatic model for the four terminal MOSFET valid in all regions of operation," *IEEE Trans. on Electron Devices*, vol. 32, pp. 2383–2391, nov 1985.

[14] H.-J. Park, P. K. Ko, and C. Hu, "A charge conserving non-quasistatic MOSFET model for SPICE transient analysis," *IEEE Trans. on Computer Aided Design*, vol. 10, pp. 629–642, may 1991.

[15] R. Gillon, J.-P. Raskin, D. Vanhoenacker, J.-P. Colinge, and G. Dambrine, "Characterisation of soi mosfets at microwave frequencies," in *Proceedings of the 8th Int. Symp. on SOI Technology and Devices* (S. Cristoloveanu, ed.), vol. 97-23, (Paris), pp. 149–154, Electrochemical Society, Inc., Aug. 31 – Sep. 5 1997.

[16] R. Gillon, *Modelling and Characterisation of the SOI MOSFET for MMIC applications.* PhD thesis, Université catholique de Louvain, Louvain-la-Neuve, june 1998. <ftp://anonymous@ftp.emic.ucl.ac.be/Pub/Incoming/RG-Public/Thesis>.

[17] E. Dubois and E. Robilliart, "Efficient non-quasi-static MOSFET's model for circuit simulation," in *Proceedings of the IEDM '95*, pp. 945–948, 1995.

[18] B. Iñìguez, L. F. Ferreira, B. Gentinne, and D. Flandre, "A physically-based C^{∞}-continuous fully-depleted SOI MOSFET

model for analog applications," *IEEE Trans. on Electron Devices*, vol. 43, pp. 568–575, Apr. 1996.

[19] C. C. McAndrew, B. K. Bhattacharya, and O. Wing, "A single-piece C^{∞}-continuous MOSFET model including subthreshold conduction," *IEEE Trans. on Electron Devices*, vol. 12, pp. 565–567, Oct. 1991.

[20] M. Chan, K. Hui, and P. K. Ko, "A robust and physical BSIM3 non-quasistatic transient and AC small-signal model for circuit simulation," *IEEE Trans. on Electron Devices*, vol. 45, pp. 834–841, 4 1998.

MICROWAVE NOISE MODELING OF CMOS TRANSISTORS

Tajinder Manku
RF Technology Group, University of Waterloo
Waterloo, Ontario, Canada

ABSTRACT

In this paper we systematically develop an understanding of the noise behavior within MOS devices operating at microwave frequencies. A bottom to top approach is taken to accomplish this – device physics to network modeling to device layout. Our hope is that the results within this paper will provide RF CMOS circuit designers with a better understanding of the noise properties of MOS devices as well as to help them design better low noise amplifiers and mixers.

1. INTRODUCTION

CMOS technology has become a viable technology choice for RF integrated components for the wireless communication market [1,2]. Because of this, radio frequency integrated circuit (RF IC) designers are pushing the need for a better RF noise model for submicron MOS transistors (e.g. [3,4]). The noise models currently available are inadequate in predicting the performance of low noise circuit components such as low noise amplifiers (LNA's) or front end mixers. Due to the poor MOSFET RF noise models currently available within most circuit simulators, "good" low noise design requires several revisions. The main difficulty in developing a "universal" noise model for MOS transistors is due to our limited understanding of the noise properties of MOS devices at RF frequencies. Although there has been some research work (e.g. [5-13]) illustrating the basic trends between microwave noise parameters and device structures, all the observations have not isolated the various noise sources within the device.

In this paper the physical origins of the various noise sources within a MOSFET are described with the help of a quasi 3-D microwave noise simulator [14,15] coupled with measurement results.

2. PHYSICAL NOISE SOURCES

Though MOSFET's and MESFET's are very different in their large signal operation, their small signal equivalent noise model are very similar. Consequently, one may use the results of the long time studies of the microwave noise behavior of MESFET's [16,17], and apply the same sort of analysis to MOSFET's. At the microwave frequencies, the intrinsic and extrinsic noise sources of a MOSFET are generated via diffusion noise. The intrinsic noise sources are:

1. **Drain Channel noise,** $\overline{i_d^2}$ - this is due to the diffusion noise generated by the carriers in the channel region[1].
2. **Gate resistance noise,** $\overline{v_{rg}^2}$ - this noise is due to the thermal noise generated by the gate resistance.
3. **Induced gate noise,** $\overline{i_g^2}$ - this noise is due to the diffusion noise generated by the carriers in the channel which capacitively couples on to the gate node as a gate current.

In order to understand the characteristics of drain channel noise and induced gate noise both numerical simulation techniques along with experimental results will be used. The numerical simulations are based on a quasi 3-D device noise simulator [14] developed at the University of Waterloo. A brief description of the physics involved with the numerical procedure is provided below – a more detailed description can be found in [14] and [15].

Using a device simulator the DC conditions of the device are numerically solved. Once this is completed, a frequency dependent source $i(\vec{r},\omega)$ is injected at position $\boldsymbol{r}$ within the device, which simulates the local noise at the

[1] Diffusion noise is a generalization of both shot and thermal noise.

position $\boldsymbol{r}$. The currents at all the terminal nodes of the device are recorded as a function of the frequency of the injected noise, as well as its position $\boldsymbol{r}$. From this information, a "spatially depend current gain function" (commonly referred to as the Green's function) is constructed; i.e.

$$G_g(\vec{r},\omega) = \frac{i_g(\vec{r},\omega)}{i(\vec{r},\omega)}; G_d(\vec{r},\omega) = \frac{i_d(\vec{r},\omega)}{i(\vec{r},\omega)} \tag{1}$$

where i_d and i_g are the recorded currents at the drain and gate, respectively; G_d and G_g are the "current gain functions" for the drain and gate, respectively. Using this information, one can obtain the power spectrum densities as well as the correlation factors. This is done using the following equation (see [14,15] and the papers within);

$$i_\alpha i_\beta^* = \int_\Omega \nabla_{\vec{r}} G_\alpha(\vec{r},\omega) \cdot \hat{K}(\vec{r},\omega) \cdot \nabla_{\vec{r}} G_\beta^*(\vec{r},\omega) d\vec{r} \tag{2}$$

where the integrand, $\nabla_{\vec{r}} G_\alpha(\vec{r},\omega) \cdot \hat{K}(\vec{r},\omega) \cdot \nabla_{\vec{r}} G_\beta^*(\vec{r},\omega)$ describes the amount of noise at position $\boldsymbol{r}$; this term is typically quoted as the "local noise density". Here, $\beta = d,g$, $*$ denotes the complex conjugate, and $\hat{K}(\vec{r},\omega)$ represents the diffusion noise tensor at point $\boldsymbol{r}$ and is given by,

$$\hat{K}(\vec{r},\omega) = 4q^2 \hat{D}(\vec{r}) n(\vec{r}) \tag{3}$$

Here, $\hat{D}(\vec{r})$ is the diffusivity tensor and $n(\vec{r})$ is the electron density at the DC operating point at position $\vec{r}$.

A 0.5μm LDD nMOS transistor was simulated using the device noise simulator. The DC IV characteristics of the simulated device were calibrated with measurement results from the LDD 0.5μm CMOS process. The maximum f_t was simulated to be ~21 GHz which agreed with our experimental results. In Fig. 1 we have plotted the drain current density I_{ds}, transconductance g_m, zero drain voltage channel conductance g_{do}, and the f_t as a function of gate-source voltage.

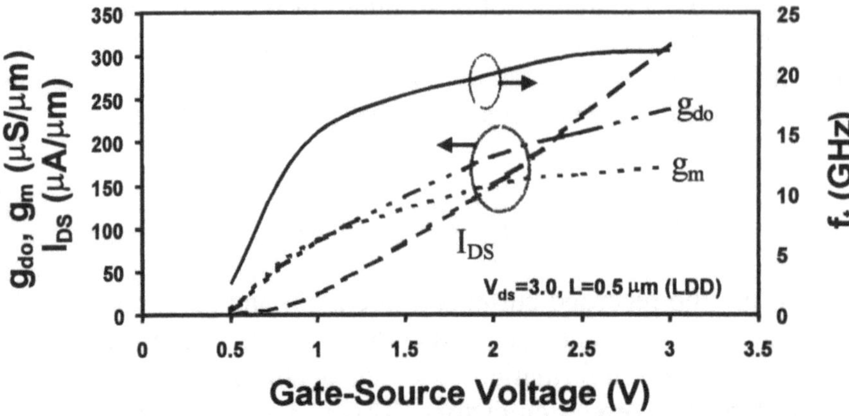

Figure 1: Drain current density, transconductance, zero drain voltage channel conductance, and the f_t as a function of gate-source voltage.

Fig. 2 illustrates typical examples of the drain and gate "local noise density"; recall that the local noise density describes the amount of noise at position ***r***, and by integrating over the device the current noise spectrum at the respective terminal is obtained. Most of the noise can be seen to be due to the channel region. The drain local noise density peaks closer to the source side whereas the gate local noise density has one peak near the source and another near the drain where the peak closer to the source side is relatively larger. The local noise density is larger near the source side because the concentration of electrons, ***n*** is larger near the source compared to the drain – see equation (2) in conjunction with (3).

The diffusion noise within the channel gives rise to both drain channel noise as well as induced gate noise. The drain channel noise is typically described by the equation,

$$\bar{i}_d^2 = 4kT\gamma g_{do}\Delta f \tag{4}$$

Here, γ is a bias dependent parameter, g_{do} is the zero drain channel conductance (this is sometimes taken to be equal to the transconductance of the device since $g_m \sim g_{do}$), k is Boltzmann's constant, T is the temperature of the carriers in the channel, and Δf is the noise bandwidth.

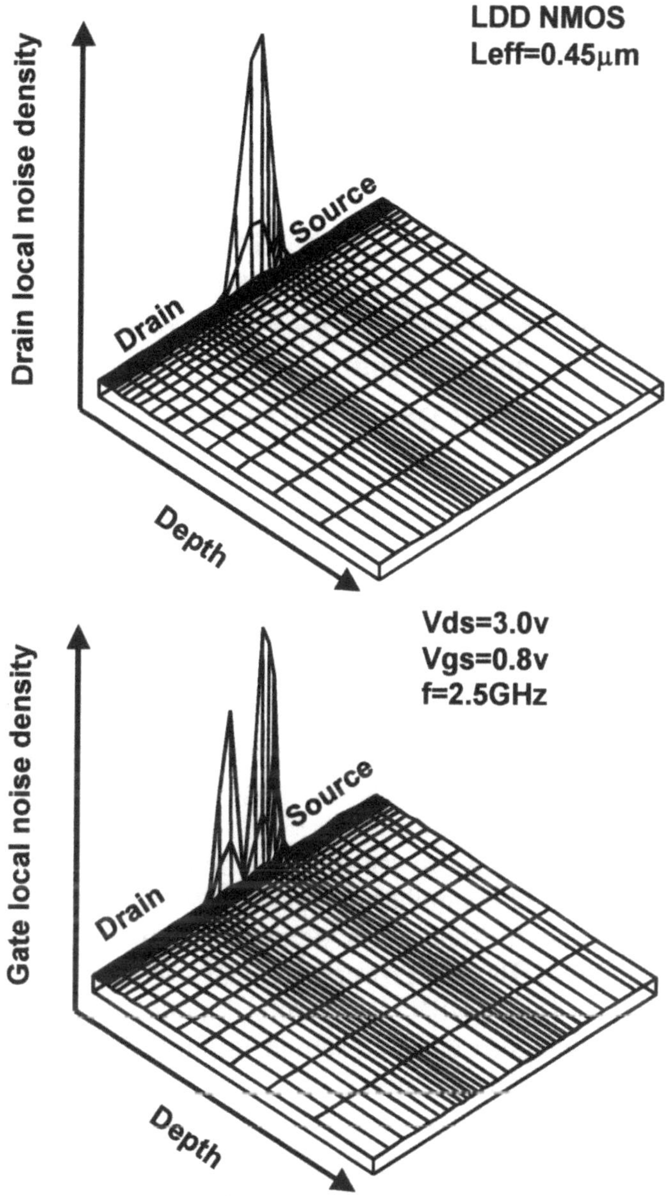

Figure 2: Drain and gate local noise density within the device.

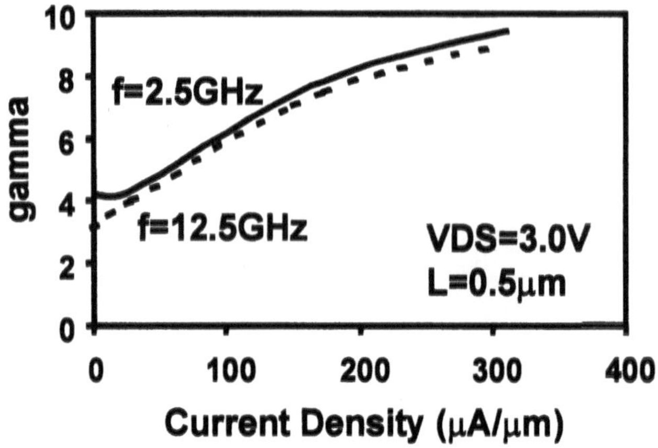

Figure 3: Gamma as a function of current per width of device.

For long channel devices it has been determined that the parameter γ is equal to *2/3*. However, for short channel devices the parameter γ has been observed to be greater than *2/3* (see [9,15]); when a device length of 10μm was simulated using the noise simulator a γ value of *0.7* was obtained. The question is 'where is this excess noise coming from in a short channel device?' From our simulation results it was determined that the hot electron effect is not significant enough to cause γ to increase up to the values reported in the literature. Our simulation results have led us to believe that this excess noise is due to shot noise near the source end due to an equivalent Drain Induced Barrier Lowering (DIBL) – a good review of DIBL can be found in [18]. DIBL is a strong function of the following processing parameters: junction depth of source, doping of source, if a LDD structure is used, and the oxide thickness.

In Fig. 3 we have plotted the parameter γ as a function of gate-source voltage. The results were extracted from the noise simulator for two frequency values. As the gate-source voltage increases the factor γ increases. This can be explained in terms of DIBL.

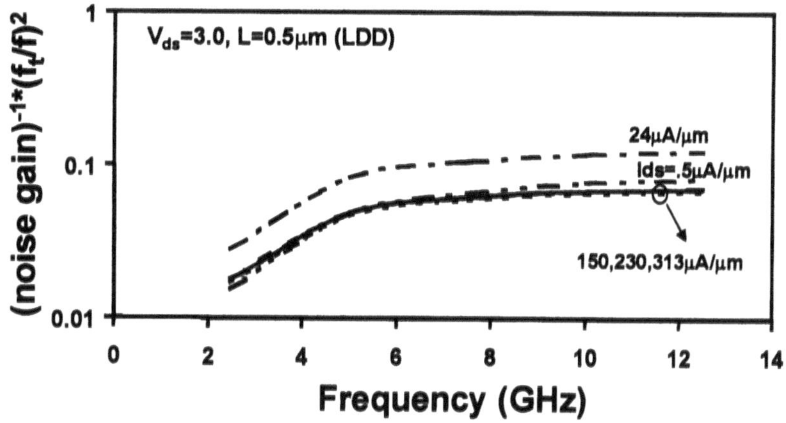

Figure 4: The contribution of the gate noise at the output compared to drain channel noise [15].

An important quantity is the relative magnitudes of $\bar{i}_g^2$ to $\bar{i}_d^2$. The ratio $\bar{i}_d^2/\bar{i}_g^2$ can be interpreted as the effective current noise gain seen from the gate port to the drain port; note that for a bipolar transistor the current noise gain is equal to the DC current gain of the device. The effective current gain noise can be approximated by comparing the average impedance the carriers see in the drain direction to the gate direction. This comparison can be expressed as $(1/\omega C_{gs})/r_i$; r_i is the effective channel resistance and $(1/\omega C_{gs})$ is the impedance of the gate-source capacitance. Therefor the effective current gain noise can be approximated by the expression,

$$\frac{\bar{i}_d^2}{\bar{i}_g^2} \cong \frac{1}{\omega^2 C_{gs}^2 r_i^2} \tag{5}$$

It has been shown by others (e.g. [19]) that r_i for a long channel device is equal to *1/(5g*$_m$*)*. By using this, equation (5) can be rewritten in the form,

$$\frac{\bar{i}_d^2}{\bar{i}_g^2} \cong 25\left(\frac{\omega_t}{\omega}\right)^2 \text{; estimated} \tag{6}$$

In Fig. 4 we have plotted $\bar{i}_g^2/\bar{i}_d^2(f_t/f)^2$ as a function of frequency using bias as a parameter. For a frequency of 2.5GHz we find that;

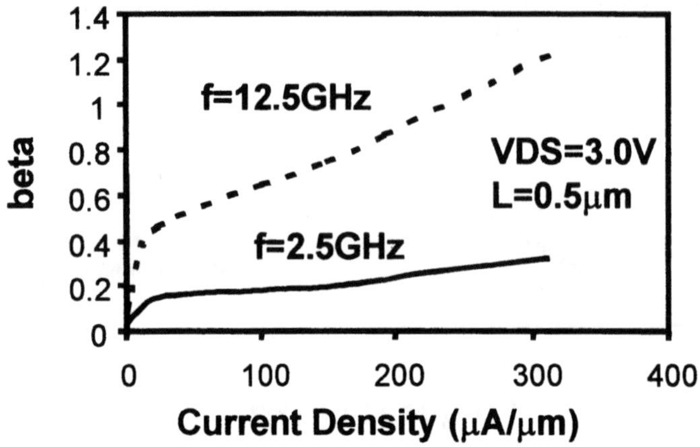

Figure 5: Beta as a function of current for two frequencies [15].

$$\frac{\bar{i}_d^2}{\bar{i}_g^2} \cong 40\left(\frac{\omega_t}{\omega}\right)^2 ;\text{2 - D simulator} \tag{7}$$

Since the current gain of a MOSFET is given by f_t / f, the effective output noise current due to the gate noise is equal to $\bar{i}_g^2 (f_t / f)^2$ (neglecting the fact the drain and gate noise are correlated). The contribution of the gate noise at the output compared to the drain channel noise is therefor $\bar{i}_g^2 / \bar{i}_d^2 (f_t / f)^2 \sim .03$ or *3%*.

The expression for induced gate noise is typically written in the form,

$$\bar{i}_g^2 = 4kT\beta \frac{\omega^2 C_{gs}^2}{g_{do}} \Delta f \tag{8}$$

where C_{gs} is the gate-source capacitance and β is a bias dependent parameter, but for a long channel device it is equal to 4/15 (~0.27). According to (8), induced gate noise decreases with V_{gs} bias. The behavior of β with bias has been extracted from the noise simulator and has been plotted in Fig. 5. Note that β is a strong function of frequency. For a lower

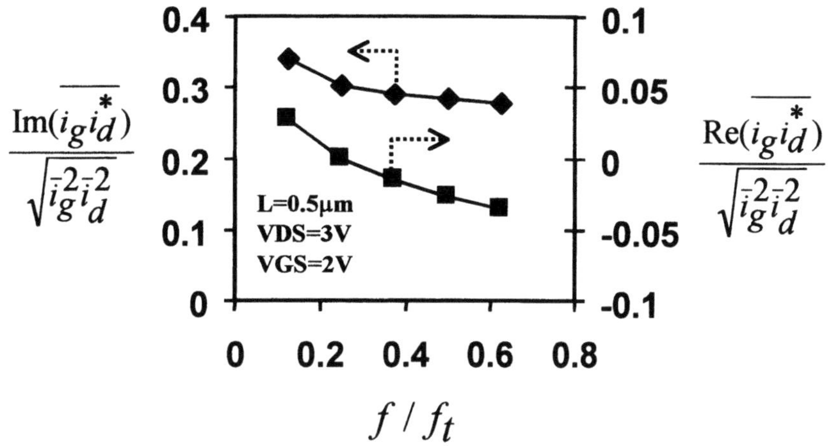

Figure 6: Normalized correlation factors between the induced gate noise and drain channel noise.

frequency value of 2.5GHz β is approximately equal to β for a long channel device (~*0.27*). For a higher frequency of 12.5GHz β is relatively higher in value.

Since the channel drain noise and induced gate noise are generated by the same noise source they are correlated. In Fig. 6 the normalized correlation between the drain current noise and the gate current noise has been plotted; i.e.,

$$B + jC = \frac{\overline{i_g i_d^*}}{\sqrt{\overline{i_g^2}\,\overline{i_d^2}}} \tag{9}$$

where B and C are defined as the real and imaginary parts of the normalized correlation factor, respectively. Simulations reveal that B is approximately equal to zero. This is to be expected since carriers travelling to the gate are displaced by 90 degrees. For low frequencies our results extrapolate to a value of $C=0.4$ which corresponds to the predicted value for long channel devices [16]. As the frequency of operation increases, C approaches *0.3*. From our simulation results we observed that the maximum correlation

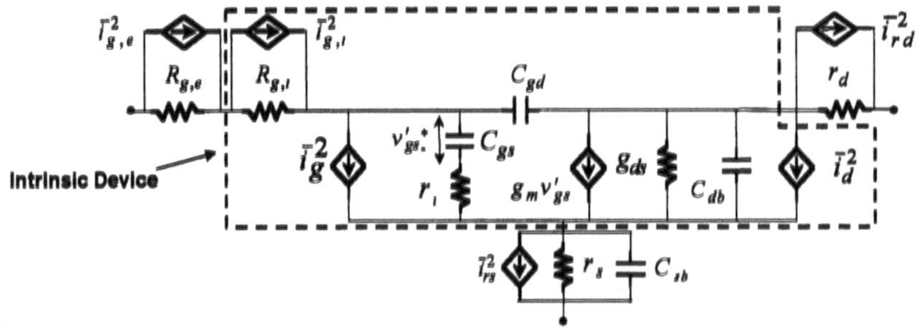

Figure 7: Noise network model of a MOSFET. The intrinsic portion of the device is encompassed in the dotted line.

between the drain's "spatially depend current gain" and the gate's "spatially depend current gain" occurs near the center of the channel region.

The gate thermal noise arises from the resistance of the gate material. It is commonly assumed that this noise source can be ignored if one uses a finger structure for the gate. However, it will be shown in the next section that the gate resistance is very important in the design of low noise devices even if a finger structure is employed – i.e. it sets the minimum noise figure of the device. The noise introduced by the intrinsic portion of the gate structure of width, W and channel length, L is given by the expression,

$$\overline{v}_{g,i}^2 = 4kT\frac{R_{g,sq}}{3}\frac{W}{L}\Delta f \tag{10}$$

where $R_{g,sq}$ is the sheet resistance of the gate material. The factor of three arises from a distributed effect [20].

4. MICROWAVE NOISE PARAMETERS – i.e. what you measure

A small signal model of a MOS transistor is shown in Fig. 7. Aside from the well known lumped elements (i.e. gate-drain capacitance (C_{gd}), gate-source capacitance (C_{gs}), output conductance (g_{ds}), output transconductance (g_m), etc…) two extra elements have been included. These include the channel charging resistance (r_i), and the gate resistance ($R_{g,i}$ (intrinsic) and $R_{g,e}$ (extrinsic)).

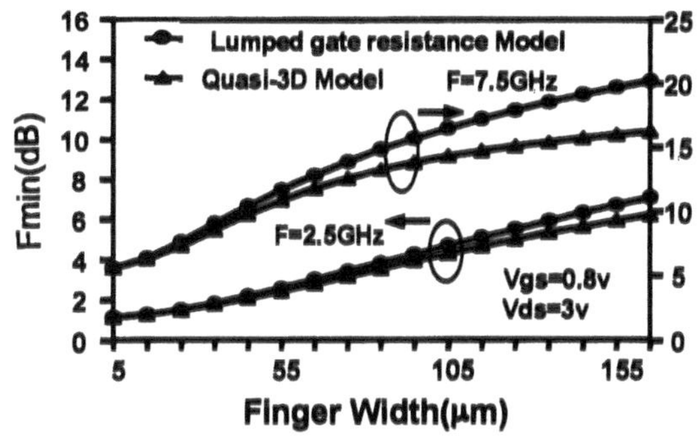

Figure 8: F_{min} as a function of finger width for two frequencies. Also illustrated is the effect of using a lumped gate resistance model in comparison to a distributed gate resistance model.

In order to arrive at a better understanding of the overall noise performance of a MOS transistor, only the intrinsic portion of the device will be considered. Furthermore, the gate drain capacitance will be neglected. Though this component should be included, our goal here is to obtain some basic trends in designing a MOS transistor for optimum noise performance, and not exact numerical values. In order to determine the noise parameters, the equivalent the input reference noise sources, v_n and i_n are first determined;

$$v_n = v_{g,i} - \frac{i_d}{g_m} + (R_{g,i} + r_i)i_g - j\frac{f}{f_T}(R_{g,i} + r_i)i_d \tag{11}$$

$$i_n = i_g - j\frac{f}{f_T}i_d \tag{12}$$

Note that the two noise sources (11) and (12) are correlated since they contain common noise terms. By using equations (11) and (12) all the noise parameters can be obtained [3]:

$$G_n = \left(\frac{f}{f_t}\right)^2 \gamma g_{do} \times F_1 \tag{13}$$

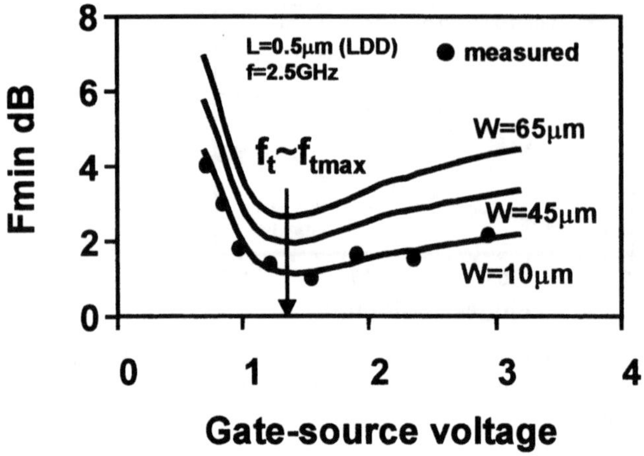

Figure 9: F_{min} as a function of gate source voltage for various gate width.

$$R_{opt} = \sqrt{(R_{g,i} + r_i)^2 + \left(\frac{f_t}{f}\right)^2 \frac{F_4 \times R_{g,i}}{\gamma g_{do}} + \frac{F_2^2 F_5}{\gamma g_{do}^2 F_1}\left(\frac{f_t}{f}\right)^2} \tag{14}$$

$$X_{opt} = \frac{F_2}{2\pi f C_{gs}} = -F_2 X_{in} \tag{15}$$

$$\begin{aligned} F_{\min} &= 1 + 2G_n(R_{opt} + R_{g,i} + r_i) \\ &\cong 1 + 2F_3 \frac{f}{f_t}\sqrt{\gamma g_{do} R_{g,i}} \end{aligned} \tag{16}$$

where G_n is the noise conductance, R_{opt} is the optimum noise resistance, X_{opt} in the optimum noise reactance, F_{min} is the minimum noise figure, and X_{in} is the input reactance of the device. The parameters F_1 to F_4 are weak functions of bias, and are less than one – the details of these parameters will not be give here, but can be derived from equations (11) and (12); see [21]. If induced gate noise is neglected (i.e. $i_g = 0$) F_1 to F_4 are equal to 1. The parameter F_5 is equal to 1 if induced gate noise is included and 0 if it is ignored (i.e. $i_g = 0$). In the bottom equation for F_{min}, it has been assumed that $R_{g,i} + r_i$ is relatively small in value.

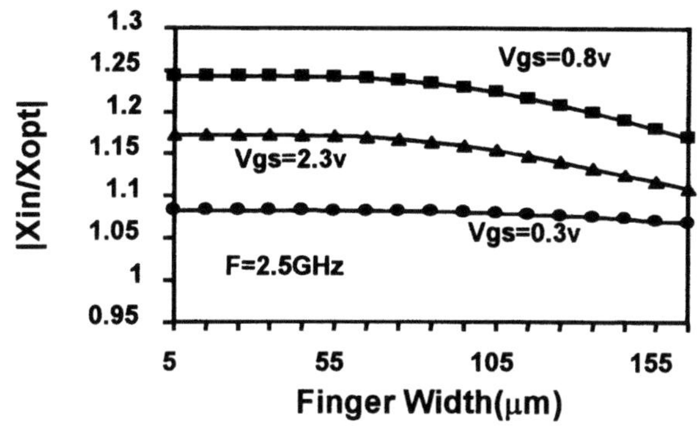

Figure 10: Xin/Xopt as a function of finger width for various bias conditions.

The minimum noise figure of the device is directly proportional to $\sqrt{R_{g,i}}$. Therefor, the larger the gate resistance the higher the minimum noise figure. The dependency of F_{min} with $R_{g,i}$ has been illustrated in Fig. 8 by plotting F_{min} as a function of finger width. For larger finger widths, the minimum noise figure increases in value due to the increased thermal noise in $R_{g,i}$. Also in Fig. 8 we have compared the lumped approximation of $R_{g,i}$ (i.e. $R_g/3$) to a distributed gate resistance model. As expected, only at higher frequencies and larger finger widths is a distributed model required. Another point to note is that the minimum noise figure is directly proportional to $1/\sqrt{g_m}$. Therefor by increasing the g_m of the device the minimum noise figure drops in value. One method to independently increase the g_m of the device is to increase the drain-source biasing current. In Fig. 9 the minimum noise figure of a MOS device as a function of biasing current has been plotted. For low values of bias current, the minimum noise figure decreases in value with increasing bias current. However, at some point an increase in the increasing bias current causes the minimum noise to increase in value. This is due to the fact g_m saturates in value (see Fig. 1), but γ increases with bias current (see Fig. 3).

The factor F_2 in the equation for X_{opt} can be shown to be equal to the following;

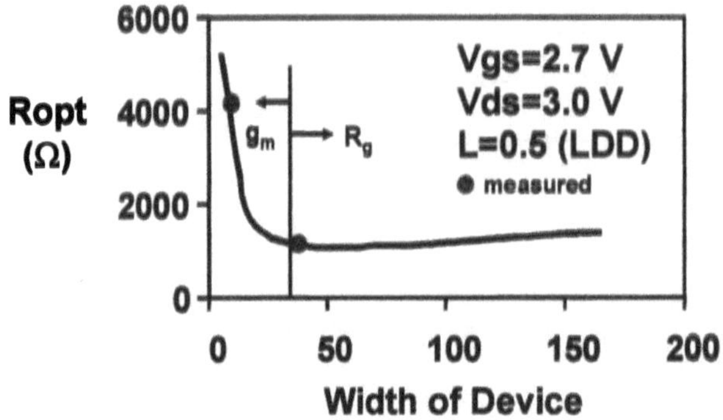

Figure 11: R_{opt} as a function of device width.

$$F_2 = \frac{1 - C(f_t/f)\sqrt{\beta^{-1}(f)}}{1 + (f_t/f)^2 \beta^{-1}(f) - 2C(f_f/f)\sqrt{\beta^{-1}(f)}} \qquad (17)$$

where $\beta(f)$ is proportional to f^{-2} and is equal to the effective current gain noise; i.e. $\bar{i}_d^2/\bar{i}_g^2$. For long channel devices, F_2 is predicted to be equal to *0.83* [16]. From the results in equation (7), F_2 is predicted to be closer to 1; recall that if the overall contribution of induced gate noise is small, $F_2 \sim 1$. This is the ideal situation since it corresponds to simultaneously matching the reactive part of the device for both power and minimum noise figure. However, if $F_2 \neq 1$, there exists an error in matching for both power and noise.

From the simulation results plotted in Fig. 10 one notes that the effect of induced gate noise with regards to the optimum noise reactance is small. That is to say, one should be able to match for both noise and power since $F_2 \approx 1$ and $X_{in} \approx -X_{opt}$. This also implies that the magnitude of induced gate noise is much less pronounced in MOS transistors as compared to MESFET's [17]. Other evident supporting $F_2 \approx 1$ can be found in [3,11].

In order to completely match for noise, the optimum noise resistance has to be equal to the driving source resistance – i.e. 50Ω. However, both measurement results and simulation results (see Fig. 11) indicate that R_{opt} is

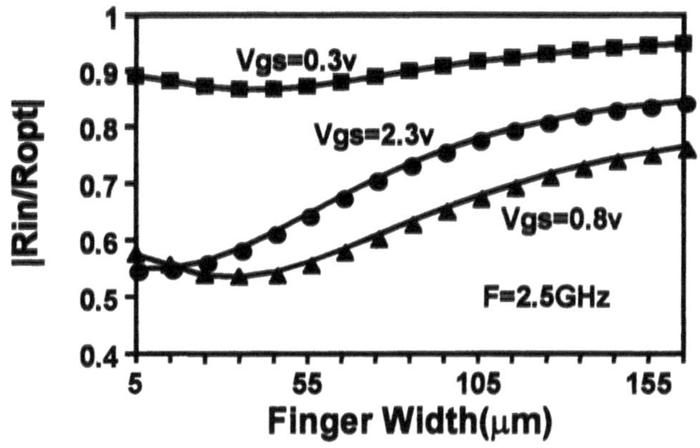

Figure 12: A comparison between of Rin with Ropt.

considerably larger than 50Ω for 0.5μm devices. In terms of finger width R_{opt} has two distinct regimes. For small values of finger widths R_{opt} is proportional to $1/g_m$ and for larger values R_{opt} is proportional to the gate resistance. In Fig. 12 a comparison is made between R_{opt} and the input resistance which is related directly to the channel resistance, r_i and the gate resistance, R_g. For short finger widths and with moderate bias $R_{in} \sim 0.55R_{opt}$. Because R_{opt} is relatively large it is difficult to match for noise by just using finger width and biasing as parameters. In any case, if one does try to use the finger width as a parameter this would in turn alter the minimum noise figure. As we shall show next, noise matching can be achieved by using the number of gate fingers as a parameter to satisfy the condition $R_{opt} = 50\Omega$.

If we assume that a single finger transistor has an equivalent input current and voltage of i_{eq} and v_{eq}, respectively, the noise figure of this transistor can be written as,

$$F = F_{\min} + \frac{G_n}{R_s}\left|R_s - Z_{opt}\right|^2 \qquad (18)$$

where F_{min} is the minimum noise figure of the single finger, G_n is the noise conductance of a single finger, R_s is the driving source resistance, and Z_{opt} is the optimum noise impedance of a single finger. By adding N fingers and by keeping the drain current density the same, the equivalent noise current and voltage can be shown to be equal to $i_{eq}\sqrt{N}$ and $v_{eq}/\sqrt{N}$, respectively. Using

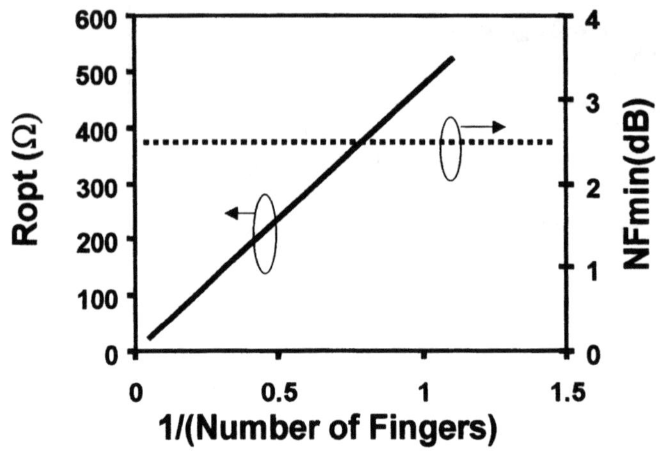

Figure 13: Typical results for Ropt and Fmin as a function 1/N.

these values for the equivalent noise current and voltage, the new expression for noise figure becomes:

$$F^{(N)} = F_{\min} + \frac{NG_n}{R_s}\left|R_s - \frac{Z_{opt}}{N}\right|^2 \qquad (19)$$

where the superscript N denoted the noise figure for the fingered device. Note that the minimum noise is independent of N, the noise conductance scales as N, and the optimum noise impedance scales as $1/N$. One of the important results here is that the minimum noise figure is independent of the number of fingers, and is only a function of one finger which is in turn a strong function of the gate resistance. To illustrate this, we have depicted typical results of F_{min} and R_{opt} as a function of the number of fingers in Fig. 13. Note that F_{min} is independent of the number of fingers and R_{opt} scales as

Parameters	Number of Fingers	Width (<30 μm)
F_{min}	indep.	~W
G_n	N	W
R_{opt}, X_{opt}	1/N,1/N	~1/W,1/W
F_{max}	indep.	~1/W
F_t	indep.	indep.

Table I: Scaling rules for various microwave parameters with device layout for a fixed current density.

1/N. The advantage of using the number of fingers as the parameter to match the optimum noise resistance to the source resistance is it does not effect the minimum noise figure.

3. CONCLUSIONS

Scaling rules for the microwave noise parameters are easily arrived at from simple analytical expressions for the elements or from the measurements of different size devices from the same foundry. Most of the scaling rules are intuitively obvious and element values are easily scaled up or down based on the total gate width and the number of fingers. In table I a summary of all the various scaling rules for the noise parameters are presented. The scaling rules are given for both gate width and the number of fingers.

4. ACKNOWLEDGEMENTS

The author would like to thank Jerry Molnar (Mitel Semiconductor) and John Nisbet (Nortel Wireless Networks) for their continual support as well as Siborg Systems Inc. The author would like to thank all the research associates and his graduate students for helping in developing the contents within this work.

5. REFERENCE

[1] L. E. Larson, "Integrated Circuit Technology Options for RFIC's – Present status and future directions," IEEE J. of Solid State Circuits, vol. 33, pp. 387-399, 1998.

[2]A. A. Abidi, "Low-power radio-frequency IC's for portable communications," Proc. of IEEE, vol. 83, no. 4, pp. 544-569, April 1995.

[3] T. Manku, "Microwave CMOS – Device Design and Physics," IEEE J. of Solid State Circuits, in print, March 1999.

[4] Qiuting Huang, Francesco Piazza, Paolo Orsatti, and Tatsuya Ohguro, "The impact of scaling down to deep submicron on CMOS RF Circuits," IEEE J. Solid State Circuits, Vol. 33, pp. 1023-1036.

[5] S. P. Voinigescu, S. W. Tarasewicz, T. MacElwee, and J. Ilowski, "An assessment of the state-of-the-art 0.5μm bulk CMOS technolog for RF applications," Proc. IEEE International Electron Devices Meeting, pp. 721-724, 1995.

[6] T. Ohguro, H. Naruse, H. Sugaya, E. Moriguji, S. Nakamura, T. Yoshitomi, T. Morimoto, H. S. Momose, Y. Katsumata, and H. Iwai, "0.18μm low voltage/low power RF CMOS with zero Vth analog MOSFET's made by undoped epitaxial channel technique," IEEE International Electron Device Meeting, December, pp. 837-840, 1997.

[7] H. S. Momose, F. Morifuji, T. Yoshitomi, T. Ohguro, M. Saito, T. Morimoto, Y. Katsuma, H. Iwai, "High frequency AC characteristics of 1.5nm gate oxide MOSFET," Proc. IEEE International Electron Device Meeting, December, pp. 105-108, 1996.

[8] R. R. J. Vanoppen, L. M. F. de Maaijer, D. B. M. Klaassen, and F. F. Tiemeijer, "RF Noise Modelling of 0.25μm CMOS and Low Power LNAs," IEEE International Electron Device Meeting, December, pp. 317-320, 1997.

[9] A. A. Abidi, "High frequency noise measurement on FET's with small dimensions," IEEE Trans. on Electron Devices, vol. 33, pp. 1801-1805, 1986.

[10] D. K. Shaeffer and T. H. Lee, "A 1.5V, 1.5GHz CMOS low noise amplifier," IEEE J. of Solid State Circuits, vol. 32, no. 5, pp. 745-759, May 1997.

[11] Qiuting Huang, Paolo Orsatti, and Francesco Piazza, "Broadband, 0.25μm CMOS LNAs with Sub-2dB NF for GSM Applications," IEEE 1998 Custom Integrated Circuits Conference, pp. 67-70, 1998.

[12] A. Rofougaran, G. Chang, J. J. Rael, J. Y. –C Chang, M. Rofougaran, P. J. Chang, M. Djafari, J. Min, E. Roth, A. A. Abidi, and H. Samueli, "A single-chip 900MHz spread-spectrum wireless transceiver in 1μm CMOS Part II: receiver design," IEEE J. of Solid State Circuits, vol. 33, pp. 535-547, 1998.

[13] J. Crols and M. S. J. Steyaert, "A single-chip 900MHz CMOS receiver front-end high performance low-IF topology," IEEE J. of Solid State Circuits, vol. 30, no. 12, pp. 1483-1492, December 1995.

[14] Yi Lin, Lan Wang, Michael Obrecht, and Tajinder Manku, "Quasi 3D Device simulations for the microwave characterization of MOS devices," Technical Digest of IEEE International Electron Device Meeting, pp. 77-80, 1998.

[15] T. Manku, M. Obrecht, and Y. Lin,"RF simulations and physics of the channel noise parameters with MOS transistors," IEEE Custom Integrated Circuit Conference, in print, 1999.

[16] A. Van der Ziel, *Noise in Solid State Devices and Circuits*, New York, Wiley, 1986.

[17] J. Michael Golio, *Microwave MESFET's & HEMT's*, Artect House, Boston, 1991.

[18] Ronald R. Troutman, "VLSI limitations from Drain-Induced Barrier Lowering," IEEE J. of Solid State Circuits, vol. 14, pp. 365-391, 1979.

[19] Y. P. Tsividis, *Operation and modeling of the MOS transistor*, McGraw Hill, 1987.

[20] E. Abou-Allam and T. Manku, "A small signal MOSFET model for radio frequency IC applications," IEEE Trans. on CAD of Integrated Circuits and System, pp. 437-447, May 1997.

[21] G. Gonzalez, *Microwave Transistor Amplifiers and Analysis and Design*, Prentice Hall, 1997.

Integrated filters and oscillators

Introduction

Increasing use of communication tools goes along with the design of analog filters and oscillators with greater flexibility and higher bandwidth.
This is reflected in the following paper.

The first and fourth papers cover high-frequency GmC filters and polyphase filters by Wim Dehaene and Asad Abidi, respectively.

Programmability is covered by the second and third paper on BiCMOS HF GmC filters by Hans Voorman and CMOS HF GmC filters by Rinaldo Castello, respectively.

Finally, very high-frequency filters and oscillators are covered by the fifth and sixth papers on passive integrated RF filters by Nick Pulsford, and Integrated GHz VCO's by Peter Kinget.

Johan Huijsing

Specific Aspects of high frequency Gm-C filters

Wim Dehaene

Alcatel Microelectronics
Excelsiorlaan 44-46
B-1930 Zaventem,Belgium

e- mail: wim.dehaene@mie.alcatel.be

ABSTRACT

This paper discusses typical problems encountered during the design of high frequency Gm-C filters. The first part handles the problem of parasitic singularities. The second part discusses tuning systems for HF Gm-C filters. The discussion is done with a a seventh order 0.05° equiripple on the phase low pass filter as an example.

1. Introduction.

In this paper some typical aspects in the realisation of high frequency Gm-C filters are discussed. For a general discussion of Gm-C filters it is referred to [Tsividis94] and the references in that paper.

The first problem discussed is that of the transconductor suited for HF Gm-C filters. The main problem there is the role of unwanted singularities in the transfer function of the transconductor. Several examples will be discussed on how to deal with this problem.

The second problem is that of the dominant role of parasitic capacitances in the response of the filter and how this affects the tuning of the filter.

An alternative tuning strategy, capable of dealing with parasitic capacitors, is demonstrated.

Throughout the discussion a specific type of filter is used as design case: a seventh order 0.05° equiripple on the phase low pass filter. This is a typical filter used as pulse equalizer in the read channel of a modern hard disk drive. The required cut-off frequency in that system reaches up to 50 MHz.

The most important specification of this filter is the ripple on its group delay. This ripple must be smaller than 2% for all frequencies f between 0.2 fc < f < 1.5 fc. Frequency fc is the cut-off frequency of the filter. This specification is motivated by the fact that in a hard disk, as in other communication or storage systems, the information is represented by the presence or the absence of a pulse in a certain time window. If the pulse is shifted in time by a component in the signal path, the information itself is affected.

2. Transconductors for HF Gm-C filters.

Linear transfer functions in general are characterised by the position of their singularities in the s-plane. The singularities of a seventh order 0.05 ° equiripple low pass phase filter are given in the table below

	normalized fo	fo for fc = 50 MHz	Q -factor
complex poles	1.148	57.4 MHz	0.681
complex poles	1.718	85.9 MHz	1.115
complex poles	2.317	115.8 MHz	2.023
real pole	0.861	43.05 MHz	-

Table 1: s-plane location of the poles of a 7th order 0.05 ° equiripple low pass phase filter

Any other singularity is called parasitic. During the design it is made sure that parasitic singularities are located “sufficiently far” away from the desired poles or zeros. In fig. 1 the influence of four parasitic poles, one per filter section, on the ripple on the group delay is illustrated. [Dehaene96]

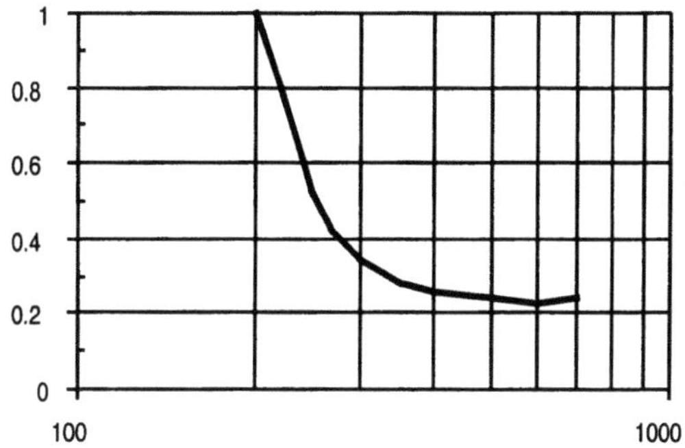

Fig. 1: influence four coincident parasitic poles on the ripple on the group delay.

Taking into account that the ideal filter has a group delay ripple of 0.27 %, the figure shows clearly that the parasitic poles must be located above 300 MHz. In practice, this boundary will be even higher because a filter section will have more than one parasitic pole.

In [Mehr97] this is calculated in terms of the excess phase, due to parasitic singularities, of a transconductor in the filter. The transconductance is approximated as:

$$Gm \approx Gm0 \cdot e^{s\tau_p}$$
$$\text{with } \tau_p = \sum_i \frac{1}{p_i} - \sum_i \frac{1}{z_i}, \qquad (1)$$

p_i and z_i are the poles and zeros of the filter

Filling this in, in the transfer function, it can be shown that the singularities are now given by:

$$s_i = s_e \cdot e^{-s\tau_p}$$

s_i is the ideal location (2)

s_e is the location affected by the excess phase

For small pole shifts this reduces to:

$$\Delta s = -s_i^2 \tau_p$$
$$\arg(s_e - s_i) = \pi + 2\arg(s_i) + \pi \frac{1-\mathrm{sgn}(\tau_p)}{2} \qquad (3)$$

This is graphically depicted in fig. 2. Remark that for the highest Q poles, there is a risk that they move in the right hand plane. The filter becomes unstable in that case.

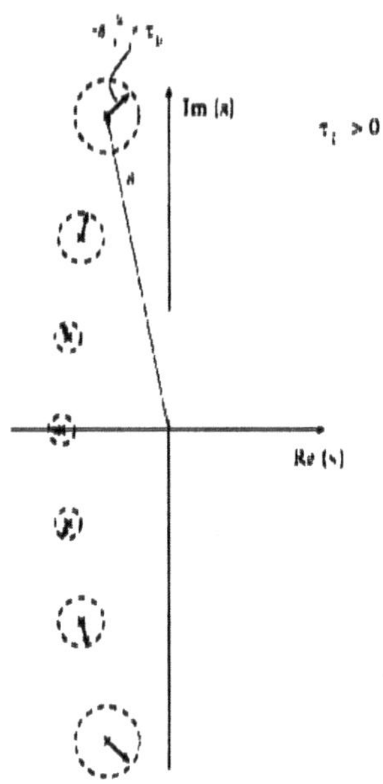

Fig. 2: pole shift due to excess phase in the transconductors in the Gm-C filter. [Mehr97]

From the discussion above it is clear that transconductors with a minimum of parasitic singularities located at very high frequencies are needed if a high frequency Gm-C filter is to be realised. Several solutions exist for this problem. Three representative ones are presented in this paper.

The first HF transconductor is shown in fig. 3 The actual transconductance is formed by a MOS differential pair operating in the linear region (M1,M2). The output impedance of the transconductor is enhanced by the bipolar cascode devices (Q1,Q2). Via the bases of these devices the drain-source voltage of M1,M2 and thus the overall transconductance is controlled. [Alini92]

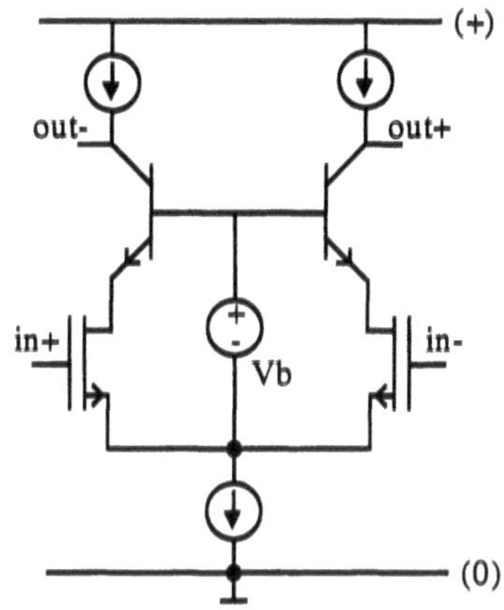

Fig. 3: high frequency BiCMOS transconductor with bipolar cascodes.

The most important parasitic singularity in the transfer function of this transconductor is the pole associated with the bipolar cascode devices. It can be approximated by:

$$pole = gm/Cpar \tag{4}$$

Cpar in this equation is the sum of all parasitic capacitance at the internal node in the signal path of the transconductor. Due to the inherently large transconductance of the bipolar transistor, and its relatively small capacitance at the source, the parasitic pole can be shifted up to frequencies in the 1 GHz range.

A second HF transconductor is described in [Mehr97]. Its schematic is given in fig. 4. In this transconductor, M1 and M2 are used to put the differential input voltage over M7, M8. These transistors operate in the linear range, translating the input voltage in a current. The current through M1, M2 is kept constant by the feedback transistors M5 and M6. In this way the linearity of the transconductor is enhanced.

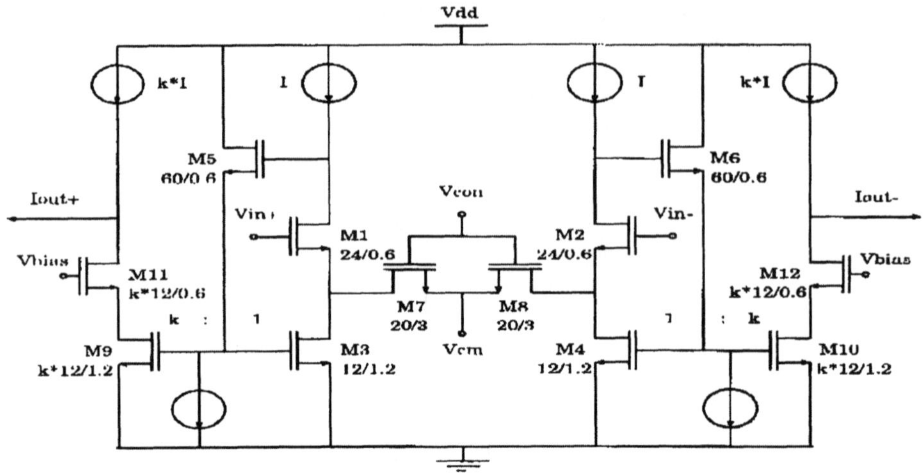

Fig. 4: Full CMOS transconductor based on MOSFET degeneration of the differential input pair. [Mehr97]

The signal current is mirrored out via the current mirrors and M4, M10 and M3, M9. The output of the mirror is cascoded (M11,M12) to increase the transconductor's output impedance.

The parasitic poles in this CMOS transconductor are caused by the two nodes in the feedback loop and the cascodes. In this case it are the transconductance of MOS transistors and their parasitic capacitance that influence the frequency behaviour. This is heavily dependent on the technology. As technology shrinks, the parasitic capacitors for a given W/L transistor become smaller. This is not only due to the smaller L but also to reduced drain and source areas that can be drawn if the technology has a smaller feature size, see fig. 5.

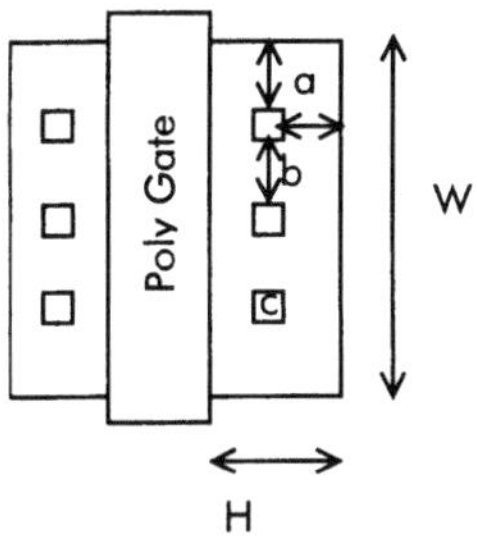

	CMOS 0.5 μ	CMOS 0.7 μ
a [μ]	0.3	0.6
b [μ]	0.8	1.2
c [μ]	0.6	1
W [μ]	4	6.6
H [μ]	1.2	2.2
Area [μ^2]	4.8	14.5

Fig. 5: feature sizes for a 0.7μ and a 0.5μ CMOS technology

A third transconductor used to realise a HF Gm-C filter is shown in fig. 6. [Dehaene96],[Dehaene97]. The actual transconductance here consists of only a differential pair. Therefore the linearity must be guaranteed by providing a relatively high overdrive voltage. The transconductor of fig. 6 is used in a CMOS technology were the implementation was only possible at the expense of excessive amounts of power.

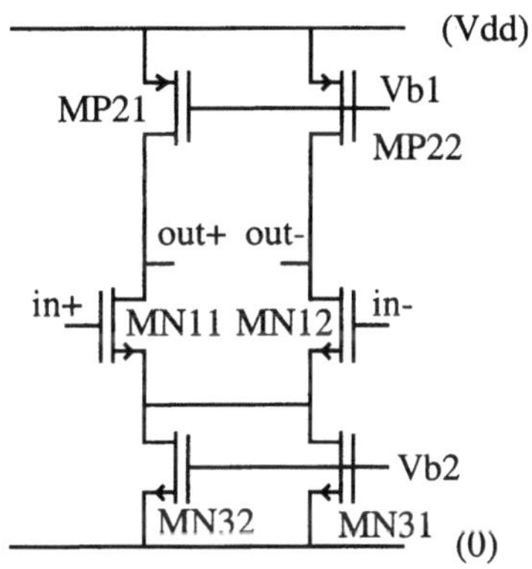

Fig. 6: basic CMOS transconductor.

In this transconductor the strategy used to deal with parasitic poles is to avoid their presence: the transconductance has no internal nodes only input nodes and output nodes. The main disadvantage of this strategy is the low output impedance of the transconductor due to the absence of cascodes. This is solved by the use of an appropriate tuning system as discussed in the next paragraph. However if the technology allows the use of cascodes this will be much more efficient.

3. Tuning HF Gm-C filters.

The first problem in tuning HF Gm-C filters is to deal with the parasitic capacitors. At the frequencies under consideration it is not possible to neglect parasitic capacitance next to the explicit capacitor that is used in the Gm-C integrator. When the filter that is actually used in the application can be tuned itself, this is not necessarily a problem. However, this is rarely the case because then the filter is unavailable for signal processing part of the time.

In most tuning systems a replica of a part of the filter is tuned. The same bias is then applied to the actual filter. The accuracy of the tuning is thus determined by the matching between filter and replica. For an accurate tuning system the ratio of parasitic capacitance to the explicit capacitor in the tuning system must match the ratio of parasitic capacitance to explicit capacitor in the filter. This is illustrated by means of the principle diagram of a charge comparison based tuning system, see fig. 7 .

The charge stored, per clock period, on the capacitor C_2, when clock ϕ is low, is equal to:

$$q_1 = C_2 \frac{I}{g} \tag{5}$$

When clock ϕ is high (switch ϕ closed) this charge is transferred to C_1. The charge extracted, per clock period, from capacitor C_1 is:

$$q_2 = kIT$$

(T is the clock period) (6)

The factor k is the ratio between the currents of the two sources in fig. 7. In equilibrium both charges are equal. Equating (5) and (6) gives:

$$\frac{g}{C_2} = \frac{1}{kT}$$

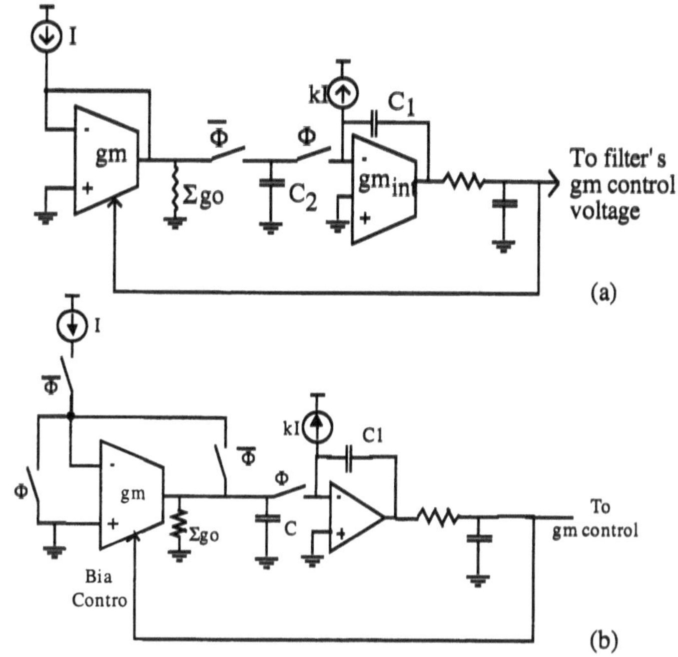

Fig. 7: principle of a charge comparison based tuning system (a). Version with more optimal switch arrangement (b)

The charge stored, per clock period, on the capacitor C_2, when clock ϕ is low, is equal to:

$$q_1 = C_2 \frac{I}{g} \tag{7}$$

When clock ϕ is high (switch ϕ closed) this charge is transferred to C_1. The charge extracted, per clock period, from capacitor C_1 is:

$$q_2 = kIT$$

(T is the clock period) (8)

The factor k is the ratio between the currents of the two sources in fig. 7a or b. In equilibrium both charges are equal. Equating (5) and (6) gives:

$$\frac{g}{C_2} = \frac{1}{kT} \tag{9}$$

The tuning system thus locks the g/C_2 ratio to a clock frequency and a current ratio.

The position of the switches in fig. 7b is more optimal from the matching point of view because the parasitic capacitor is now also discharged.

For a low frequency filter the capacitor in the tuning system can be made larger than the capacitors in the filter as long as matching is conserved. This allows to operate the tuning system at a relative low frequency. When a large capacitor is used to tune a HF Gm-C filter this has the effect that the parasitic capacitor in the tuning system is negligible while this is not the case in the filter itself. This leads to a considerable tuning error. Therefore the frequency of the tuning system must be high. This is the second problem in tuning HF Gm-C filters: the high tuning clock leads to tuning errors caused by clock feed through and charge loss in the switches. An alternative to operate the tuning system at a low frequency using the same capacitors as in the filter is to use a large ratio between the currents in fig. 7. However this makes one of the currents very small. This again compromises the accuracy. As conclusion it can be said that it is not possible to put the tuning clock frequency much lower than one order of magnitude below the frequency of the filter's singularities.

However, Alternative tuning systems are possible if the following considerations are taken into account. The tuning systems considered actually lock the time constants in the filter to a reference. This is not sufficient to guarantee a certain cut off frequency, nor does it make the filter's shape insensitive to technology variations. The reason for this is that nothing is done about the Q-factor of complex poles. Furthermore, tuning the absolute values of the time constants is not always necessary.

In a lot of applications where HF Gm-C filters are used, the function of the filter is double. First it has to equalize the channel response. The second function is anti-aliasing. In this case the phase response, the "shape", of the filter is much more important than the absolute value of its cut-off frequency. This "shape" is not determined by the absolute value of the poles but by their relative position and by the Q-factor of complex pole pairs. This is shown in fig. 8. As long as the ratio of the time constants and the Q factor of the different sections remains constant, the relative position of the curves does not change. In that case the shape of the group delay characteristic does not change neither.

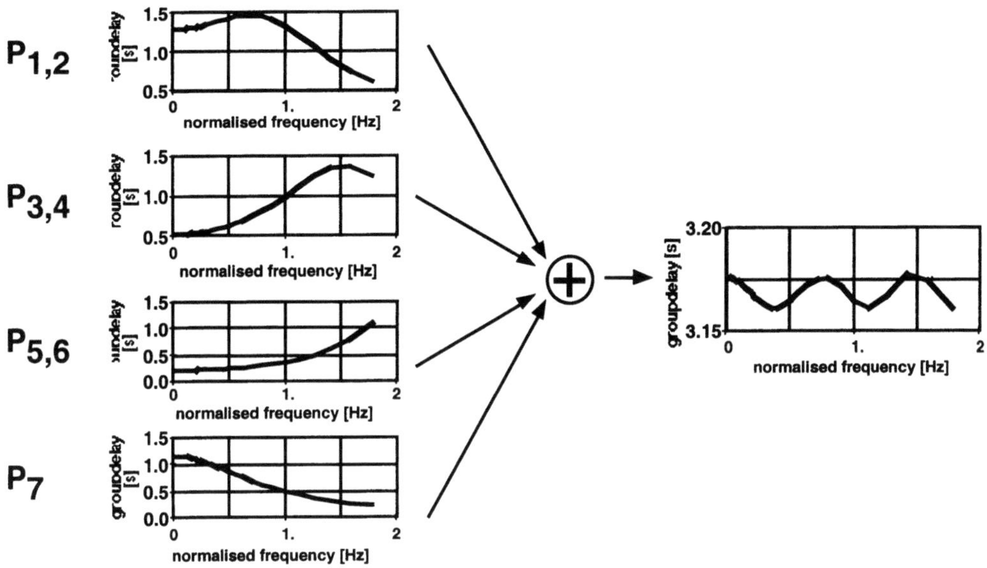

Fig. 8: filter group delay as a combination of the group delay of the different sections.

Advantage can be taken from this to come to a more accurate tuning system operating at a lower frequency. This is illustrated with a design example [Dehaene96],[Dehaene97]. The filter in this design is built with biquadratic sections as shown in fig. 9. (for the single pole section a similar strategy is used) This biquad is symmetrical with respect to the two integrating nodes. Therefore the impedance seen at this nodes is equal. This allows to omit the explicit capacitors. Only the sum of all parasitic capacitors is used as integrating capacitor. With a non symmetrical biquad this would not be possible because it is not feasible to control the ratio of this sum of parasitics accurately otherwise than to make them equal. The transconductors used in the filter were already shown in fig. 6. Using only the parasitics as capacitor is also a low power approach because the parasitic capacitance is the minimum capacitance possible at every node

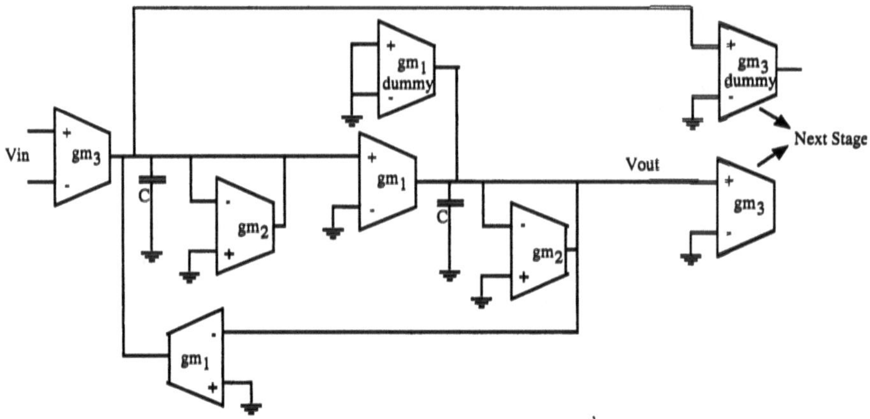

$$H(s) = A \frac{\omega_o^2}{\omega_o^2 + \frac{\omega_o s}{Q} + s^2}$$

$$A = \frac{gm_3}{gm_1(\gamma^2 + 1)}$$

$$Q = \frac{\sqrt{\gamma^2 + 1}}{2\gamma}$$

$$\omega_o = \frac{gm_1\sqrt{\gamma^2 + 1}}{C} = \frac{1}{\tau}\frac{\sqrt{\gamma^2 + 1}}{\gamma}$$

$$\tau = \frac{C}{g}$$

$$\gamma = \frac{g}{gm_1}$$

C is the capacacitance seen at node 1, 2 of the biquad :

$$C = \sum_i C_{p,i}$$

g is the conductance seen at node 1, 2 of the biquad :

$$g = gm_2 + \sum_i g_{o,i}$$

($C_{p,i}$: the parasitic capacitances of the OTAs)

($g_{o,i}$: the output conductances of the OTAs)

Fig. 9: schematic diagram of the symmetrical biquad

The quantities tuned in this filter are the ratio of the natural frequency of every section relative to the first section and the Q-factor of the complex pole pairs.

Tuning the Q-factor is relatively simple because it only depends on a ratio of conductances. the principle diagram of the tuning system is given in fig. 10.

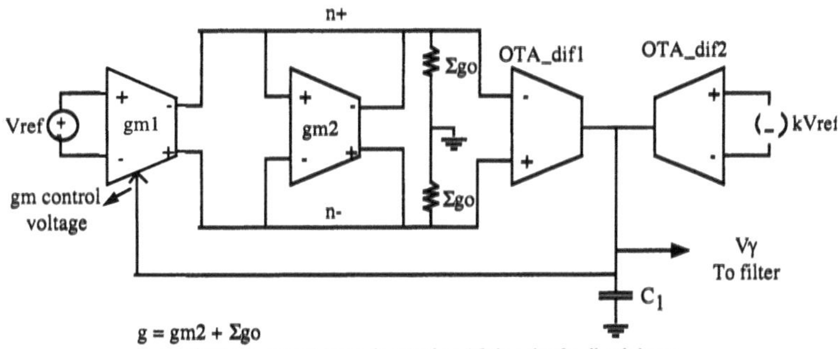

Fig. 10: principle diagram of the Q-tuning system.

The principle diagram of the tuning system for time constant ratios is shown in fig. 11. During clock phase Φ capacitors C1 and C2 are charged to a reference voltage Vref. When the clock goes low, these capacitors are discharged over the corresponding parallel conductance. The resulting waveform is also shown in fig. 11.

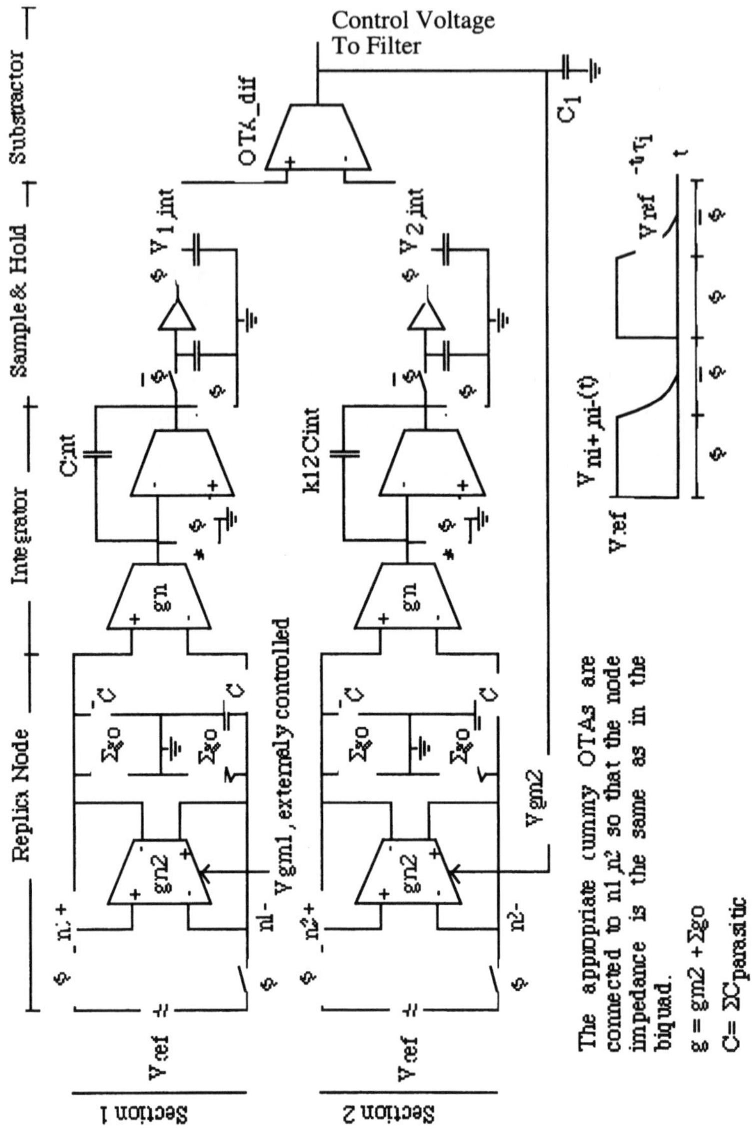

Fig. 11: principle diagram of the time constant ratio tuning system.

The discharge part of the waveform is given by:

$$v_1(t) = Vref\ e^{-\frac{t}{\tau_1}}$$
$$v_2(t) = Vref\ e^{-\frac{t}{\tau_2}} \qquad (10)$$

This voltage is integrated and sampled. Under the assumption that the integration time is much larger than the time constants, the output of the sample and hold blocks is given by:

$$v_{1,int}(t) = \frac{-gm}{Cint}\int_0^{\infty} Vref\ e^{-\frac{t}{\tau_1}}dt = \frac{gmVref\ \tau_1}{Cint}$$
$$v_{2,int}(t) = \frac{-gm}{k12\ Cint}\int_0^{\infty} Vref\ e^{-\frac{t}{\tau_2}}dt = \frac{gmVref\ \tau_2}{k12\ Cint} \qquad (11)$$

The voltages in (11) are subtracted from each other and filtered. The resulting voltage is used to close the feedback loop so that $v_{1,int}$ equals $v_{2,int}$ when the loop is settled:

$$v_{1,int}(t) = v_{2,int}(t)$$
$$\frac{gmVref\ \tau_1}{Cint} = \frac{gmVref\ \tau_2}{k12\ Cint} \qquad (12)$$
$$k12 = \frac{\tau_2}{\tau_1}$$

This shows that the tuning system locks the ratio of the time constants to a ratio of capacitors. Capacitor ratios can be very accurately controlled. The clock period does not determine the accuracy of the tuning. It must even be sufficiently lower than the filter time constants.

The two proposed tuning systems still have to be completed with a tuning system for the absolute accuracy of one time constant. In a lot of applications this can be a fairly simple, low accuracy (10 %) tuning system. Even if high accuracy is required this needs to be implemented only once with the proposed strategy. The other time constants can be tuned relative to the accurate one.

The measured response and group delay of a filter with the tuning systems discussed above is shown in fig. 12. The specification of 2% ripple on the group delay is met.

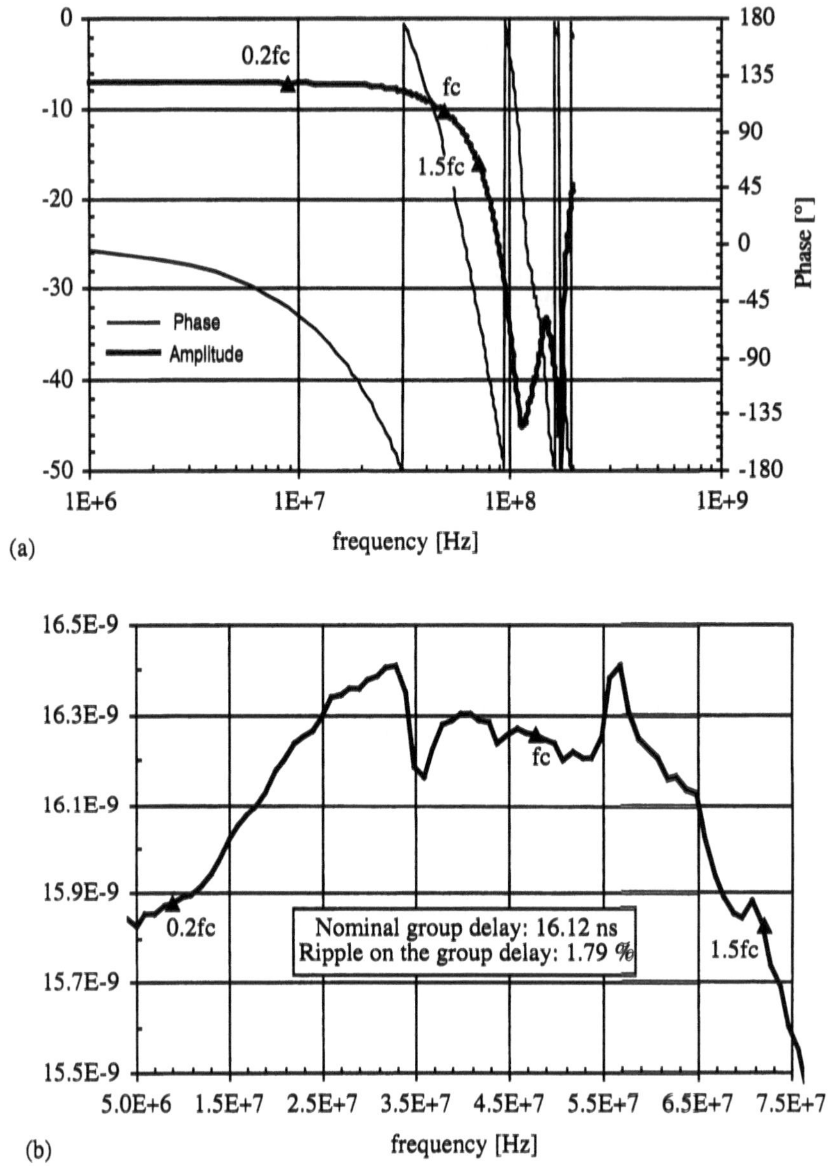

Fig. 12: (a) measured transfer characteristic and (b) group delay of the filter.

The conclusion from this is that it is possible to deal with the parasitic capacitors. It is even possible to let them play a useful role in the system. Furthermore, it has been shown that the use of HF clocks can be avoided.

4. References

[Alini92] Alini R., Baschirotto A., Castello R., "Tunable BiCMOS continuous-time filter for high frequency applications", IEEE Journal of Solid-State circuits, Vol. 27, no. 12, pp. 1905-1915, December 1992.

[Dehaene96] Dehaene W., "CMOS integrated circuits for analog signal processing in hard disk systems", Ph. D. dissertation, K.U.Leuven, 1996.

[Dehaene97] Dehaene W., Steyeart M., Sansen W., "A 50 MHz standard CMOS pulse equalizer for hard disk read channels", IEEE Journal of Solid-State circuits,Vol. 32, no. 7, pp. 977-988, July 1997.

[Mehr97] Mehr I., Welland D, "A CMOS continuous-time Gm-C filter for PRML read cahnnel applications at 150 Mb/s and beyond", IEEE Journal of Solid-State circuits,Vol. 32, no. 4, pp. 499-513, April 1997.

[Tsividis94] Tsividis Y., "Integrated continuous time filter design – an overview', IEEE Journal of Solid State Circuits, Vol. 29, no.3, pp. 166-176, March, 1994.

Tunable bipolar and biCMOS Gm-C filters for high-frequencies

Hans Voorman and Hugo Veenstra
Philips Research Laboratories
Prof. Holstlaan 4 (WAY 51)
5656 AA Eindhoven, The Netherlands

Abstract

We consider known, modified and new transistor circuits for transconductors as well as their application in integrated tunable Gm-C filters for high frequencies, that is for hundreds of MHz until more than 1 GHz. The transconductor circuits use NPN bipolar transistors and in most of the examples also vertical PNP transistors. The low delay and low input capacitance of the new circuits makes them particularly suitable for application at very high frequencies. Their high power efficiency and low noise lead to an unusually low power consumption in combination with a high signal-to-noise ratio. Their circuit simplicity yields low cost.

Introduction

The filter types, which we consider, are continuous-time analog high-frequency tunable electronic Gm-C filters for integration in bipolar or biCMOS IC processes. This means that the filters treat the signals in a continuous-time analog manner, in contrast to switched-capacitor filters, for example, which sample the signals or operate on sampled signals. By high frequencies we mean hundreds of MHz including the very low GHz region. Although the filter circuits can also be used for low frequencies, down to the kHz region even, we emphasize design aspects which are related to applications at high frequencies, such as equalizers for bit detectors for magnetic and optical recording (for harddisk, CD, DVD, etc.), selective amplifiers for receivers and filters for high speed baseband data links, for example between different IC's.

The functions of the filters are: noise suppression (shaping), interference rejection and pulse shaping. In general, there are gain as well as group-delay requirements. Noise suppression requires a steep high-frequency roll-off. Low frequencies are rejected to eliminate DC-offsets, thermal asperities and other baseline variations and to suppress the low-frequency noise of the mains and of the motor of a diskdrive, for example.

The filters are Gm-C filters. Other filter types use integrators or are based on log-domain filtering. For a survey on continuous-time filters we refer to Voorman [1], for a more recent application to De Veirman, et al. [2] and for log-domain filters for high frequencies to Frey [3].

We consider resistive elements, which are integrated as transistor circuits. Matching and resistance control via the bias current are exploited for programming, tuning and adaptation of the filters. The capacitors are either junction capacitors or dedicated oxide, nitride or oxi-nitride capacitors or capacitors, which use the MOS gate oxide as a dielectric.

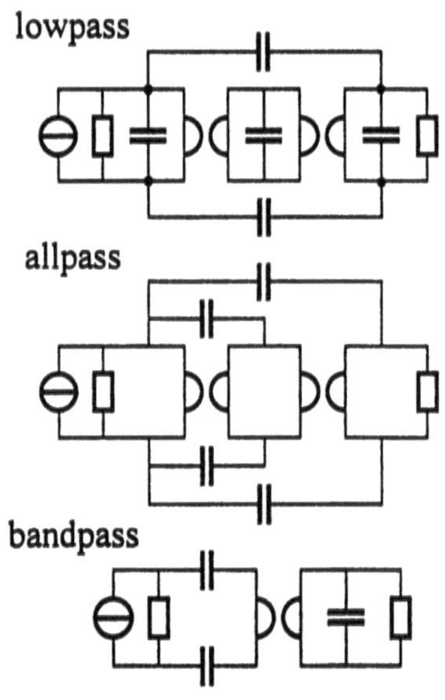

Figure 1. Some typical gyrator filters: a lowpass, an allpass and a bandpass, all three of second order. Each filter consists of a source - in these examples a current source - with a(n internal) resistance, a lossless two-port filter network and a load resistance. When properly designed, these filters show a low gain sensitivity to element value variations. For high frequencies their low number of nodes is important. The filters are implemented as fully balanced circuits.

In many cases Gm-C filters are derived from gyrator filters.
Some typical examples of gyrator filters have been depicted in fig. 1. The circuits are of the type: lossless two-ports between resistive terminations. Gm-C filters are implemented using two element types: Gm elements or transconductors or voltage-controlled current sources (VCCS elements) as resistive elements and capacitors (C). Source and load resistors are implemented as input-port-to-output-port short-circuited transconductors. Gyrators are made as combinations of two cross-coupled transconductors. Two Gm-C (transconductor-capacitor) filter examples have been depicted in fig. 2.

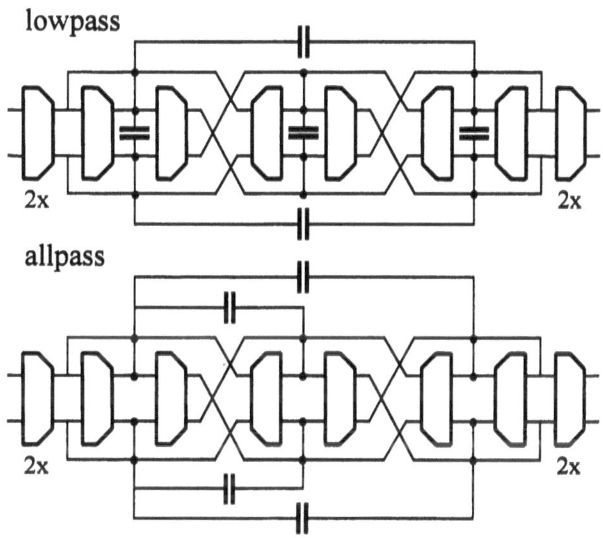

Figure 2. Two examples of Gm-C or transconductor-capacitor filters: a lowpass and an allpass, both of second order. From the left to the right we see: an input transconductor which has the function of the current source, a transconductor-resistor which is a transconductor with input and output port interconnected, a pair of transconductors which represents a gyrator, etc. At the right we see the load transconductor-resistor and a transconductor which couples to the next filter stage or to the filter output. In general, the capacitors have different values and the transconductances of the transconductors are equal, with the exception of those which have been indicated by 2x. They have a two times higher transconductance to obtain an overall 0 dB low-frequency gain.

For high frequencies filter designers use first-, second- and third-order filter sections, which are cascaded in an optimum order. In many cases signal enhancement in a second-order filter section is avoided by making a combination with a first-order filter section. This is the main reason for the use of third-order filter sections. Higher-order sections are generally too inaccurate, because the losses of the simple high-frequency circuits start to interfere with the 'lossless between resistive terminations' character of the filter sections. In practice, allpass or delay sections are given a high-frequency roll-off (extra pole) already during the optimization of the pole-zero pattern with respect to system requirements.

Low-Q

Generally speaking, high-frequency Gm-C filters are low-Q filters.
This statement is supported by the following two arguments.

If we consider filters as consisting of coupled resonance circuits, parasitic phase-shifts of the constituting elements influence the quality factors. A gyrator-capacitor resonance circuit consists at least of a gyrator or two transconductors and two capacitive elements. In most cases spurious phase shifts caused by losses of the capacitors can be kept low by proper layout design. Phase shifts in the transconductors:

$$d\varphi = -\arctan(f/fc) \approx -f/fc\ , \qquad (1)$$

where fc is the 3-dB cut-off frequency of the transconductors, are non-negligible at high frequencies. The two transconductors influence the quality factor Qo of a resonance circuit with resonance frequency fo according to

$$1/(Qo+dQo) \approx 1/Qo + 2\ d\varphi \approx 1/Qo - 2\ fo/fc\ . \qquad (2)$$

Hence,

$$dQo/Qo \approx 2\ Qo\ fo/fc\ . \qquad (3)$$

For example, with Qo = 10, dQo/Qo = 10 % and fo = 500 MHz, fc should be 100 GHz. This is not very realistic.
In practice the phase shift of the transconductors limits the quality factor Qo to relatively low values. Control loops - which add complexity (and cost) - can give some relief.

The second argument refers to the signal-to-noise ratio and the power consumption. By way of example let us consider a symmetrical gyrator capacitor resonance circuit. For Qo >> 1 the noise voltage is given by

$$\langle Vn^2 \rangle \approx 2.F.Qo\, k.T/C\,, \tag{4}$$

where F is a noise factor and C is the value of the capacitors, see Blom and Voorman [4] or Voorman [5], for example.
The signal-to-noise ratio is

$$SNR = Vs^2 / \langle Vn^2 \rangle\,, \tag{5}$$

where Vs is the effective signal voltage at the output of the filter.
Because of the enhancement in the resonance circuit, the maximum internal signal voltage Vs[max] = Qo.Vs, approx.. Further, Vs = Rg.Is and Vs[max] = Rg.Is[max], where Rg is the gyration resistance.
Let the power efficiency of the gyrators be

$$\eta = Vs[max].Is[max]/P, \tag{6}$$

where P is the power consumption of the gyrator.
With resonance frequency $\omega o = 2.\pi.fo = 1/(Rg.C)$ we obtain

$$P = SNR\;\; 4.\pi\, k.T\;\; Qo^3\, fo\;\; F/\eta\,, \tag{7}$$

$$= 5.10^{-18}\;\; SNR\;\; Qo^3\, fo\,, \tag{8}$$

where we have assumed realistic values F = 1 and η = 1%.
For a signal-to-noise ratio SNR = 50 dB, Qo = 10 and fo = 500 MHz we obtain P = 250 mW.
This is a high value for a single resonance circuit in a filter and it increases with the cube of the quality factor.
One should observe that there is an exception to this rule: in resonance amplifiers the voltages on the resonance circuits are directly fed to the next stage: in that case Vs[max] = Vs and the power consumption is a linear function of Qo (instead of a cube function). In resonance amplifiers the signal enhancement Qo is included in the gain. In filters the signal enhancement is purely internal.

In conclusion, because of the phase shifts in the transconductors and to keep the power consumption of the filters low, the quality factor of the resonance circuits, and hence of the filter poles, should be kept low. At high frequencies for experiments high-Q Gm-C filters can be made and tuned, but for products we use only relatively wideband Gm-C filters. Luckily, most modern high-frequency systems do not require the implementation of narrowband analog filters.

Imperfections, impedance level and IC process

We construct the filters from transconductors and capacitors. Practical imperfections are low input and low output resistances, stray capacitances and delays for transconductors, series resistances and stray capacitances for filter capacitors and stray capacitances of the interconnect. For accurate biasing voltage drop across the resistance of interconnect should be accounted for (in the circuits and in the lay-out). As the high input and output resistances for the transconductors are less important for low-Q filters than for high-Q filters, we can use relatively simple transconductor circuits (without buffers). This keeps internal signal delays low.

Parasitic capacitances can often be included in the filter capacitances or they can be added to the filter transfer function before the filter optimization is carried out. For accurate filters they should not be too large. In an integrated circuit stray capacitances to a filter node are typically between 0.1 pF and 1 pF. Filter capacitances are an order of magnitude larger. A typical filter capacitor is 3 pF. This value gives an indication for the maximum impedance level of the filter: $1/(2.\pi.f.C)$. It is typically 50 kOhm at 1 MHz, 10 kOhm at video frequencies (5 MHz), 500 Ohm at 100 MHz and 50 Ohm at 1 GHz.

Realisation of these low impedance levels without introducing large stray capacitances is easier when using bipolar transistors, than when using MOS transistors. This, and the lower noise and further the large control range (several orders of magnitude in practice), motivates our choice for transconductors from bipolar transistors for high-frequency filters.

The circuits contain high-ohmic nodes. For this reason they need complementary transistors (with a few exceptions). For most of the circuits we assume the availability of high-frequency PNP transistors (vertical PNP or VPNP transistors) in addition to high-frequency NPN's. The filters can be integrated in bipolar and (C)bi(C)MOS processes.

Linearized transconductors

There are two basic circuits for linearized transconductors from bipolar transistors, see fig. 3 and Voorman [6].

The first is a parallel circuit of two long-tailed pairs with emitter ratios 1 : m and m : 1. Let the collector current Ic of a bipolar transistor be $Ic = Io.exp(Vbe/VT)$, where Io is the saturation current and $VT = k.T/q$. For the first long-tailed pair with emitter ratio (= ratio of saturation currents) 1 : m, a tail bias current I/2 (half of the total bias current) and a differential input voltage V the differential output current

$$dI1 = (Ic1 - Ic2)/2 = I/4 \,.\, \tanh((V/VT - \ln(m))/2) \,. \quad (9)$$

For the two pairs the output current

$$dI = I/4 \,.\, \tanh((V/VT - \ln(m))/2) + I/4 \,.\, \tanh((V/VT + \ln(m))/2) \,, \quad (10)$$

$$= I/2 \,.\, (\exp(u) - \exp(-u)) \,/\, (a + \exp(u) + \exp(-u)) \,, \quad (11)$$

where $u = V/VT$ and $a = m+1/m$.
The zero voltage ($V = 0$) transconductance Gm is

$$Gm = (I/VT) \,/\, (2+m+1/m) \,. \quad (12)$$

The transconductance as a function of the input voltage is maximally flat for $m = 2 + \sqrt{3} = 3.732\ldots$, which corresponds with $Gm = (I/VT) / 6$. In practice we choose $m = 4$ with a slight overshoot of the characteristic. In that case $Gm = (I/VT)\ 4/25$.

For low nonlinear distortion signals should not be larger than 50 mV(eff). To increase the maximum signal voltage, the circuit has been extended in different ways with more parallel stages, however, the simple 1 : 4, 4 : 1 transconductor has shown to be by far the most useful circuit.

The second circuit type has a resistor string between the two input terminals and the bases of the transistors are connected to different points of the resistor string in order to obtain a linearized transfer function. All emitters are connected to the bias current source I. The simplest of these circuits and also the most useful one, has been depicted in fig. 3 at the bottom.

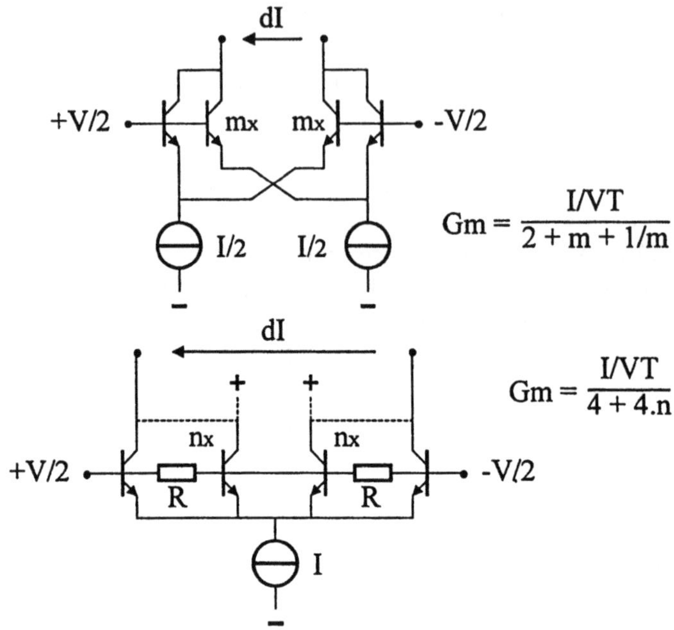

Fig. 3. Two basic linearized transconductor types.
Input voltage as well as output current are taken differential: Ileft = I/2 + dI, Iright = I/2 − dI. The characteristics are of the same type. The circuit at the bottom can handle a factor two larger signal voltage, which is better for the signal-to-noise ratio for a given total bias current. A disadvantage is that its resistor string reduces the input resistance of the transconductor. Practical parameter values are m = 4 (some overshoot) and n = 2 (maximally flat). Both circuit types can be generalized, for example with more stages, more resistors and more transistors, in order to increase the maximum input voltage
(for a given distortion level).

The two resistors have equal values. Let the transistor ratio be 1 : n : n : 1. The n-emitter transistors are in the center. They carry equal currents, which can be drawn from the outputs without effect on the transfer, as the output current is differential. In this case the sum current from the outputs is invariant and equal to I. The currents can also be drawn directly from the supply: class A/B operation. In that case the sum current is variable, but we have less noise.

With anti-symmetric excitation the center node is virtual ground.
From left to right we denote the collector currents by Ix, Ic, Ic, Iy.
For the transistor pair at the left V/2 = VT ln(n.Ix/Ic).

Hence,

$$Ix = Ic/n \exp(+V/(2.VT)) . \tag{13}$$

Similarly,

$$Iy = Ic/n \exp(-V/(2.VT)) . \tag{14}$$

From $Ix + 2.Ic + Iy = I$ we derive

$$Ic = n.I / (2.n + \exp(V/(2.VT)) + \exp(-V/(2.VT))) . \tag{15}$$

Expression (15) can be used in (13) and (12).
Subtraction yields for the differential output current $dI = (Ix - Iy)/2$

$$dI = I/2 \, (\exp(u/2) - \exp(-u/2)) / (2.n + \exp(u/2) + \exp(-u/2)) , \tag{16}$$

where, as before, $u = V/VT$.
Comparison with expression (11) shows that we have the same characteristic for a two times larger voltage and $2.n = m + 1/m$.
For the same bias current the transconductance or Gm value is a factor two lower: $Gm = (I/VT) / (4 + 4.n)$.
We have a maximally flat behaviour for $m = 2 + \sqrt{3}$ and for $n = 2$.
Maximum signal voltages are 100 mV(eff).
The fact that this transconductor can handle a two times larger signal voltage is important for low signal distortion and large SNR. Its relatively low output capacitance, in particular when the transconductor is used in class A/B, keeps stray capacitances low. The main disadvantage of this transconductor is its low input resistance, because of its resistor string.

Load circuit

The output currents of the transconductors consist of a bias part, which is common mode and a signal part, which is differential. To get rid of the bias current we use a load circuit. For a simple example, see fig. 4. The circuit is low-ohmic for common-mode current: it should pass the bias current and it is high-ohmic for differential current: it should not damp the resonance circuits in the filters. The combination of transconductor and load circuit can directly be used in Gm-C filters. In cases where the output currents of several transconductors combine, we need one load circuit only. In practice we need half the number of transconductors as number of load circuits.

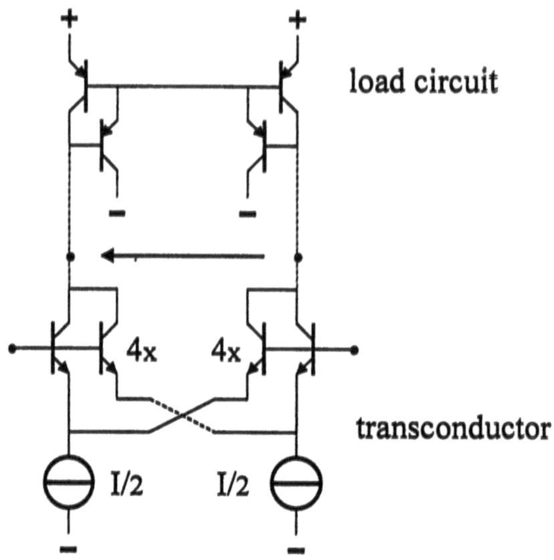

Fig. 4. A 1:4, 4:1 transconductor with load circuit.
To get rid of the bias current of the transconductor we need a load circuit. The circuit should be low-ohmic for bias current (= common-mode current) and high-ohmic for signal current (= differential current). The figure shows a simple circuit example, which can directly be used in transconductor-capacitor filters. In the filter parallel load circuits can be replaced by a single one.

All-NPN circuits

Fig. 5 shows the 1:4, 4:1 transconductor with an all-NPN type load circuit. With this combination and capacitors we can design tunable all-NPN Gm-C filters. For the load circuit each of the base-emitter voltages of the long-tailed pair in the center is equal to a diode voltage (at the top). Hence the base-emitter voltages of the outer transistors are equal as are their collector currents. The circuit is high-ohmic for differential currents and low-ohmic for common-mode currents. In filters, parallel load circuits can again be replaced by a single one. The rest of the filter design is straightforward.

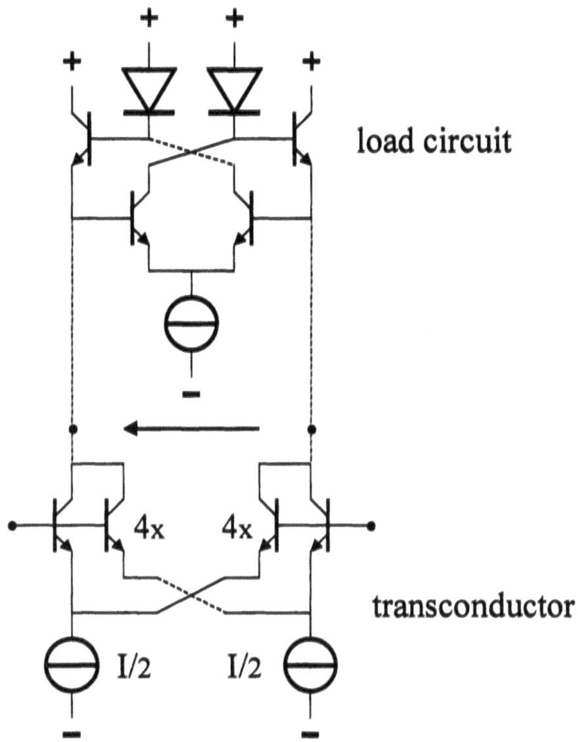

Fig. 5. An NPN 1:4, 4:1 type transconductor with an all-NPN load circuit. In the outer left and right transistors of the load circuit we have equal currents. The load circuit is high-ohmic for differential signal currents and low-ohmic for common-mode bias currents. The arrow represents capacitive currents. The circuits can be used for all-NPN tunable Gm-C filters. The minimum supply voltage is 3V, approx.

The performance is limited by the high-frequency performance of the load circuit and also by its noise. Because of the loop in the load circuit, the high-frequency performance is inferior to the performance of the transconductor circuit. Further, the ring of base-emitter diodes is fairly noisy. Of course, we could use resistors to reduce the noise, see for example [2], but this is not compatible with free tunability.

Complementary transconductor

A significant improvement can be obtained by using complementary transconductors, see fig. 6. We use NPN and PNP type transconductors, which share the same bias current. In this case the load circuit can be combined with a transconductor, see the right part of fig. 6. This can be done at the PNP side - as shown - as well as at the NPN side. In the first case all filter nodes are referenced to the positive supply. This can be an advantage when input and output signals are also referenced to the positive supply. In that case it saves level-shift circuits. The use of load circuits at the NPN side references all filter nodes to the negative supply, that is to the substrate. In general, this gives a better power supply rejection as most of the stray capacitances are to the substrate.

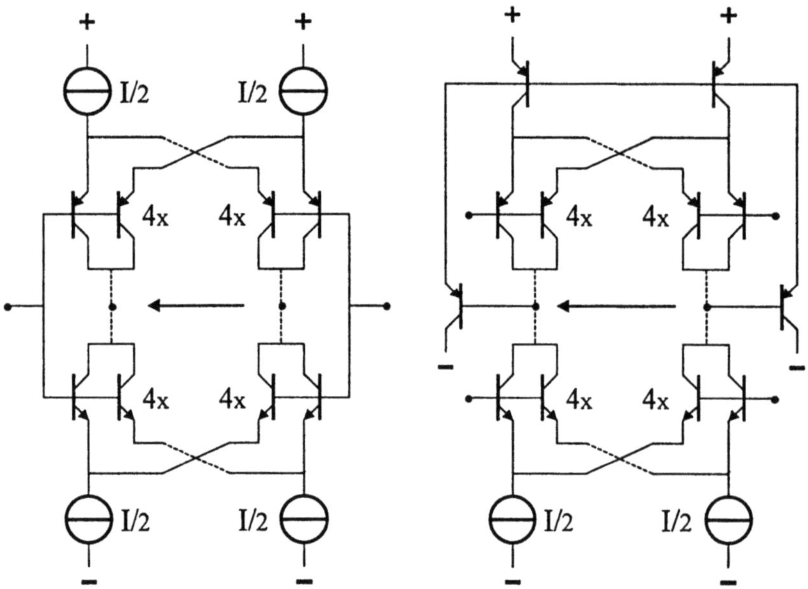

Fig. 6. Complementary 1:4, 4:1 transconductors.
At the left we see a transconductor with NPN transistors and one with PNP transistors. If we interconnect inputs and outputs of the two transconductors, we get a transconductor with twice the transconductance, as needed for the coupling between different filter stages. Also gyrators can be made from an NPN and a PNP type transconductor. NPN and PNP transconductors share the same bias current, which results in a low power consumption. At the right side we show a PNP type load circuit combined with a PNP type transconductor.

A first advantage of the approach with complementary circuits is that we obtain a reduction of 50% of the bias current, because all bias currents are shared by two transconductors. Furthermore, the noise is lower. The load circuit itself is less noisy and its noise current is not entirely single-ended, but partly common mode. Because we have only a part of the signal voltage across the load circuit, its output impedance has a reduced influence on the filter performance.
In fig. 7 we show several aspects of Gm-C filter design with these complementary transconductors and NPN type load circuits.

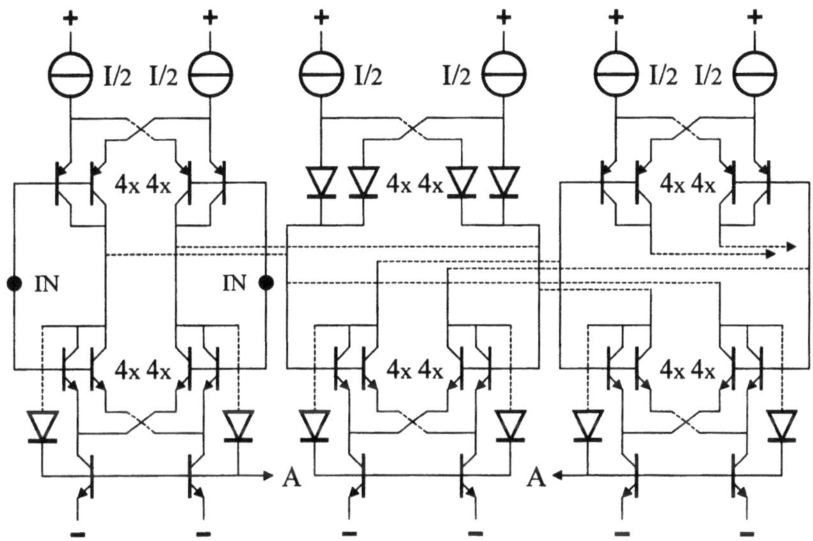

Fig. 7. Aspects of filter design with complementary 1:4, 4:1 transconductors. The figure shows the input part of a filter. Capacitors have not been indicated. The load circuits are NPN in type. All filter nodes are referred to the negative supply (ground and/or substrate) as are most of the stray capacitances. This keeps the crosstalk from the positive supply low. We can connect the wells of the VPNP transistors of the transconductors to the corresponding emitters to further reduce the crosstalk from the supply. The 2x input transconductor is at the left. It is a fully complementary circuit, which acts as a current source. The source resistance has been depicted in the center at the top. The input-output short-circuited PNP-type transconductor contains PNP transistors with base and collector short-circuited: PNP type diodes, which have been replaced by NPN-type diodes to improve the high-frequency performance. The wells and bases of the diodes can be merged pairwise, which results in multi-emitter structures (to reduce the stray capacitances further). The NPN-transconductors in the center and at the right side constitute a first gyrator, etc., etc. Nodes A can be connected and one of the diode pairs to A can be omitted (the pairs are in parallel).

New transconductor

To the simplest of the second transconductor type - the transconductor type with resistor string – we add a similar transconductor from PNP transistors with input port and output port connected, see fig. 8, left part. With anti-symmetric excitation the midpoints of the two resistor strings are at virtual ground. The currents in the four 2-emitter transistors are all equal. We can pass these currents through. If we implement the 2-emitter transistors as diodes, see the right part of fig. 8, the resistor strings are not needed any more. The voltage at the midpoint remains the same.

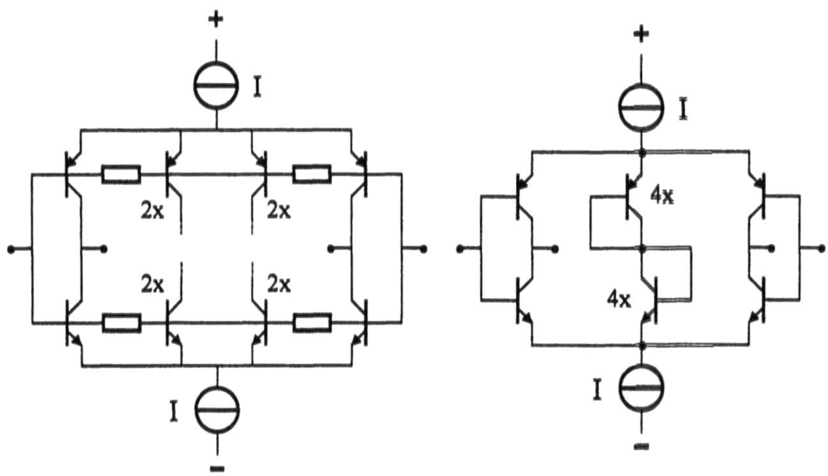

Fig. 8. We take a linearized transconductor with PNP transistors and one with NPN transistors. We put them in parallel (left part of the figure) and let them share the same bias current. In the case of anti-symmetric excitation the midpoints of the resistor strings are virtual ground. All 2-emitter transistors carry the same current. Connection of these transistors as diodes (collector and base short circuited) and connecting the center points of the resistor strings has no influence on currents and voltages. Even, omission of the resistor strings does not change the current in the transistors or the voltage at the midpoint. The resulting transconductor has been depicted in the right part of the figure.

The combination eliminates the resistor string and hence it eliminates the low input resistance of the transconductor. It doubles the output signal current for the same input voltage and bias current. Hence, it doubles the power efficiency (compared with the power efficiency of the non-complementary circuits). All bias current sources become common mode, which results in a significant reduction of the noise current. Finally, the input capacitance of the transconductor is low. We should observe that the transistors in the transconductor cannot be very small, because the base series resistance should be kept low (for low noise) and DC offsets should not become too large. The input capacitance is from the input transistors (1-emitter transistors). The larger transistors with larger capacitance are in the center. They are common mode, where they do no harm to the high-frequency performance. Also in this respect we obtain a significant improvement, in particular compared with the 1 : 4, 4 : 1 transconductors, where the large 4-emitter transistors are connected to the filter nodes.

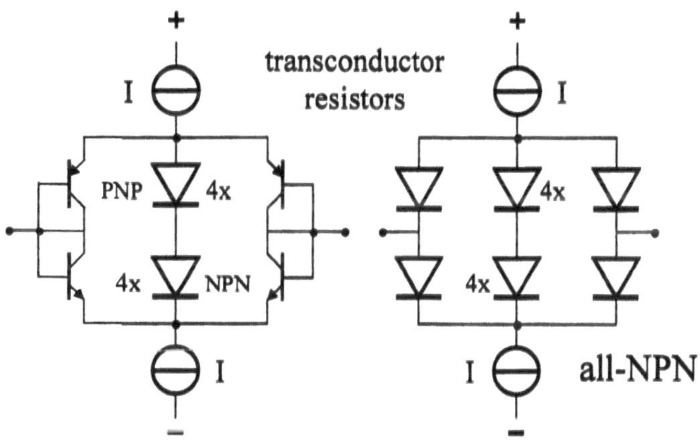

Fig. 9. Transconductor resistors as combinations of diodes.
If we interconnect input port and output port of a transconductor, all transistors become diodes (left). All PNP diodes can be replaced by NPN diodes (right) and we obtain an all-NPN tunable resistor. In almost all cases the bias current sources on either the PNP or the NPN side can be replaced by diodes. The circuits can be used for the source and load resistors in the Gm-C filters.

A further simplification can be obtained for the source and load resistor circuits in the filters, see fig. 9. In fig. 10 we show a gyrator with its biasing. The cross-coupling is needed to get a DC-stable circuit. Further aspects of Gm-C filter design with the new transconductors have been indicated in fig. 11.

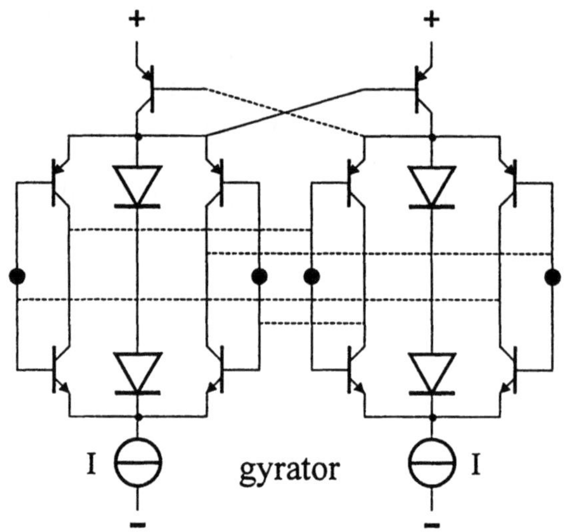

Fig. 10. In the figure we show two transconductors, which have been interconnected as a gyrator, together with PNP biasing transistors. The cross coupling for the biasing transistors is needed for negative feedback (for DC stability).

Additional performance aspects

As the mobility of holes is lower than the mobility of electrons, NPN transistors have a substantially higher fT than (V)PNP transistors. The fT is a yardstick for the performance of the transistors in cases that they are current controlled. In the transconductors the transistors are 'voltage' controlled and the cut-off frequency fV applies. For high fV we need a low base series resistance, which results for the NPN transistors in a transistor geometry BEBC (Base-Emitter-Base-Collector), that is in an emitter with two bases alongside.

In general the dope in the base of the VPNP transistor is much higher than the dope in the base of the NPN. This means that the pinch resistance and hence the base series resistance of the VPNP is much lower for the same geometry. This allows us to use a different geometry CEBEC for the VPNP, which has a lower collector series resistance and can handle more current. Doing this, the high frequency performance of the NPN and VPNP transistors for these filter applications differs less that one would expect from the ratios of the fT's. It is our experience that not only concerning power consumption, SNR and simplicity, but also in terms of speed these transconductors are very competitive.

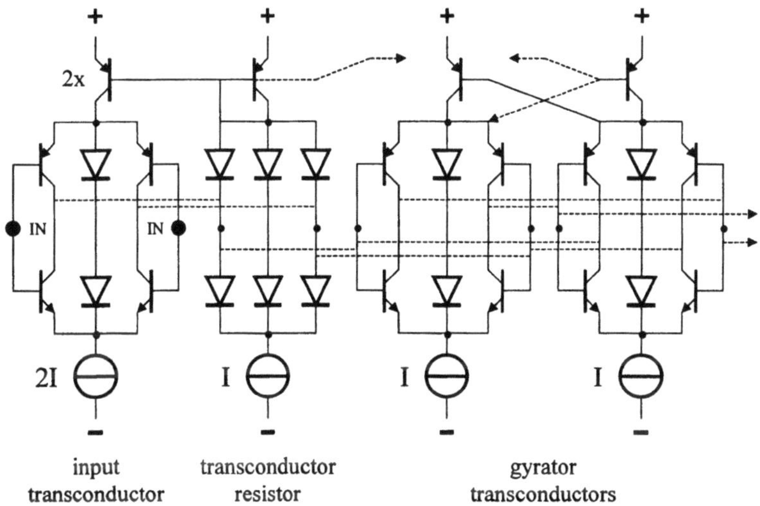

Fig. 11. Aspects of filter design with the new transconductor type.
We have depicted the input part of a filter section. Capacitors have been omitted. It reveals all important aspects of the filter circuit design. At the left we have the 2x input transconductor. Its output signal current is fed to the source resistor. The source resistor has been constructed from NPN diodes. The right-hand transconductors constitute a gyrator. In this case the cross-coupled biasing of the gyrator cannot be applied, because this would result in a translinear circuit with 1 NPN and 3 PNP transistors (in the transconductor resistor we have replaced the PNP diodes by NPN diodes). Instead, we can connect the base of the right-hand PNP to the bases of the PNP diode and the 2x PNP bias current source.

Experimental work

This work on high-frequency Gm-C filters tracks back to the design of video lowpass filters and further Laguerre filters and analog tapped delay lines for adaptive filters, such as ghost cancellers for television signals. The 5 MHz designs were made in a 2.5 μm IC process with fT = 3 GHz for the NPN and 800 MHz for the VPNP (for references, see: Voorman [1]). Later we have implemented CMOS and bipolar equalizers (PR4 cosine filters) for digital video tape recording for 20 Mb/s. For our work on Hard Disk Drive IC's, see, for example, Bergmans [7], we have successfully designed first- and second-order allpass filters and lowpass filters for the feedforward part of the equalizer of our Dual-DFE channel IC, see fig. 12. Particularly interesting was the good agreement between simulation and measurement regarding overflow levels of the filter sections. Overflow in the lowpass section causes mainly amplitude distortion. In the delay (allpass) sections zero-crossings of the signals are shifted due to the nonlinear distortion. The allpass sections had a low frequency delay of more than 3 bits. A large filter delay is particularly difficult for high-frequency filters, because gain and group-delay imperfections tend to increase with increasing filter delay, see [8], for example. The first version of the IC was for a maximum user data rate of 240 Mb/s and a second for 360 Mb/s both in a 1 μm biCMOS process with fT = 6 GHz for the NPN transistors and 1 GHz for the VPNP's. Evaluation of the corresponding silicon showed good performance up to 300 Mb/s. We worked on a re-design for 700 Mb/s in a 0.6 μm biCMOS process with fT = 10 GHz for the NPN and 3 GHz for the VPNP when this project was stopped. Circuit simulations had already shown that the filters would certainly not be a bottleneck for the design of the new channel IC. Taking into account these experiences and our more recent ideas, which have been included in this text, and the fact that there are much faster IC processes gives us the firm belief that the methods can readily be applied for bandwidths of hundreds of MHz until somewhat over 1 GHz.

Fig. 12. Chip photo of the analog equalizer of our DDFE channel IC for HDD (HDD = Hard-Disk Drive). The DDFE or Dual DFE uses two DFE's (DFE = Decision Feedback Equalizer), with a survivor path for erasures (erasures are uncertain decisions) in order to enhance the Bit-Error Rate (BER). The DFE's are preceded by an analog equalizer, which in our case is of order 7. From the left to the right we distinguish a group of 5 transconductors: an input transconductor and a first-order allpass section and a group of 7 transconductors: a coupling transconductor and a second-order all-pass section. The allpass sections are needed to shape the pulses towards the minimum-phase target response, that is they remove leading ISI (Inter-Symbol Interference). The decision feedback part removes trailing ISI. The next group of transconductors is the lowpass part of the filter which consists of two second-order filter sections. They are used to remove out-of-band noise. The equalizer is used in a bitrate synchronous fashion for bitrates from 60 – 360 Mb/s. Between the two lowpass sections (right-hand part of the photo) we add with adaptively controlled weights the filter signal and its second derivative in order to continuously correct the spectral balance: high versus low frequencies. This provides a continuous correction of the response to flying height variations of the read head of the disk drive.

Conclusions

We have presented improvements and new bipolar transistor circuits for transconductors for high-frequency Gm-C filters. The circuits are based on two fundamental types of linearized transistor circuits (see fig. 3). Introduction of complementary circuits (see figures 6 and 7) gives a power efficiency improvement of a factor two and leads in a natural way to a new transconductor circuit, which can handle a factor two higher signal voltages and has a lower input capacitance (see figures 8 – 11). The circuits can be used in 3V (and higher) applications up to very high frequencies: hundreds of MHz up to 1 GHz. They are extremely power efficient, have a low excess noise and a low circuit complexity (low cost).

References

[1] 'Continuous-time analog integrated filters' by J.O. Voorman in 'Integrated continuous-time filters' edited by Y.P. Tsividis and J.O. Voorman, pp. 15-45, 1993.

[2] 'A 3.0 V 40 Mb/s Hard Disk Drive read channel IC' by G.A. De Veirman, et al., IEEE Journal of Solid-State Circuits 30, pp. 788-799, 1995.

[3] 'Log domain filtering for RF applications' by D.R. Frey, IEEE Journal of Solid-State Circuits 31, pp. 1468-1475, 1996.

[4] 'Noise and dissipation of electronic gyrators' by D. Blom and J.O. Voorman, Philips Research Reports 26, pp. 103-113, 1971.

[5] See [1]: page 28, fig. 27.

[6] See [1]: pages 32, figures 33 and 34 and 188, figures 4 and 5.

[7] 'Dual-DFE read/write channel IC for hard-disk drives' by J.W.M. Bergmans, et al., IEEE Transactions on Magnetics 34, pp. 172-177, 1998.

[8] See [1]: page 43, fig. 54.

High Speed Analog Filters Using Scaled CMOS Technology

R. CASTELLO[1], I. BIETTI[2] and F. SVELTO[3]

1-Dipartimento di Elettronica - Universita' di Pavia,
Pavia, Italy
2-STMicroelectronics- Data Storage Division
Assago - Milan, Italy
3-Dipartimento di Ingegneria - Universita' di Bergamo,
Bergamo, Italy

ABSTRACT

Data on a 7th order programmable gm-C filter in 0.35μm CMOS with 160MHz maximum pole frequency are provided. Techniques to increase the bandwidth of gm-C filters in further scaled technologies, while preserving dynamic range and accuracy, are presented. GHz passive LC filters, based on master slave tuning, are introduced.

1. INTRODUCTION

High frequency analog filters are used in two areas. First, as pre-processing blocks in front of an A/D or as post processing blocks after a D/A. Examples are antialias filters for digital TVs, equalizers in digital modems and preconditioning filters in data recording channels (HDD). Second, as building blocks in the front end of a ΣΔ converter. Although sample domain filters have been used up to a few hundred MHz clocks this paper will concentrate on continuous time filters. Among them we will consider both active and passive ones. Since filters are generally part of complex digital systems in CMOS technology, this paper deals exclusively with CMOS circuits.

As IC technologies scale, the frequency of integrated filters increases. This is key to sustain the progress of systems like disk drive channels. In addition, it may allow to integrate higher speed blocks e.g. image or IF filters in wireless transceivers. On the other hand, submicron CMOS implies a reduced supply voltage and a lower gain due to the fine line lithography.

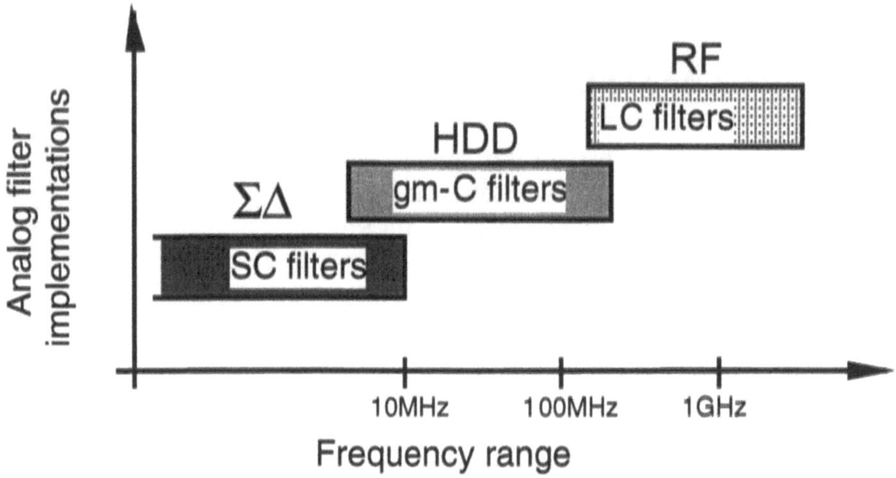

Figure 1. Analog filters applications.

Applications and implementation of anolog continuous time filters are shown in Fig 1.

2. Gm-C Filters

Frequency range from tens to a few hundred MHz is covered by gm-C filters; however, they offer relatively poor Dynamic Range (DR) and precision. The building block of a gm-C filter is a V-to-I converter (transconductor). Maximum DR corresponds to maximum voltage swing (VMAX) and gm/I ratio and minimum excess noise factor (F) of the transconductor according to the following expression:

$$DR = \beta\sqrt{\frac{nI}{KTBw}}\sqrt{\frac{g_m}{FI}}V_{MAX} \quad \text{[eq. 1].}$$

where n is the filter order, Bw its bandwidth and β depends on its transfer function.

In MOS technology a transconductor can be realized in different ways [1] each with a different DR. In all cases there is a bias condition that gives the maximum DR for a given power consumption. Traditionally all spreads in capacitors and gm have been compensated by changing the bias current and the same has been done to tune the filter bandwidth. This technique moves the bias point away from its optimum value causing DR degradation and changes the gain by a factor closely related to the overall tuning (t). Bandwidth adjustment can also be done with a digitally controlled capacitor array[2]. Due to the phase error caused by the finite on-resistance of the switches, this method is used at moderate frequencies or requires error calibration. Technology scaling and layout optimization can extend the range of this technique[3].

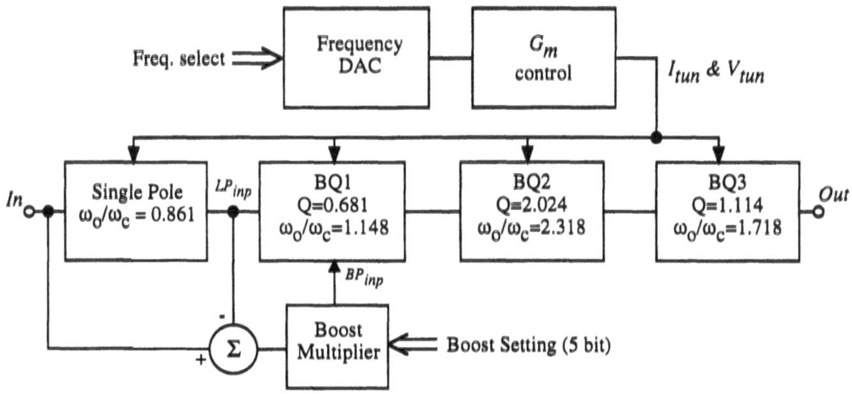

Figure 2. Overall filter block diagram.

Filter implementation	7th order LPF phase-equiripple
Temperature Range	0-120° C
Supply Voltage	3.3V ± 10%
Area	0.15mm^2
Power Consumption (fc=70MHz Vdd=3.3V)	55 mW Typ. (Filter + control circuits)
Boost Programmability	0-13dB (5 bit DAC)
Differential output for THD=-40dB (worst)	400mV$_{pp}$
Integrated output noise (fc=70MHz)	1.15mV$_{RMS}$
Dynamic Range (Wst/Bst)	42dB / 50dB
Tuning Range	14-70MHz

Table 1. Performances of a Gm-C filter for HDD applications.

In a recent 3V design for a 300 Mbit/sec PRML HDD using .35μm CMOS technology a discrete gm tuning has been used to change the cut-off frequency while current tuning has been used to compensate the spreads. The 5:1 bandwidth programmability has been broken in different regions programming the transconductor size. Fig. 2 shows the filter structure with Q and bandwidth of each biquad while the overall performance achieved is shown in Table 1. The small area ($0.15mm^2$) is important both for cost reasons and reduced substrate noise pick up (the filter was tested isolated and with a 300 MHz DSP core without appreciable difference). On the other hand, the bias point (worst and best case) had to be changed by 5 to 1 causing a DR reduction of about 10 dB and a DC-gain variation by a factor of more than 2. The latter effect limits the frequency response precision. A microphotograph of this discretely tuned filter is shown in Fig. 3.

Figure 3. 7th order LPF for HDD applications in 0.35μm

Fig. 4 shows the transconductor used in this design (4a) together with two other possible ones [4-5]. Assuming a very small tuning requirement all the circuits have similar DR. However, the one shown in Fig. 4b allows the largest swing while that of Fig. 4c has the smallest swing and maximum DC-gain variation.

Although DR is the key target for a stand alone filter, for filters embedded within large digital systems crosstalk immunity may even be more important. A significant figure of merit in defining the immunity of an analog filter from external disturbances is the product of the maximum voltage swing times the admittance level at the various nodes normalized to the current consumption according to the following equation.

$$NI = \frac{Gm}{I} V_{MAX} \text{ [eq. 2]}$$

This is because the majority of on chip disturbances are given by current noise. The amount of voltage noise that results is proportional to the noise current itself and to the impedance of the node where it is injected.

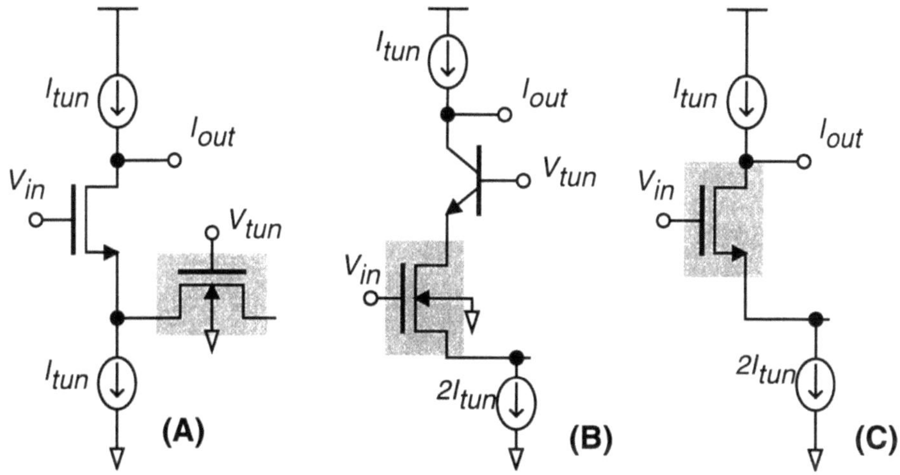

Figure 4. Possible Transconductor Topologies

Even in terms of this new figure of merit, the circuit of Fig. 4c has the worst performance for t greater than 2. The other two circuit are comparable although the one of Fig. 4b is better at low tuning while the one of Fig. 4a is better at high. For the applications described above, having to insure a tuning of more than 4 to 1, the circuit of Fig. 4c was not usable. The topology adopted was the one of Fig. 4a mostly because the technology used was pure CMOS. In this case, in fact, the circuit of Fig.4b becomes much less competitive. In a BiCMOS implementation, on the other hand, this latter topology would probably have been the best choice.

Increasing filter bandwidth requires to use further scaled technologies with a further reduction of the DR and an increased variation of the transconductor DC-gain. To maintain acceptable performances the DR should be kept as close as possible to the theoretical maximum and the gain as constant as possible. Conceptually this requires to separately adjust the load capacitors and the transconductor gm. Furthermore discrete tuning has to be used also to compensate temperature and technology variation. In this way the parameter variation is reduced proportionally to the number of discrete elements that can be switched in or out of the signal path. In addition, the DC-gain variation decreases drastically and can be compensated for.

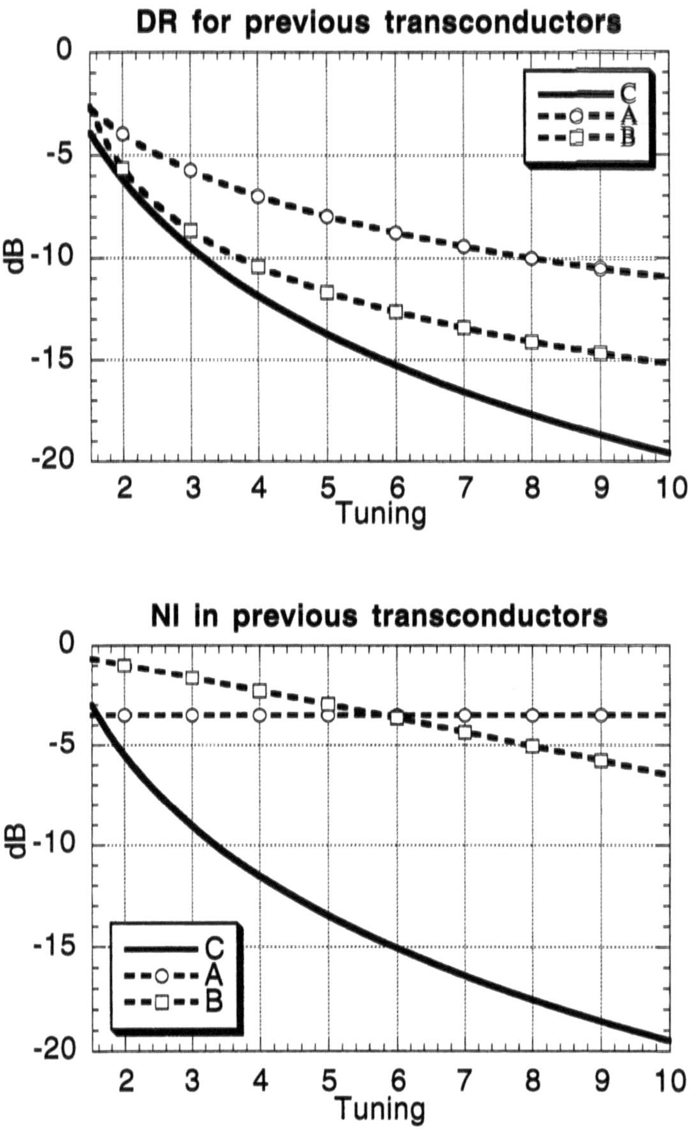

Figures 5a 5b. DR and NI for different transconductor types.

If a very large number of discrete tuning elements is used this approach makes the tuning ratio (t) to approach 1. In this situation the relative performance of the three transconductors of Fig. 4 changes. This is because while for the circuit of Fig. 4c both the DR and NI decrease linearly with t and the DC gain variation increases linearly with t, for the other two the DR decreases approximately as $t^{1/2}$, the gain variation is either constant (Fig. 4b) or increases as $t^{1/2}$ (Fig. 4a) and finally NI stays to the first order constant [1]. As t is made to approach 1 the circuit of Fig. 4c improves its performance much more than the other two to the point to become comparable with them. In addition being the simplest one, and potentially the fastest, becomes the topology of choice. DR and NI for the 3 transconductor types are plot in Fig. 5a and 5b.

The circuit schematic of a transconductor based on saturated MOS devices that implements discrete gm tuning is shown in Fig. 6. The input devices are made up of elementary units with gates and drains directly connected in parallel and sources separated. An array of elementary current sources can be connected through a set of switches to the various device sources. To program the filter bandwidth some of the unit current sources are either sent to an elementary differential pair or diverted to the supply. The unit differential pairs which are not connected to the common source node are left floating. To compensate for technology and temperature spreads the total current through the transconductor is kept constant while the number of biased (floating) pairs is changed. Using this configuration no switch is placed in the signal path and the total parasitic capacitance at the filter nodes stays constant while tuning (to first order).

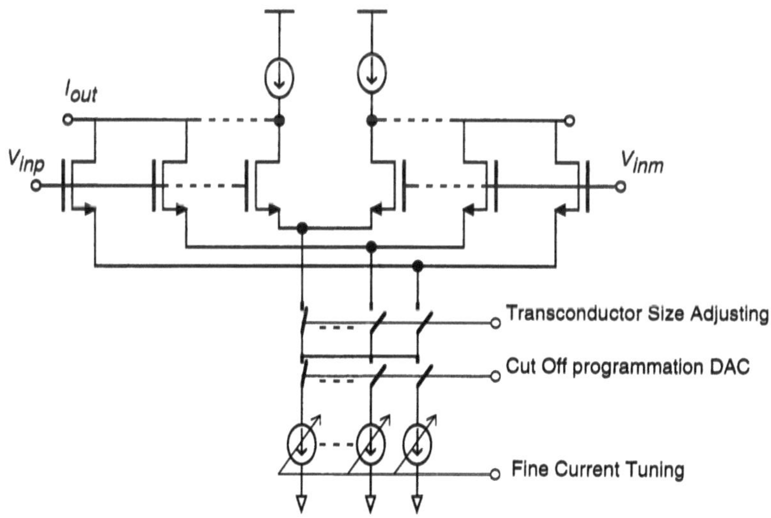

Figure 6. Schematic of a transconductor with variable size.

To reduce the number of differential pairs used, coarse tuning can be done by transconductor size adjusting while fine tuning can still be implemented by changing the reference bias current (see Fig. 6). Table 2 shows the parameters spread in a single saturated differential pair transconductor tuned just to compensate temperature and technology variations (no cut-off programming). In this case all the tuning is done changing the reference current. On the same table is also reported the parameters spread for a transconductor made up of 4 differential pairs tuned both to compensate variations and to program by 2:1 the cut-off. In this case coarse and fine tuning approach is used. This data refers to .25μm CMOS technology and to max. input differential signal of 500mV peak to peak.

Parameter	Single diff. pair	Multiple diff. pair
Current Tuning	7 ÷ 1	2.5÷1
Overdrive Variation	4 ÷ 1	1.5÷1
DCgain spread	1.8 ÷ 1	1.3÷1
Min DCgain	50	70
Min. Power supply	2.0 V	1.6V

Table 2. Comparison between single pair and multiple pairs transconductor.

Discrete gm tuning could be implemented using the approach proposed by Khorramabadi et al. [6] for active RC filters. The filter is calibrated using an A/D converter and the computing power of a DSP generally available. Coarse calibration is performed at power-on and periodically during operation (when filtering is not required). In these phases the bias current is set to the nominal value. Between calibration cycles the number of elementary unit used is stored in a register and kept constant, while the bias adjustment loop is turned on. Using this technique and assuming a 20 degree temperature variation between cycles, the extrapolated performance achievable for a 1.8 V 500 MHz 4th order low pass filter in a .18 μm technology is shown in Table 3.

When using a pure digital CMOS technology without double poly option the problem of how to implement the capacitors arises. In this case an interesting possibility is to use MOS capacitors. The main problem in this case is the non linearity caused by the voltage dependence of an MOS capacitor. Nonetheless this approach can often be feasible due to three reasons. First the relatively limited voltage swing achievable in low voltage continuous time filters. Second the fact that often differential solutions are used. This creates a partial cancellation of the capacitor voltage variation. Third the typical availability in deep scaled down

CMOS technologies of a natural device (zero threshold) at no extra cost. Using this device to built the MOS capacitor gives a reduced voltage dependence of the capacitance. This is because a zero threshold device can be easily kept in strong inversion across the entire voltage swing.

Filter implementation	4th order LPF
Cut-Off Frequency	500MHz
Supply Voltage	1.8V
Area	<< $0.1mm^2$
Power Consumption (Vdd=1.8V)	50 mW
Differential output -40dB THD	$500mV_{RMS}$
Integrated output noise	$200uV_{RMS}$
Dynamic Range	60dB

Table 3. Extrapolated performances for 0.18μm Gm-C filter.

An advantage of this solution is the partial correlation that exists between the value of gm and that of the MOS capacitor in the presence of technology variations. A consequence of this is the total tuning required to compensate technology and temperature variations is reduced from about 2.5 to 1 to just slightly more than 2 to 1. It follows that a smaller number of discrete elements is required to achieve the same performance level. In this case the tuning strategy that compensate for all the spreads while maximizing DR and NI in a saturated MOS transconductor (i.e. maximizing the V_{gs} -VT of the input device) is slightly different. The proper tuning strategy will not compensate the capacitor spread acting on the capacitor itself (as in the previous case) but by changing the reference current. The required percentage change of current is the same as the variation in the capacitor value. The residual gm variation will instead have to be compensated acting on the device size keeping the reference current constant (at least in a simplified ideal model).

The use of a BiCMOS technology potentially solves the problem of tuning in a very straightforward way. This occurs when a purely bipolar transconductor is used whose performance is almost independent of the amount of tuning required up to very large values of t [9]. A limitation of this approach is the small voltage swing (only a few thermal voltages) that can be achieved. This may be less of a problem at low supply voltage where the voltage swing is necessarily limited.

3. LC Filters

For higher frequencies a possible approach is to use LC based filters. LC resonators are used in low phase-noise oscillators and selective filters in the GHz range. Unfortunately inductors in standard digital CMOS technology have Q values well below 10. Q-enhancement techniques can be used but this approach adds noise since it uses extra active elements in the signal path. The extra noise can, however, be acceptably small in certain cases. For example, in band pass amplifiers, it is possible to make the output noise dominated by the input devices. Furthermore, both output noise and gain increase with the amount of Q enhancement giving an equivalent input noise fairly independent of the required Q. The key problem, when using Q enhancement, is its accurate control. This problem has not been solved in a reliable way and remains a crucial point for the future of high Q high frequency integrated filters.

The other key problem is the tuning of the center frequency. High Q filters require precise center frequency control. A promising approach_makes use of a voltage controlled reactive element (varactor). Varactors have been implemented with p-n junctions but recently MOS varactors have been proposed[6-8]. They offer a potentially larger tuning range and a more linear capacitance versus voltage characteristic. A measured CV plot for an MOS varactor is shown in Fig. 7[8].

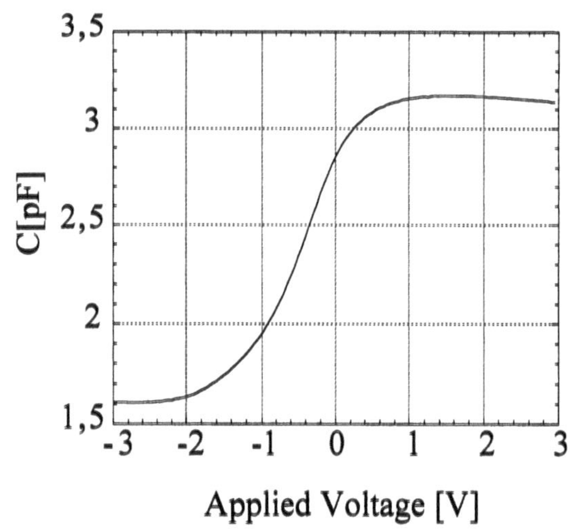

Figure 7. Capacitance versus applied voltage for an MOS varactor.

This new type of varactors are intended to be used in the resonators of an integrated VCO. Using the same reactive element also in the band pass filter enables a master-slave tuning with the VCO as the master. This concept is shown, in a simplified form, in Fig.8 and has the following potential advantages: tuning is done without active elements with no extra noise and no extra power consumption, the controlling voltage required for the tuning is already available (assuming a frequency synthesizer is part of the system). There are two possible limitations of this technique, the achievable linearity of the tank and the achievable matching between the VCO resonator and the filter tank. In both cases the limiting factor comes from the non linearity in the C-V characteristic of the varactor.

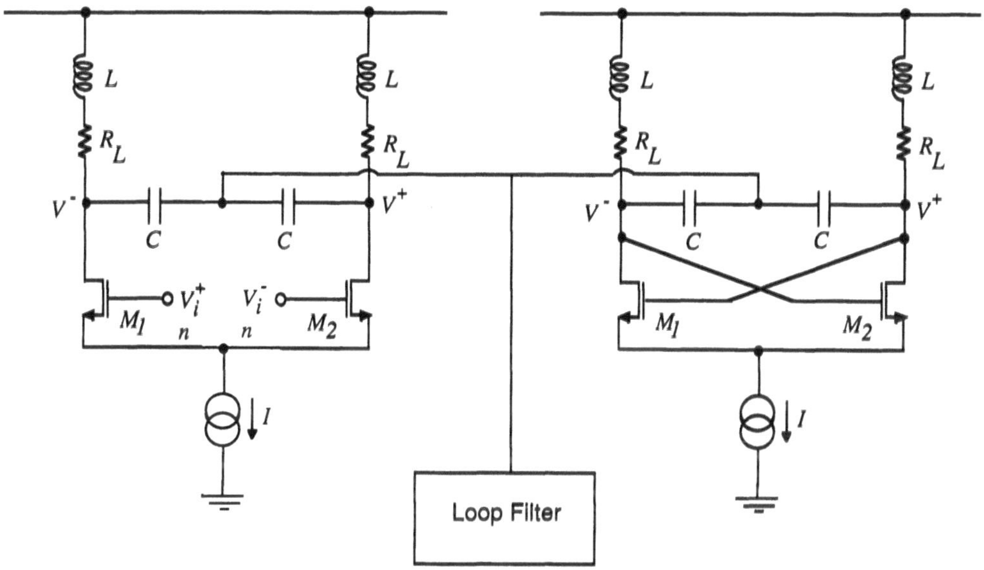

Figure 8. Master - Slave tuning concept. The VCO is used as the master.

With respect to the first problem, the varactor will introduce third order intermodulation due to its non-linear characteristic. To verify how the varactor impacts on the IIP3 of a narrowband LNA, an ideal LNA input stage (i.e. infinitely linear) has been simulated together with an LC load using the MOS varactor. Fig. 9 shows the IIP3 of the LNA, set to give a gain of 20dB together with the normalized capacitance of the MOS varactor. The IIP3 is always higher than 0 dBm, provided the control voltage is higher than –0.7V. This corresponds to a tuning range of ±20% around the nominal value.

With respect to the second problem, the achievable matching between the VCO oscillation frequency and the narrowband LNA center frequency is limited by the following three factors:

1. The intrinsic matching between inductance and capacitance of the two tanks.

2. The matching between the varactor-to-parasitic ratio in the two circuits.

3. The equivalent value of the varactor capacitance in the VCO due to its non-linearity in the presence of large signal swing, as compared with its small signal value.

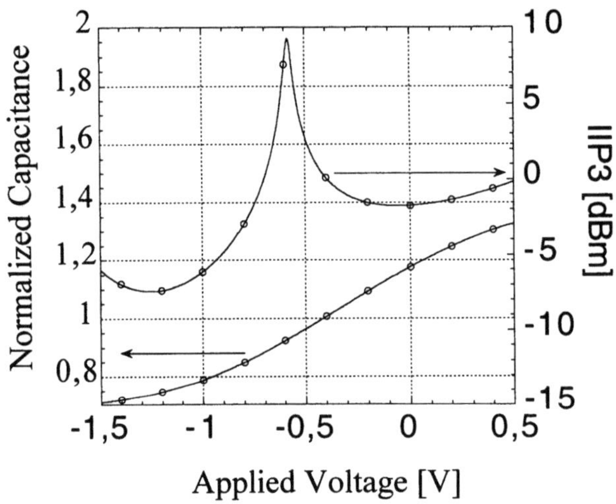

Figure 9. IIP3 of an ideal LNA input stage using the MOS varactor in the LC load. The C-V characteristic of the varactor is also shown.

To evaluate the achievable matching, in a realistic situation, an LNA, a mixer and a VCO have been designed. The target specifications correspond to a 1.6 GHz GPS system. A Q of 8 for the inductors and a 2.7V supply have been assumed. The entire front-end system has been optimized taking into account loading effects. The complete circuit is shown in figure 10, where the total parasitic capacitance at each node has been indicated. Based on this design, the effect of both the varactor non linearity and the parasitic capacitance at the different nodes can be computed. In the optimized design, the varactor-to-parasitic-ratios for the LNA and VCO have very close values and only a very small adjustment had to be done to compensate the difference. This adjustment produces a negligible degradation in the performance, compared with the optimum design. The simulated performance for the various blocks, assuming a ± 20% varactor tuning, is reported in table 4.Notice the low NF and good IIP3 for the entire front-end together with a good VCO Phase Noise.

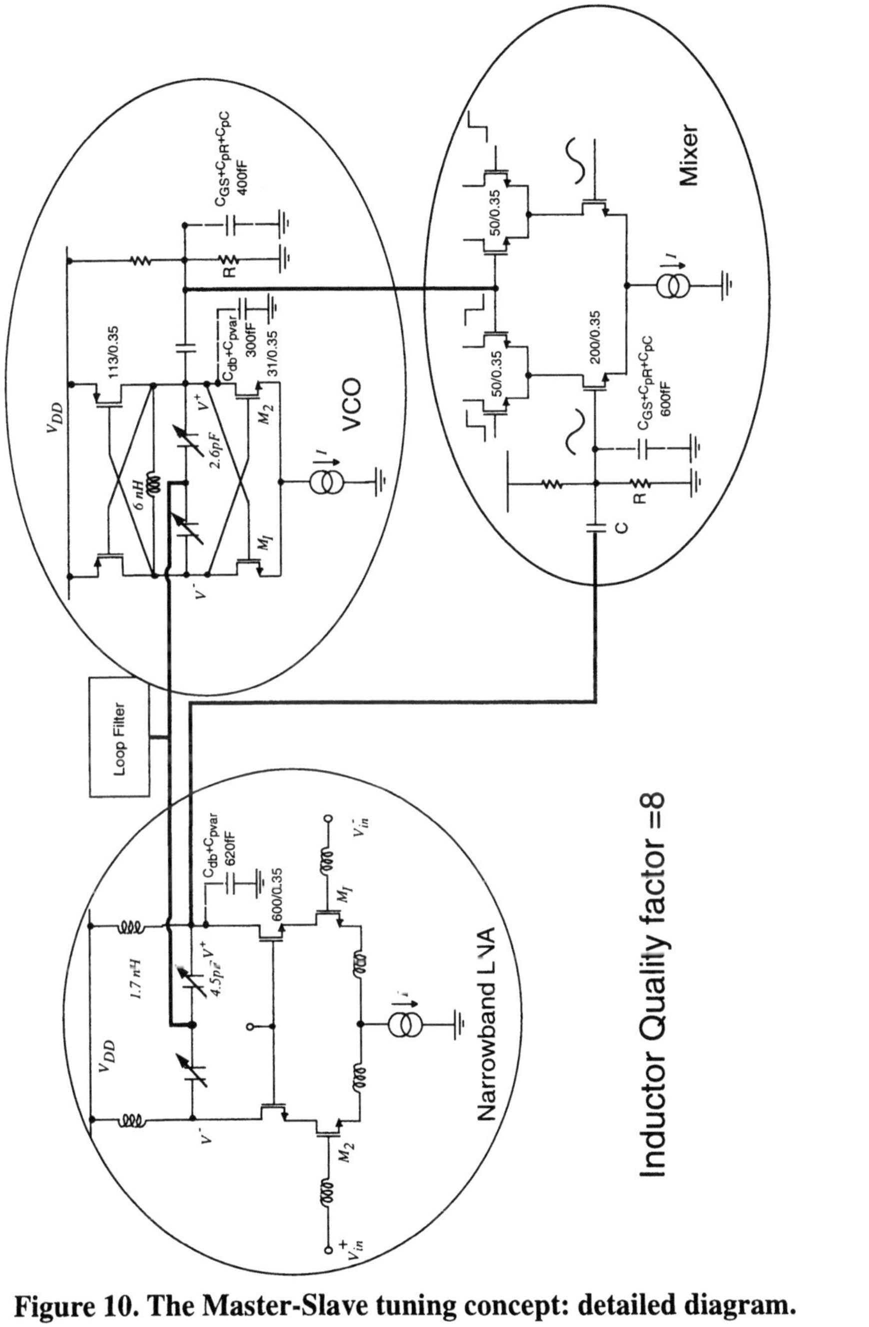

Figure 10. The Master-Slave tuning concept: detailed diagram.

Once the varactor-to-parasitic ratio is equalized, the matching between the VCO resonator and the LNA tank depends on the parasitic matching (about 15%) and the varactor matching (about 2%). The mismatch between the VCO and the LNA varactor, due to the large signal oscillation of the VCO (2 Vpk-pk differential) was computed to be about 3%.

Taking all the above into account, the expected center frequency matching is about 3%. This is a 3 to 4 improvement compared with the untuned case for spiral inductors and even more for bond wire ones. Using a more linear varactor (under investigation) should enable a matching close to 2%.

Carrier Frequency	1.6 GHz
LNA Gain	20dB
LNA NF	3 dB
LNA IIP3	-2dBm
Mixer NF	14 dB
Mixer IIP3	0 dBm
VCO Phase Noise @ 100 kHZ	-103dBc/Hz
VCO 1/f Phase Noise Corner	30 kHz
Current Consumption	14 mA
f0 matching	3%

Table 4 . Performance of the circuit of Fig.10

4. Conclusions

Technology scaling offers the potential to increase the frequency range of application of different analog filter techniques. Analog performance like DR or precision all tend to degrade as the feature size is reduced. Ways to overcome some of these limitations with the objective to preserve performance while increasing speed have been explored. In particular the concept of discrete tuning in gm-C filters and of master-slave tuning in LC selective amplifiers have been introduced. Extrapolated performance for realistic design targets indicate the fundamental feasibility of these concepts. Many practical problems e.g. Q variation with discrete tuning in gm-C filters and cross-talk in master-slave LC filters have not been addressed in any detail in this paper. More work on these areas is still required to change these concepts into actual products.

References

[1]R. Castello "Low voltage continuous time filters" in Advances on Analog Circuit Design, editors W. Sansen, H. Huijsing, R. van der Plassche, Kluver Academic Publishing.

[2]Y. Mehr, D.R. Welland "A CMOS Continuous-Time Gm-C Filter for PRML Read Channel Applications at 150 Mb/s and Beyond" in IEEE *J. Solid-State Circuits*, vol. 32, pp. 499-513, April 1997.

[3]A. Kral, F. Behbahani, A. Abidi "RF-CMOS Oscillators With Switched Tuning" in IEEE *CICC Dig of Tech. Papers*,, pp. 555-558, 1998, Santa Clara.

[4]R. Alini, A. Baschirotto and R. Castello "Tunable BiCMOS Continuous-Time Filters for High-Frequency Application" in IEEE *J. Solid-State Circuits*, vol. 27, pp. 1905-1915, December 1992.

[5]W. Dehaene, M.S.J. Steyaert and W. Sansen "A-50MHz Standard CMOS Pulse Equalizer for Hard Disk Read Channels" in IEEE *J. Solid-State Circuits*, vol. 32, no 7, pp. 977-988, July 1997.

[6]H. Khorramabadi, M.J. Tarsia, N.S. Woo "Baseband Filters for IS-95 CDMA Receiver Applications Featuring Digital Automatic Frequency Tuning" in *ISSCC Dig of Tech. Papers*, Feb. 1996, pp. 172-173.

[7]R. Castello, P. Erratico, S. Manzini, F. Svelto, "A ±30% Tuning range varactor compatible with future scaled technologies," in *VLSI Symp. on circuits, Dig. Papers*, pp.34-35.

[8]T. Soorpanth, C.P. Yue, D. Shaeffer, T. Lee, S. Wong, "Analysis and optimization of accumulation mode varactor for RF ICs," in *VLSI Symp. on circuits, Dig. Papers*, pp.32-33.

[9]N. Rao, V Balan, R. Contreras, "A 3V 10-100MHz Continuous-Time Seventh-Order 0.05° Equiripple Linear-Phase Filter" in *ISSCC Dig of Tech. Papers*, Feb. 1999, pp. 44-45.

Analog *RC* Polyphase Filter and Mixer Design for Large Image Rejection

F. Behbahani and A. A. Abidi

Electrical Engineering Department
University of California
Los Angeles, CA 90095-1594

Introduction

Achieving adequate image rejection poses the major obstacle to full integration of a superheterodyne wireless receiver (RX). Zero-IF RX's are attractive for full integration because the image of one half of a channel is the other half of the same channel. Thus, it is sufficient to reject the image by, say, 15 dB more than the final SNR_{min} required. Also, the receiver power dissipation is potentially lower because the subsequent amplification and filtering is at baseband. However, zero-IF suffers from several drawbacks such as DC offset and flicker noise, which cannot be easily filtered if the modulated spectrum at zero IF contains sensitive information at DC. The low-IF architecture solves this problem by downconverting the desired channel to frequencies beyond the flicker noise corner. The IF amplifiers and filters operate at frequencies substantially the same as in a zero-IF receiver, but now the image consists of some other unrelated channel two times IF away in frequency, which may be substantially larger than the desired channel. All of a sudden, an image rejection of up to 60 dB becomes necessary.

The image may be rejected by filtering prior to downconversion, or by signal cancellation. Active filters with sufficient selectivity and dynamic range are impractical at high frequencies. Alternatively, the image is cancelled by mixing quadrature phases of RF with the LO, or *vice-versa*, followed by phase shift at the IF. The ultimate image rejection is limited by the *quadrature accuracy* of the mixer input phases, *gain matching* of the mixers, and the accuracy of the *IF phase shift*. The ultimate repeatable image rejection of a simple quadrature mixer is limited to about 40 dB without tuning.

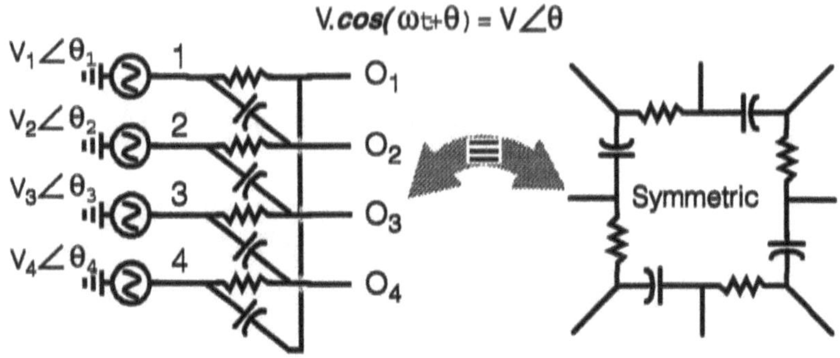

Fig.1. The passive *RC* polyphase filter.

An interesting extension of the basic structure is the ***double-quadrature architecture*** [1]. It uses four mixers with both RF and LO inputs in quadrature phases.This structure is shown to be much ***less sensitive*** to the unbalance in the phase and amplitude of the RF and LO inputs of the mixers. The image rejection is now limited by the gain mismatch between mixers, and by inaccuracies in the IF phase shifter. It will be shown that a carefully designed polyphase can provide IF phase shift with sufficient accuracy to obtain 60 dB image rejection repeatably.

Passive Analog Polyphase Filters: Principle of Operation

The polyphase filter is a symmetric *RC* structure with inputs and outputs symmetrically disposed in relative phases. Figure 1 shows the example of a four input-output polyphase. Consider four input sinewaves at the same frequency but with arbitrary amplitudes and phases, represented by four phasors as shown in Figure 2. The input phasor set can be decomposed into four orthogonal components [2]: quadrature clockwise, quadrature counter-clockwise, collinear differential, and the collinear in-phase. As shown later, the collinear components appear as common mode at the differential outputs and are cancelled.

Figure 3 shows the response of the filter to the two sets of phasors. If the input frequency is at the *RC* pole, the input vectors experience +45° and -45° phase shift. The outputs of the branches add constructively (destructively) for counter-clockwise (clockwise) quadrature inputs. Thus, the polyphase filter rejects the counter-clockwise, collinear differential, and collinear-inphase vector sets and only passes the clockwise set at the *RC* fre-

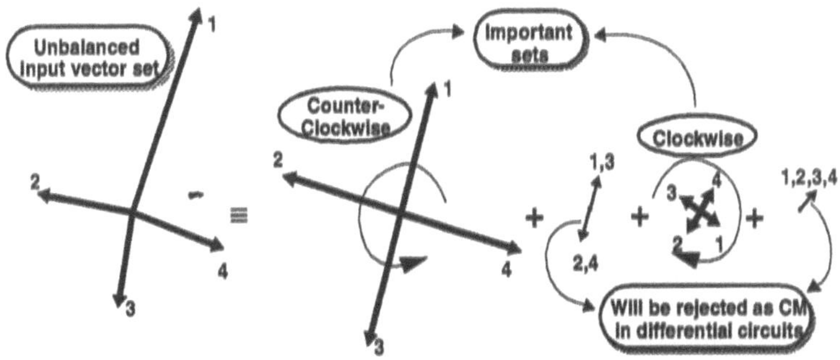

Fig.2. Decomposing four arbitrary phasors into two quadrature sequences, and two collinear sequences.

quency. The collinear phasors are rejected at all frequencies, whereas at frequencies other than the *RC* pole rejection of the counter-clockwise sequence degrades.

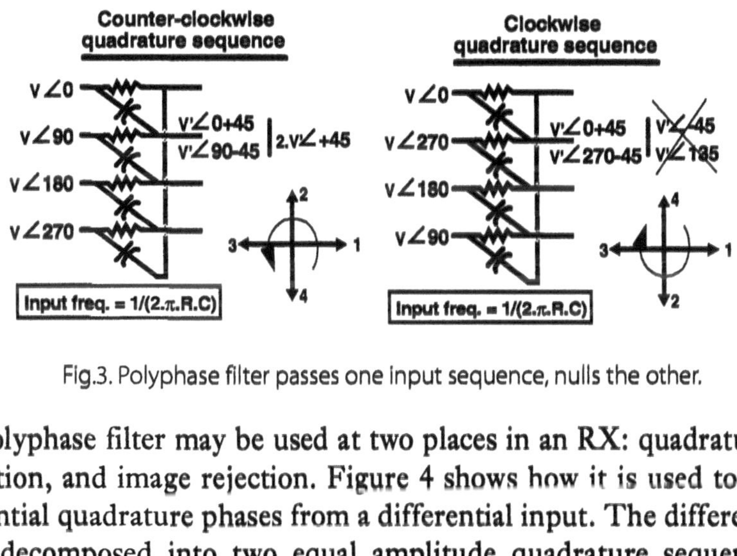

Fig.3. Polyphase filter passes one input sequence, nulls the other.

The polyphase filter may be used at two places in an RX: quadrature phase generation, and image rejection. Figure 4 shows how it is used to generate differential quadrature phases from a differential input. The differential signal is decomposed into two equal amplitude quadrature sequences, one clockwise and the other counter-clockwise. Assuming the input signal is at the *RC* pole, the polyphase rejects the clockwise sequence and only the counter-clockwise sequence, which is in perfect quadrature, survives at the output.

To implement image rejection , the downconversion mixer maps the image and desired signals into two opposite sequences, and after passing the

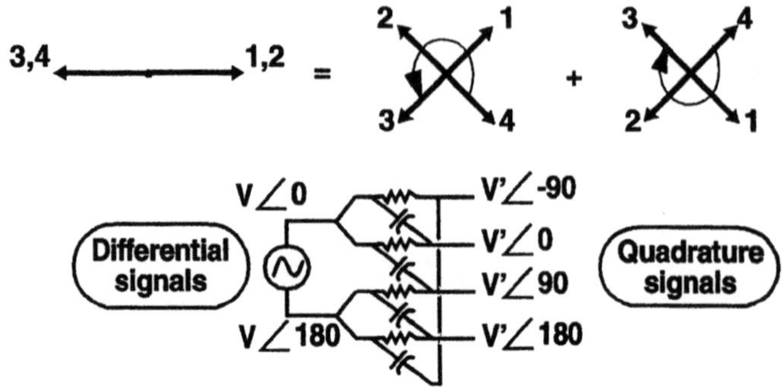

Fig.4. Quadrature generation using RC polyphase filter.

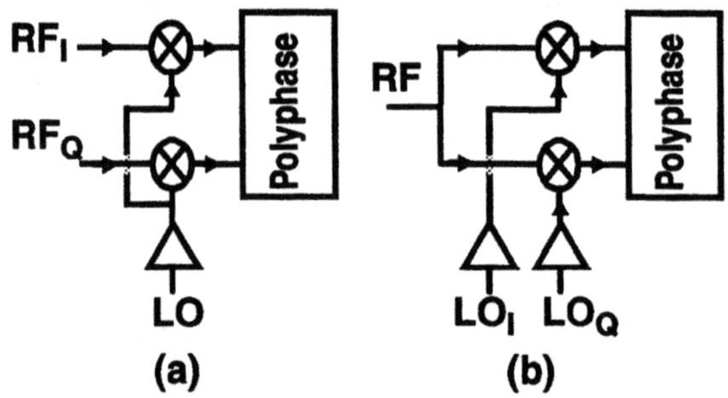

Fig.5. Two arrangements for single-quadrature downconversion mixers.

polyphase filter the sequence carrying the desired signal survives. Figure 5 shows two conventional single-quadrature down-conversion block. Using either quadrature RF signal (Figure 5a) or quadrature LO signal (Figure 5b), the downconverted IF signal would be in opposite quadrature sequences for the desired and image signal. The image signal can be rejected by the polyphase.

Practical Design Considerations

This section addresses several practical design issues for an RX polyphase filter.

a) Bandwidth: The polyphase filter perfectly rejects an input sequence at only the *RC* pole frequency. Several stagger-tuned stages of the polyphase filter may be cascaded for wideband rejection. Figure 6 shows an example of

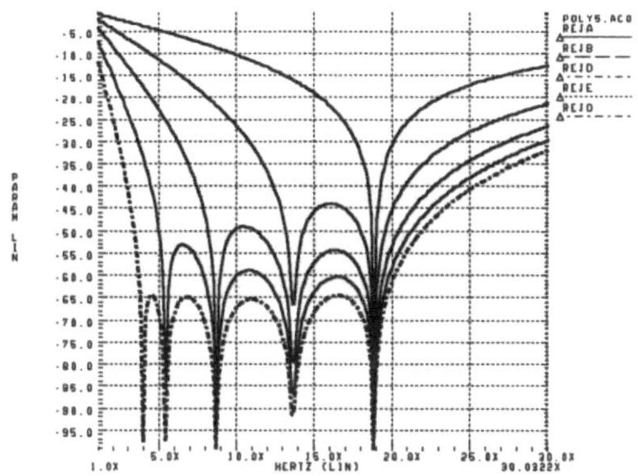

Fig.6. Example of five-stage stagger-tuned polyphase response to rejected sequence.

image rejection through a five-stage polyphase filter. Logarithmically spaced pole frequencies produce an equiripple response [3].

The larger the desired image rejection, or the higher the ratio of maximum to minimum signal frequency, the larger the number of polyphase stages needed (Figure 7). In practice, the on-chip RC time constant can vary ±25%

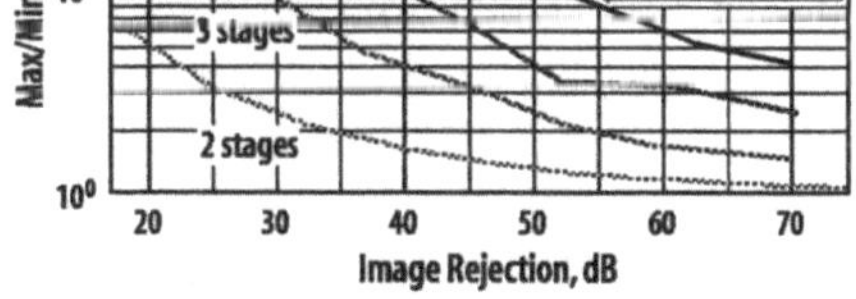

Fig.7. Frequency band of image suppression, versus number of logarithmically stagger-tuned stages cascaded in polyphase filter.

from lot to lot. Thus, the required bandwidth of the polyphase filter is actually the signal bandwidth with ±25% margin.

b) Component matching: Perturbations in the transfer function of each branch of the polyphase filter means that the phasors representing the image signal no longer cancel exactly. It can be proven that

$$\sigma(1/\text{Image Rejection}) \approx 0.25 \times \sigma(R, C) \quad (1)$$

Thus, to obtain 60 dB image rejection with 3σ yield, 0.13% matching is required among the filter resistors and among the capacitors. The variance of adjacent resistors and capacitors [4] on-chip is proportional to the inverse of their physical area.

c) Resistor cut-off frequency: On-chip resistors suffer from distributed capacitance to the substrate (Figure 8a). Associated with the resistor is a cutoff frequency

$$f_{Cutoff} = \frac{1}{2\pi L^2 \rho C_{OX}} \quad (2)$$

where ρ is the sheet resistance, L the resistor length, and C_{ox} the substrate capacitance per unit area. As a result the signal experiences phase shift

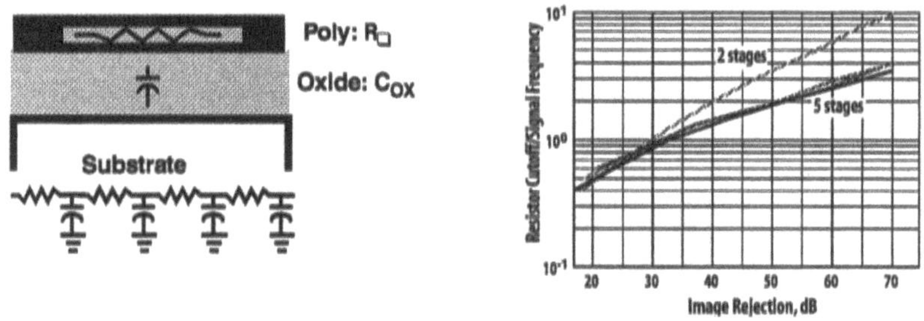

Fig.8 (a) Physical structure of polysilicon resistor. (b) Limits to image rejection as signal frequency approaches resistor cutoff.

within the resistor at high frequencies. This disturbs the intended phase shift through the filter branches and limits the achievable image rejection. Figure 8b shows the minimum ratio between the cut-off frequency to the highest pole frequency and the corresponding bandwidth for 2 to 5 stages of polyphase filters. From (2), the required cut-off frequency sets the maximum resistor length.

d) Polyphase voltage gain and loss and frequency response: Depending on the loading of the following stage, the desired signal traversing the polyphase experiences either gain or loss. As Figure 9 shows, with no loading and at the

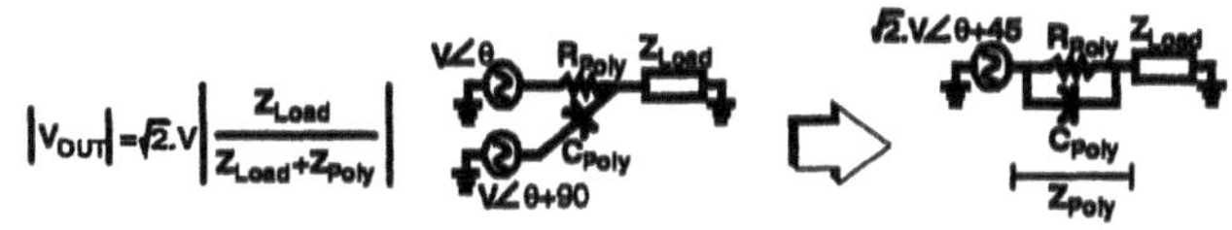

Fig.9. Signal summation at polyphase filter output.

RC pole frequency, signals reaching each output node are in-phase and add together in voltage (6 dB gain). However, within each branch the input signal experiences 3 dB loss through the *RC* or *CR* series combination, which results in a net voltage gain from each input node to the corresponding output node of 3 dB. When an identical stage is cascaded with no buffering, the resulting voltage division modifies the 3 dB of gain to 3 dB loss. The lower the load impedance, the higher the loss the desired signal suffers per stage.

On the other hand the relative image suppression is independent of the load, as long as the load is balanced at all four outputs. This can be simply deduced from the fact that a transmission zero is responsible for the null in the image sequence, and this zero is independent of the load.

If there are no grounded loads at the polyphase filter outputs, either the resistors or the capacitors transmit the input to the output with no loss at DC and very high frequencies, respectively, which results in 0 dB voltage gain. However, in practice, one filter section is loaded by either another stagger-tuned filter section, or the capacitive load of a differential pair, and always by the bottom plate parasitic associated with the filter capacitors. In either case there is no resistive path to ground, which means that the DC voltage gain is still 0 dB. However at higher frequencies there is voltage division between the polyphase and the load capacitor which produces loss.

The cascade loss of a multi-stage polyphase filter to the desired signal must be minimized. The total loss is set by tapering the resistance of each stage (and the capacitors accordingly to maintain the same pole frequencies), thus limiting the load of each stage on the previous stage. To reduce the total loss, the stages use progressively larger resistors, with the largest value in the last stage. This lowers the loading of each stage on the previous one. However, for a given resistance in the final stage, the lower resistance in the first stage of a long cascade will more heavily load the input source.

e) Polyphase input impedance: The input impedance of a polyphase filter is strongly frequency dependent. This poses a problem when it is driven by a non-zero source resistance.

In a multi-stage polyphase with capacitive loads at the output, there is no path to ground at DC, which implies a very high input impedance. At the pole frequency of the first stage, the input impedance is $R\|(1/j\omega C)$. At very high frequencies, the load capacitance to ground in series with the polyphase impedances determines the input impedance. Thus, the input impedance transitions from infinite, to $R/(1+j)$, to capacitive. When a transconductance stage, such as a Gilbert mixer, drives the filter with a wideband input, low frequency signals experience a much larger voltage gain than the desired signal which lies at the pole frequency. In a low-IF RX the low frequency signals may be much stronger interferers than the desired signal at the pole frequency. The large interferers subject to the large gain can quickly saturate the receiver. Therefore, the polyphase filter must be driven by a low source impedance to obtain a relatively flat frequency response which does not unnecessarily boost off-channel signals.

f) Polyphase filter noise: All resistors in a multi-stage polyphase filter contribute output noise. However, just as the desired signal, noise originating in the first few stages experiences loss traversing the filter. If all stages use equal resistors (and different capacitors for stagger tuning), the last stage would dominate the output noise. Therefore, the lower the final stage resistance the lower the noise, but the higher the load on the previous stage.

g) Polyphase filter output: The four outputs of the polyphase filter are differential quadrature phases (I_0, Q_0, $\overline{I}_0$, $\overline{Q}_0$).. However, in practice, the input of the circuit following the polyphase filter, usually an amplifier, is differential The output differential signal can be obtained from the polyphase output in two possible ways, shown in Figure 10. The first method is to short I_0 to Q_0 and $\overline{I}_0$ to $\overline{Q}_0$. The second is to take the output signal just from I_0 and $\overline{I}_0$. In the former, quadrature signals from the I and Q branches add together (3 dB voltage gain), but each branch loads the one connected to it (6 dB voltage loss). This results in a net loss of 3 dB. The second method, by contrast, does not suffer this loss but the quadrature filter outputs are loaded asymmetrically. This imbalances the load on the last filter stages, degrading image

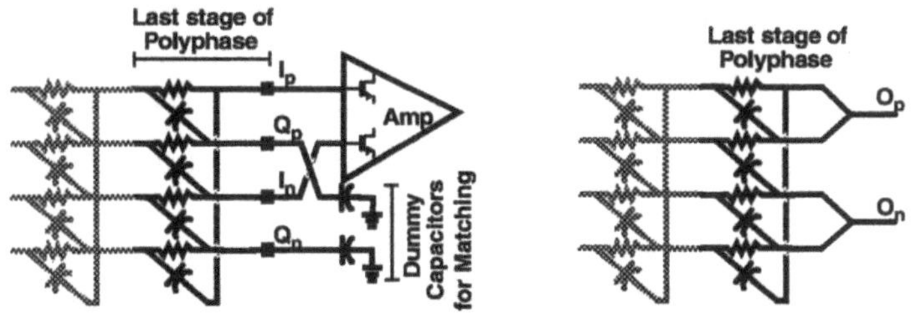

Fig.10. Two possible ways of terminating a polyphase filter.,

rejection. Dummy capacitors might be used for to balance the output load, but the matching will not be perfect in practice.

Achievable image rejection is limited by mismatch in load capacitance, depending on the ratio between the load and polyphase capacitance. If the load capacitance is smaller than the polyphase capacitance, the spread in *ratio* of the load capacitance and the polyphase capacitance must obey Eq 1. If it is comparable or larger, the fractional spread of the load capacitors *alone* should obey Eq 1. In general, the configuration of Figure 10b is much less sensitive to the load capacitance mismatch than Figure 10a.

Polyphase Filter Design Guide

This section enumerates steps in the design of the polyphase filter to obtain a given rejection response.

Step 1) The number of stagger-tuned stages required in the polyphase filter is calculated from the target depth of image rejection and the fractional bandwidth over which this rejection is desired.

Step 2) As an initial guess, the two lowest and highest poles are set at the boundaries of the rejection band. The remaining poles are equally spaced on a logarithmic axis. The actual pole locations may be fine tuned with simulations.

Step 3) The desired image rejection determines the required matching between the resistors and between the capacitors. The variance in passive components is inversely proportional to their physical area.

Step 4) Large resistors should be used to minimize the power consumption of the amplifier driving the polyphase filter input. The resistance is limited by the maximum noise at the output and the cut-off frequency. The former is the limiting factor in low noise or low frequency circuits, and the latter for high noise or high frequency polyphase filters.

Step 5) In polyphase filters with a large number of stages in cascade, the resistance should be tapered down towards the input to lower cascaded loss. Loading on the driving amplifier limits this.

Step 6) Design the driving amplifier, and if the cascade loss is too high, inter-stage amplifiers as well to preserve signal dynamic range within the polyphase filter.

Downconversion Mixers

So far the discussion has centered on the polyphase filter. Now the mixer circuits responsible for orienting the desired signal and the image in opposite sequences are considered.

As the output of the mixer must be in vector format to distinguish the down-converted desired signal from the image, either the RF input to the mixer must be a vector (in quadrature phases) or the LO input. Figures 11 (a) and (b) show two well-known downconversion architectures, labelled single-

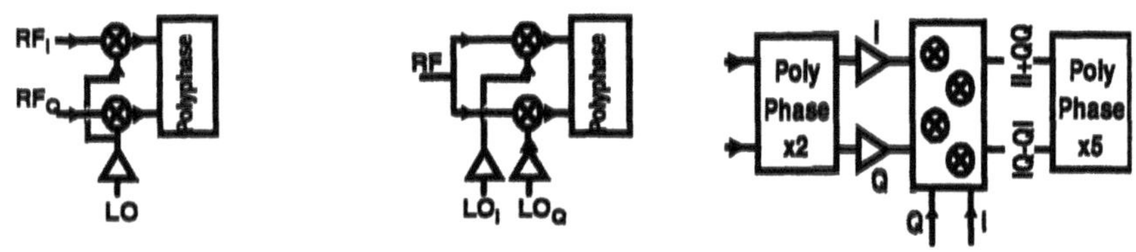

Fig.11. Single-quadrature mixers (a) and (b); Double quadrature mixer (c).

quadrature, and in Figure 11 (c) an extension which is termed double-quadrature. Multiplying a differential set of phasors with a quadrature set of phasors produces another quadrature set at the translated frequency. After downconversion, the desired signal and the image produce opposite sequences. This is shown in Figure 12 . For quadrature LO, differential RF (Figures 11b,12b), the output sequence is opposite of (the same as) the input sequence, if the LO is high (low)-side injected. For quadrature RF and dif-

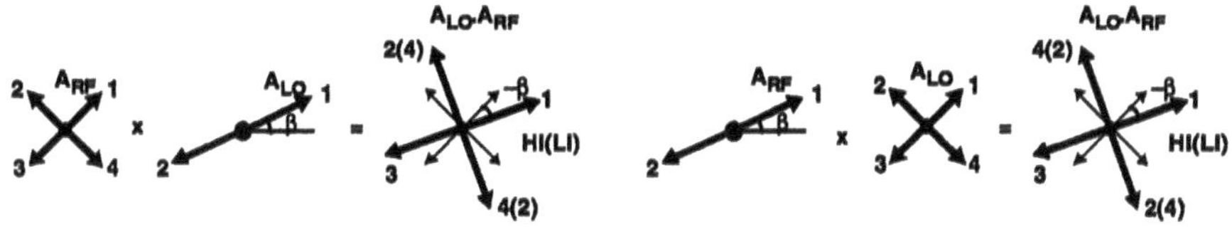

Fig.12 Phasor sequences at the inputs and outputs of the two single-quadrature mixers.

ferential LO (Figure 11a,12a), the opposite is true. Therefore, a polyphase filter with appropriate input connections following the mixer selects the desired signal and suppresses the image.

Now consider the effect of imperfections on the image. Suppose that either the RF input or the applied LO, whichever is appropriate, is not in exact quadrature. The actual (non-quadrature) phases can be decomposed into an exact desired sequence, and a small amplitude opposite sequence representing the error. Whether the error is in the LO or in the image RF input, after mixing the *error* sequence in the *image* will appear in the *same* sequence as the downconverted *desired* signal. As a result, this portion of the downconverted image will feed through the polyphase filter with the desired signal. The image rejection is therefore the ratio of the amplitude of the desired sequence to the error sequence. Exactly the same analysis applies if the amplitude of the quadrature input is unbalanced.

At high frequencies, the effect of the parasitics and the resistor cut-off frequency makes it difficult to supply the mixer with quadrature accuracy better that 1%, which limits image rejection to 40 dB. Simple *RC-CR* circuits, polyphase filters, quadrature oscillators, and divide-by-2 stages are the most common blocks used to generate quadrature LO, however only the first two may convert RF to quadrature. In all cases, the required matching between the mixer gains is the same as the required accuracy for the input quadrature signals. This is because mismatch between mixers may be modeled as perfectly matched mixers and inaccurate quadrature input signals to the mixers.

For image rejection higher than 40 dB, the double-quadrature down-conversion architecture is a better choice. This architecture uses four mixers, where both LO and RF input signals are in quadrature, and the outputs of the mixers are combined as $IF_I=RF_I.LO_I+RF_Q.LO_Q$ and $IF_Q=RF_I.LO_Q-RF_Q.LO_I$. It has been noted before, and is proven below, that quadrature inaccuracy in the LO and RF limits image rejection to the 2nd-order. For instance, to achieve 60 dB image rejection only 3% quadrature accuracy is required for the RF and LO inputs instead of 0.1% in the case of single quadrature down conversion.

To understand this property, it should be first recognized that this combination yields *two single-sideband mixers* whose IF outputs are in relative quadrature. If the mixer pairs match in gain, they produce an IF output at *either* the sum *or* the difference frequency in response to a particular combination of input sequences. It is reasonable to assume that the desired and the image signals, because they are nearby in frequency ($\omega_{IF} << \omega_{LO}$), after passing through some phase shifter take on the same input sequence (RF_{Id}, LO_{Id}). Then it is straightforward to see that after downconversion they assume opposite sequences, one of which passes the subsequent polyphase filter and the other does not. So far, this is exactly the same property as the single quadrature downconversion mixer. What is different is how errors in input quadrature or amplitude are handled. In both cases, the error is represented by the opposite sequence (RF_{Er}, LO_{Er}) (Figure 13). It is a property of the

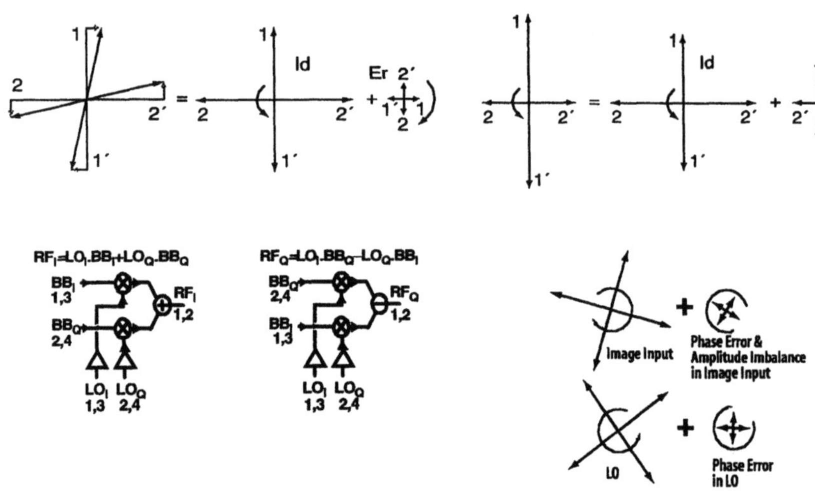

Fig.13. (a) Phase or (b) amplitude errors represented as superposition of ideal and error sequences. (c) Decomposition of double quadratuure (DQ) mixer. (d) Ideal and error sequences at the RF and LO inputs of DQ mixer.

single-sideband mixer that reversing the sequence at a given input frequency shifts the output from the difference frequency to the sum frequency, or *vice-versa*. Thus in a downconversion mixer, the error sequence in the RF input mixed with the correct sequence of the LO, or *vice-versa*, induce a mixer output at the sum frequency, which is rejected by filtering. Finally, what remains is the 2nd-order product of the error sequence in RF and the error sequence in the LO. This induces a mixer response at the difference frequency, in the same sequence as the desired signal. Therefore the image transmission through the polyphase filter is proportional to the ***product*** of the quadrature plus amplitude errors in the RF and LO inputs.

$$\text{Image Rejection} = RF_{Er}.LO_{Er}/RF_{Id}.LO_{Id} = (RF_{Er}/RF_{Id}) \times (LO_{Er}/LO_{Id}).$$

So far it was assumed that the four mixers match perfectly. Gain (or delay) mismatch in the mixers will unbalance the output (IF) sequence, which too can be decomposed into ideal and error sequences. As the error sequence for the image is oriented the same way as the ideal sequence for the desired signal, it will pass the subsequent polyphase filter. Thus, the image rejection in the double quadrature architecture is ultimately limited by matching of the four mixers.

Mixer Blocks

The choice of mixer circuit topology depends on the required image rejection and the RF. At high RF, it is difficult to match differential pairs without excessive power consumption. Therefore, for large image-rejection passive MOS switch mixers may be superior to the Gilbert cell type. The size of the switches may be chosen large enough that the on-resistance of the switch is much smaller than the polyphase resistor, with which the switch is in series. When the switches are on, the total mixer matching is then determined by the well-matched series resistors. However, in the transition from "off" to "on" the switch resistance dominates. Over the transition mismatches in threshold voltage and switch channel resistance will show as mismatch in the average mixer conversion gain. However, an LO waveform with sharp edges alleviates this problem because the mismatches appear over a smaller duty cycle. This is easily implemented at lower LO frequencies by driving the mixers with fast CMOS inverters.

A FET switch is non-unilateral. This has an important practical consequence in the double quadrature mixer. Assuming square LO waveforms, at any time two waveforms among LO_I, LO_Q, $\overline{LO_I}$, and $\overline{LO_Q}$ are high. Without loss of generality, consider LO_I, LO_Q both high as in Figure 14. Then

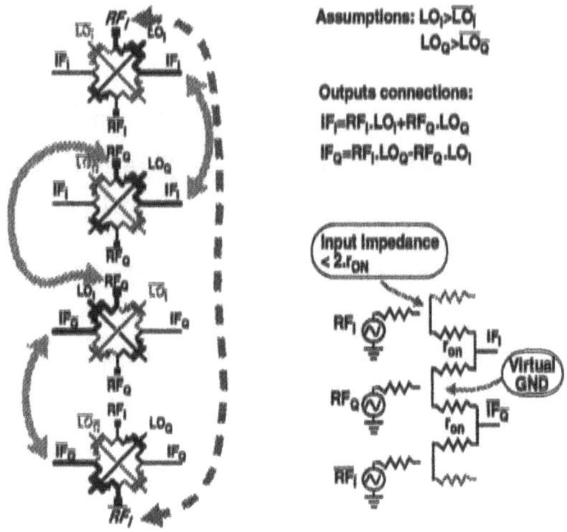

Fig.14. Passive FET mixers short two inputs to the double quadrature mixer at all times.

the two differential inputs, RF_I and $\overline{RF_I}$ are shorted through four switches. This means that each input source is shunted by $2r_{ON}$, which will almost always seriously load the driving stage and cause a very large conversion loss. An effective way to overcome this problem is to separate the four mixers into two sets of single-quadrature downconverters, divide the polyphase filter into two identical halves on a common centroid, and then connect each half of the multi-stage polyphase to each mixer output. The outputs of the two half circuits are finally connected together after the last polyphase filter section.

Amplification within Polyphase Filter

The branches of the polyphase filter match best when they only comprise resistors and capacitors. However, in a wideband multi-stage polyphase filter the cascade loss may accumulate to the point that interstage amplifiers must boost the signal if it is not to be overcome by noise. The amplifiers in each branch must match as well as the polyphase filter components. This requires large area MOSFETs and careful layout. For low frequency amplifiers the area may be enlarged by using long-channel FETs, while keeping W/L constant. However, in high frequency amplifiers the required f_T for the devices may constrain L to the minimum feature size, and FET area can only be enlarged at the expense of W/L and thus higher current drain.

For obvious reasons the amplifier must have a wide dynamic range, but to drive the highly frequency-dependent input impedance of the polyphase filter its output impedance must also be small. Figure 15 shows an amplifier

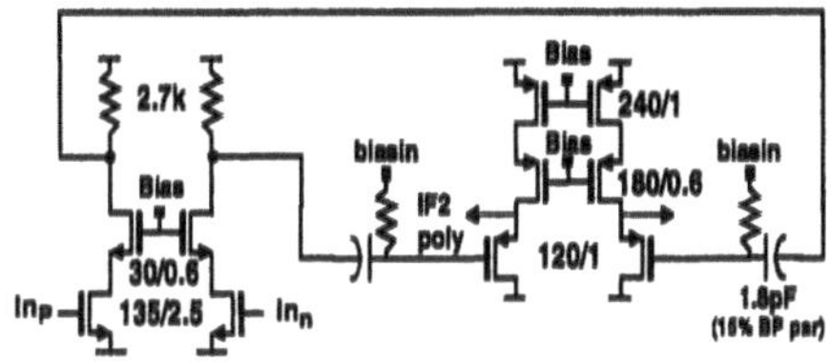

Fig.15. Interstage amplifier inserted within polyphase filter to compensate cascade loss.

circuit well-suited to this purpose. The grounded sources guarantee wide dynamic range for the input devices. The cascodes suppress the dominant nonlinearity of the MOS output impedance, and the source follower provides the required low output impedance to drive the amplifier.

The capacitor coupling between stages serves a larger purpose than merely a level shift. Due to the bottom-plate parasitic capacitance associated with the branch capacitor in the polyphase filter, the transfer function to the desired input sequence droops at high frequencies. For instance, in a 5 stage polyphase whose bandwidth extends from 4 to 20 MHz, the passband droop can be as large as 6 dB. The zero introduced by the inter-stage coupling capacitor partly compensates this droop.

Layout Considerations

Careful layout is very important to match the polyphase filter elements to the required degree. Physical area defines the matching among the adjacent resistors and the capacitors. All the R's and C's in one stage are laid out in the same orientation. Dummy resistors and capacitors covering at least a 50-μm distance are laid-out at the sides of the polyphase filter (Figure 16) to shield against edge effects during fabrication. Good matching for large image rejection mandates use of physically large resistors and capacitors, which widens the layout of the entire polyphase filter. This region can now suffer from processing gradients. To alleviate linear gradients, the polyphase

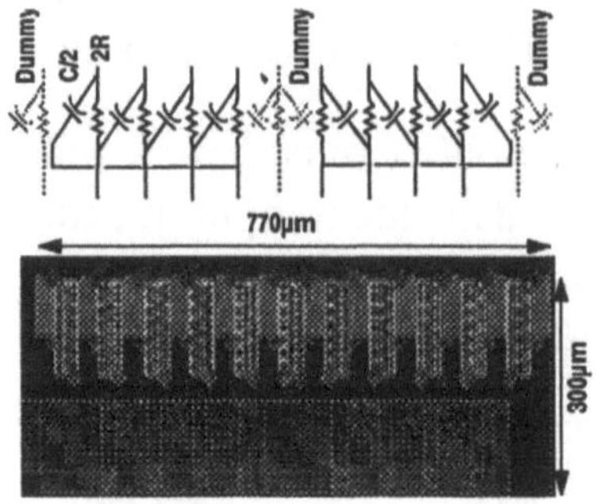

Fig.16. Layout of each stage of the polyphase filter on a common-centroid, to protect against gradient effects.

filters are laid out in common-centroid, which however makes the interconnections complicated and long, with many lines crossing each other. In general the interconnect lines are of different length, and they cross each other asymmetrically. To balance contributions of line resistance, the interconnect lengths are equalized. As shown in Figure 17a, one solution is to use serpen-

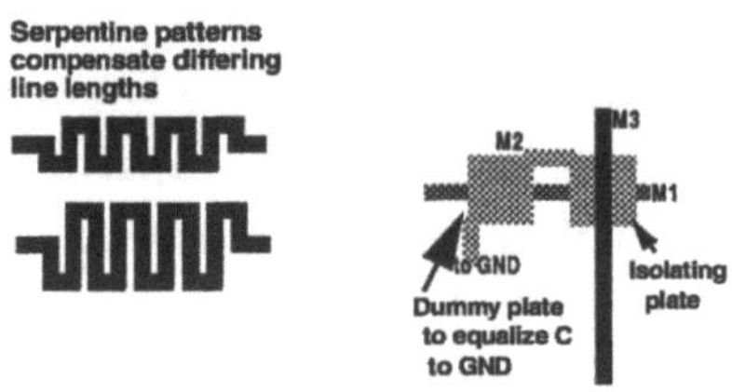

Fig.17. Interconnect used in polyphase filter layout. Serpentine patterns balance wire length, shields terminate crossover capacitance to ground.

tine shapes as part of all the parallel interconnections. By adjusting the height of the serpentine, parallel interconnection lengths may be equalized while keeping the same number of corners. Another problem is in the asymmetry in crossover capacitances. If the process has three or more metal layers, a grounded isolating metal plate can be used between the signal lines, as shown in Figure 17b. This converts the asymmetric inter-line capacitance to symmetric line-to-ground capacitance. Two parallel lines should be placed far enough apart so that the interline capacitance is low enough.

One gradient-compensated stage of the polyphase filter has 8 capacitors and 8 resistors with complicated interconnection. Considering the required distance between the parallel lines, the area of the interconnection connecting two sides of the polyphase could be more than two times of the component area. It is much more area-efficient to connect two halves of the polyphase filters at the input of the first stage and finally the output of the last stage (Figure 18).

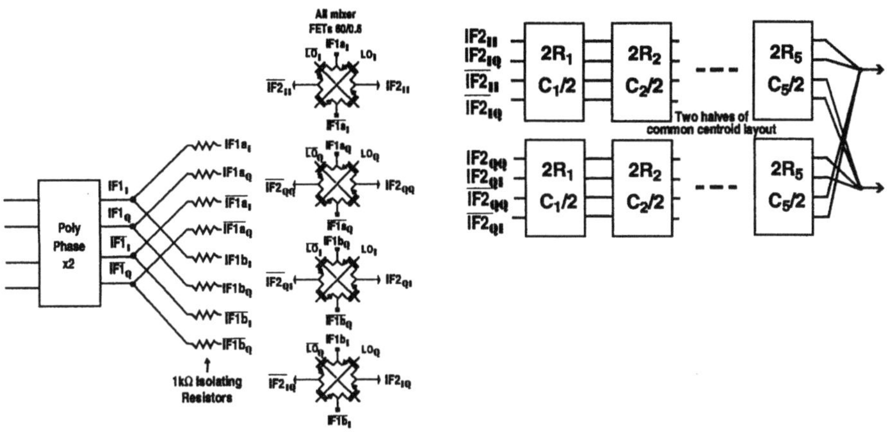

Fig.18. Actual signal path through mixers and polyphase filter.

Design Example

To verify these concepts, a standalone downconversion prototype with an input frequency of 240 MHz and an IF ranging from 5 to 15 MHz has been designed, with a target image-rejection of 60 dB across the full 10 MHz IF band (Figure 19). Following the discussion above, it uses the double-quadra-

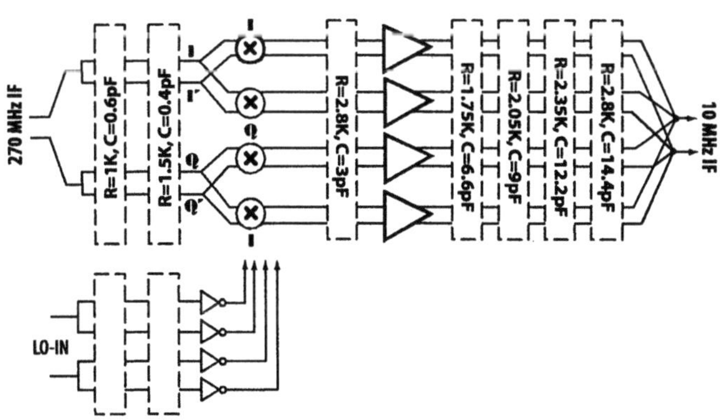

Fig.19. Block Diagram of Image Reject Downconverter prototype.

ture architecture to reduce sensitivity to quadrature errors in the RF and LO mixer inputs. For 60 dB image rejection, 3% accuracy is enough. These quadrature signals are generated in two-stage polyphase filters tuned to 240 MHz. A simple non-common centroid layout with small area components would have been appropriate, but as no information was available regarding on-chip process gradients, a common-centroid layout was used here as well.

The downconversion mixers and the IF polyphase filter must match to 0.1%. The passive 4-FET mixers are padded with polysilicon resistors. When hard-switched with very short transition time, the passive resistors determine mixer matching rather than the switch resistance. Fast inverters with small rise- and fall-time used as LO buffers guarantee sharp transitions.

Five stagger-tuned stages of the polyphase filter meet the requirement for image rejection across the 10 MHz-wide band. To allow for ±25% process variation, the image-rejection bandwidth is extended from 3.5 to 20 MHz. Test resistors were first fabricated to measure matching characteristics, and it was found that the resistor area must be 2400 μm^2 to match to 0.1%. The capacitors must be at least 0.5 pF to match to the same accuracy.

The desired signal traversing the many cascaded lossy blocks of the passive RF polyphase filter, the padded passive mixers, and the passive IF polyphase filter suffers a 25 dB loss. An amplifier is inserted to make up for part of this loss. The optimum position of the amplifier is when it cuts the loss chain in the middle, that is, after the first stage of the IF polyphase filter. The impedance level of the four remaining stages of this filter is tapered to reduce the post-amplifier loss to 10 dB, and the resistors in the last stage set the differential output noise to 7 $nV/\sqrt{Hz}$. Using Monte-Carlo simulations with published data on FET matching [5], the transistor sizes in the amplifier are determined to obtain the required matching.

The final chip layout is shown in Figure 20(a). Figure 20(b) plots the image-rejection measurement results for various samples. It is notable that the image rejection is very close to the simulated response. The average image rejection is 58.5 dB. The differential input referred noise is 26 $nV/\sqrt{Hz}$ with IIP3 of 25 dBm. The bias current of the CS, SF, and LO driver buffers are 6, 3.5, 4 mA respectively, from 3.3 V. In this prototype, these are the only active components.

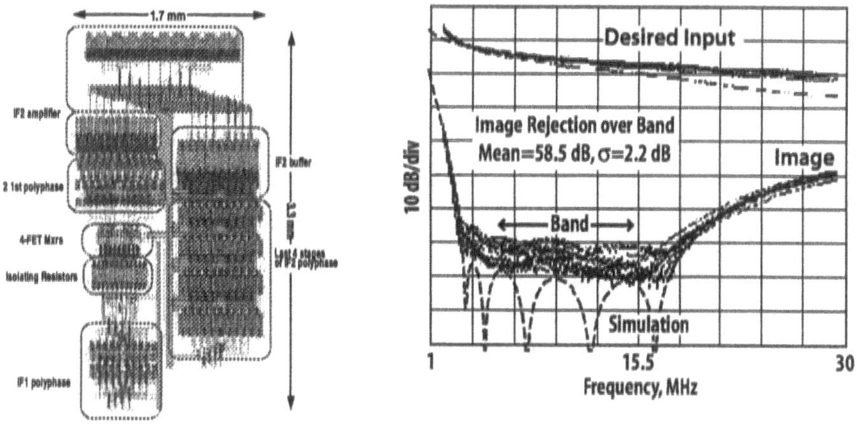

Fig.20. (a) Image Reject Downconverter prototype. (b) Measured Image Rejection in prototype, including composite results from 12 test chips.

Conclusions

The operation of passive analog polyphase filters and downconversion mixers has been described in easy to understand terms as phasor sequences. The preferred downconversion mixer architecture has been identified, and the practical limitations to the image rejection in the combination of the polyphase filter and mixer are explained. The basics are put to test in a 0.6 μm prototype wideband image reject mixer, which is able to repeatedly achieve an unprecedented 58 dB image rejection across a 10 MHz-wide bandwidth centered at an IF of 10 MHz.

References

[1] J.Crols, M.S.J.Steyaert, "A single-chip 900 MHz CMOS receiver front-end with a high performance low-IF topology", *IEEE Journal of Solid-State Circuits*, vol.30, no.12, pp. 1483-92, 1995.

[2] M.J. Gingell, "Single sideband modulation using sequence asymmetric polyphase networks", *Electrical Communication*, vol.48, no. 1-2, pp. 21-25, 1973.

[3] R.C.V. Macario, I.D. Mejallie, "The phasing method for sideband selection in broadcast receivers," *EBU Review (Technical part)*, no.181, pp. 119-25, 1980.

[4] M.J. McNutt, S. LeMarquis, and J.L. Dunkley, "Systematic capacitance matching errors and corrective layout procedures", *IEEE Journal of Solid-State Circuits,* Vol.29, no.5, pp.611-616, 1994.

[5] M. J. M. Pelgrom, A. C. J. Duinmaijer, and A. P. G. Welbers, "Matching Properties of MOS Transistors," *IEEE J. of Solid State Circuits*, vol. 24, no. 5, pp. 1433-1440, 1989.

PASSIVE INTEGRATED RF FILTERS

Nicolas J. Pulsford

Philips Research Laboratories
Prof. Holstlaan 4, 5656 AA Eindhoven, The Netherlands

ABSTRACT

The practical application of RF integration in the front end of cellular telephones is discussed. The varying requirements of an integration technology are identified based on generic system architectures and examples of commercially available integrated components are given. A thin film passive integration technology on Si is presented and an integrated 1.3 GHz phase locked loop module is evaluated.

1. INTRODUCTION

In recent years, competition in the cellular telephone market has forced original equipment manufacturers to reduce the size, cost and weight of their products while at the same time improving the performance and features, which the telephones can offer [1]. In addition to advancements in the telephone design and architecture, the component suppliers to the cellular telephone market have also responded to the continuous pressure for innovation. In the RF front end of a cellular telephone, this innovation has been driven foremost by integration and improvements in active Si and the reduction in the circuit complexity, which this brings. For example, whereas a few years ago set makers required 3 or 4 IC's in the front end of a GSM telephone to convert the signal between the baseband and the carrier frequency, they are now able to employ a single transceiver IC to perform the same function [2]. With low noise amplifier (LNA), mixer, synthesizer and oscillator (active part) functions integrated on chip, the set maker has to employ fewer external passive components to interconnect the active IC's together and ensure trouble free operation of his handset. This process is set to continue in the near future with the advent of multiple bands of operation (for example, GSM and DCS1800) and use of near zero-IF architectures to eliminate IF components in the telephone.

This abstract studies the integration of the RF components in a cellular telephone between the antenna and the (first) transceiver IC. It evaluates the requirements of an integration technology based on the performance levels of the various generic component types found in the RF front end. As an example of this approach, a technology for the integration of passive components on Si is presented. Some examples of integration are described.

INTEGRATION REQUIREMENTS FOR RF COMPONENTS

RF component integration is attractive for cellular telephones because it inherently scores in reducing the component count and normally brings with it an associated reduction in size and weight. In addition, the integration should also aim to improve on performance and cost. These 5 parameters: cost, size, weight, performance and component count provide a useful measure for integration success.

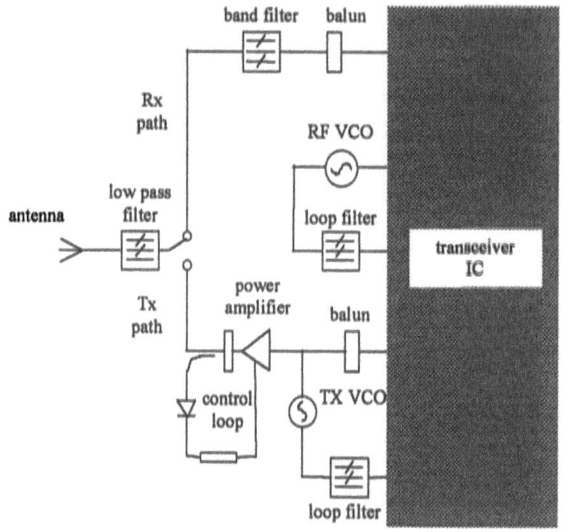

Figure 1. Generic block diagram showing the RF front end of a cellular telephone

Figure 1 shows a generic block diagram of the RF front end of a single band cellular telephone between the antenna and the (first) transceiver IC. This generic diagram is similar for both TDMA and CDMA systems as well as both IF and (near) zero-IF architectures. Telephones operating on multiple bands (e.g. GSM and DCS1800) typically have parallel Rx and Tx paths for each frequency band with a duplexer filter in front of the antenna. The generic RF functions in figure 1 are used to highlight the requirements of an integration technology for the various component functions. What is immediately apparent when working through the generic functional blocks is the wide range of different performance requirements, which they present and this in turn presents a natural barrier to integration by a single unifying technology. In general, the presence or absence of IF components are not crucial to the integration discussion because IF and RF components are from the same group of technologies.

In the receive (Rx) path, low loss, sharp filters are required to preserve the signal integrity. Telephones use filters based multilayer on ceramic and / or surface acoustic wave (SAW) technologies in the Rx path. They achieve filter characteristics, which can not be matched by discrete passive or active filters for bandwidth, insertion loss and accuracy. The balun or impedance matching network, which is often present to interface the input to the low noise amplifier (LNA) on the transceiver IC, can be made with discrete (L,C) SMD components.

In the transmit (Tx) path, efficient power modules are required to generate, amplify and control the transmit signal to the antenna. The two key functions - the power VCO (voltage controlled oscillator) and power amplifier (PA) - are typically hybrid modules containing a combination of active Si dies, (L,C,R) RF passive components, ceramic multilayer components and high value capacitors for decoupling. A balun or impedance matching network may again be present to interface to the transceiver IC.

Around the transceiver IC itself, high value capacitors are required for power supply decoupling. A RF VCO is required for the frequency mixing of the Rx and Tx signals.

In summary, a cluster of technological requirements can be defined which need to be covered to integrate a complete RF front end:

- multilayer ceramic
- SAW
- active Si
- CMC (high value C)
- NP0 C (low value C)
- L (high-Q)
- R

It is clear that no one integration technology can hope to cover all of the above technological requirements. In practice, two approaches are followed which offer alternative, but not incompatible, solutions.

The first approach is to integrate limited RF component combinations in one technology. Several examples of this type of integration are commercially available:

- SAW + balun (e.g. Murata, Matsushita [3])

SAW technology can give a balanced output eliminating the need for a separate balun function in the Rx path between the SAW filter and LNA input on the transceiver IC. This is a very efficient integration because it requires only a modest change in the SAW filter design.

- coupler + ceramic multilayer filter (e.g. Murata [4])

In addition to low loss filters, resonator, matching and coupler functions can be designed in the ceramic multilayer technology. A combination in the Tx path is the PA output coupler and filter in one component.

- R, C on active Si (most IC companies)

Active Si technologies incorporate (R,C) components which can be included on chip for simple matching and biasing functions. This is applied for reducing the number of components in VCO and PA modules.

The second approach is to apply hybrid assembly techniques to combine different technologies in the same module. This can range from simple component combinations to complete module functionality:

- low pass filter + switch (Murata [4])

A diode switch is mounted on a ceramic multilayer filter in a single package. A disadvantage of this product is the height (2.1mm) of the module.

- Low temperature co-fired ceramic (LTCC) substrates (e.g. Dupont, Kyocera, Hereaus, Motorola [5])

A high density of passive components is integrated in a multilayer substrate, which both carries active IC's and discrete SMD's and itself acts as a SMD package substrate. Generic modules include PA + matching + filtering and VCO + frequency synthesizing. Of all the commercially available technologies, LTCC offers the most complete integration possibilities.

As an alternative to LTCC substrates, research work has also focussed on the thin film integration of passive functions on Si, glass and ceramic [6-8]. The aim is to exploit the material and layer definition advantages of thin film over thick film technology for RF modules. The rest of this abstract describes work done on thin film passive integration on Si for RF modules at Philips Research.

PASSIVE INTEGRATION TECHNOLOGY ON HIGH OHMIC SI

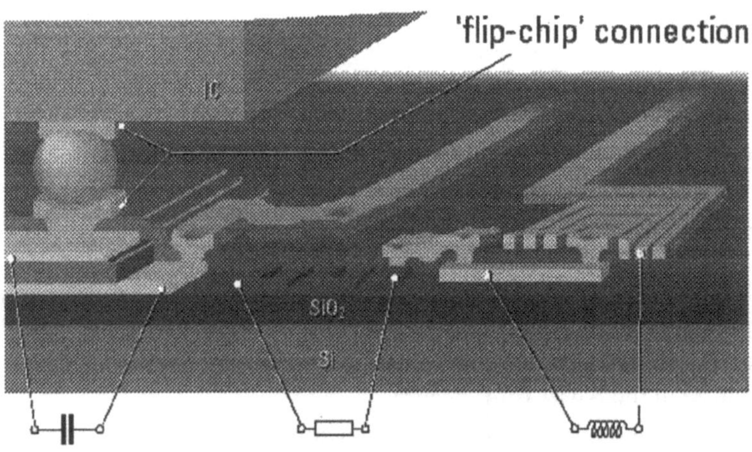

Figure 2. Cross-section of the passive integration technology on Si showing the layer build-up to form the integrated passive components.

A passive integration technology has been developed and optimized for RF applications. The basic structure of the technology is shown in figure 2. The integrated component substrate consists of a metal-insulator-metal (MIM) capacitor stack, integrated with thin film resistors. The two layers make it possible to realize not only planar inductors with a jumper (shown in cross-section), but also interconnect for the integrated components, to active components mounted on the substrate (an example of a flip-chip bump is shown) and to the outside world.

In the process flow [9], the starting substrate is high ohmic Si (ρ>3kΩcm) which has a layer of 2μm thermal oxide to isolate the passive components from the semi-insulating Si. The resistive layer is TiW(N), the bottom metal is 200nm Al, the dielectric is 200nm SiN_x and the top metal is 7μm Au grown by electrochemical deposition. The passive components are formed in these layers with the Si substrate having no active function. Indeed Si is chosen as the substrate material only for ease of processing and compatibility of packaging with active IC's. The lack of active layers in the Si also gives the freedom to optimize the technology for the passive components alone. In particular, the combination of a thick top metal (7μm Au) and a high ohmic Si substrate allows integrated inductors with a relatively high Q-factor to be designed compared to conventional Si technologies [10]. Figure 3 shows the layout and measured Q-factor of two inductors in the passive integration technology. The peak in the Q-factor with a value Q>40 close to f=1GHz makes the technology suitable for cellular telephone applications.

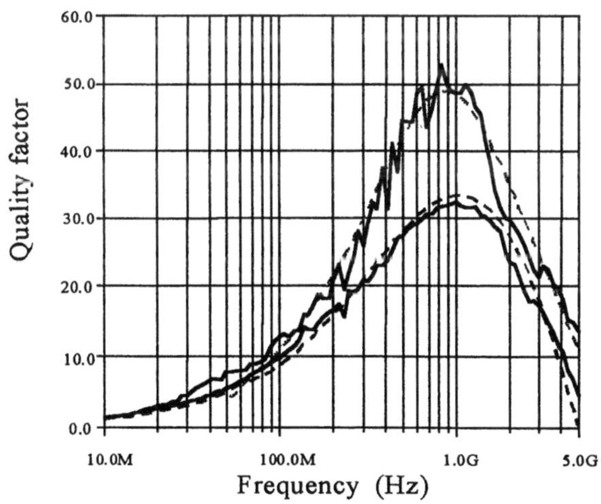

Figure 3. Measured (solid lines) and simulated (dashed lines) Q-factors for typical integrated inductors. The upper curve is for a single turn L=5nH inductor and the lower curve is for a two turn L=7nH inductor.

The dashed lines in figure 3 show the simulated Q-factor for two inductors inductors using the Philips in-house electromagnetic simulation software tool *Fasterix* [11]. *Fasterix* employs a boundary element method solution for the Maxwell equations to generate a Spice-compatible equivalent circuit for the physical layout of the integrated inductors. Extra terms are added at the equivalent circuit level to account for both the skin effect and the silicon substrate losses. There is good agreement between the measured and simulated Q-factor over a wide frequency range demonstrating the applicability of the software. A powerful advantage of Fasterix for modules combining integrated passives and active Si dies is that the equivalent circuit models for the passives can be added to the active Spice models for the active IC's to simulate the complete module performance.

PHASE LOCKED LOOP MODULE

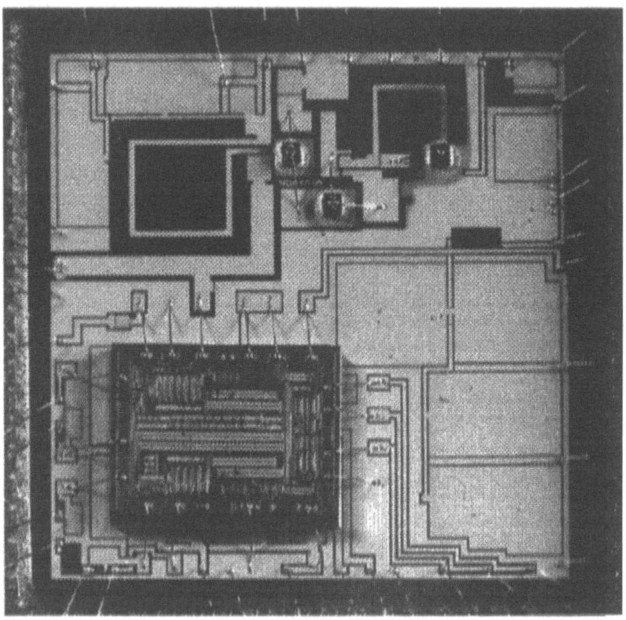

Figure 4. Photograph of the 1.3 GHz phase locked loop module

In terms of module design, the key advantages of the passive integration technology on Si are the high quality factor of the integrated (L,C) passives and the compatibility with Si packaging. As an example of the application of the technology, figure 4 shows a photograph of a 1.3 GHz phase locked loop module in an open ceramic package where passive integration and active Si dies are combined together. The passive integration substrate (8.3 x 8.3 mm) contains 55 integrated passive components. The bare dies mounted on top are a synthesizer IC and two RF transistors and a varicap. The synthesizer IC is attached with non-conducting epoxy and the three discrete dies with conducting silver glue or eutectic bonding. A number of the integrated passive functions can be identified. In the VCO (upper half of module), two single turn inductors are visible. The biasing resistors and resonant capacitors for the VCO are all integrated together with decoupling capacitors for the power supply. In the low pass filter (bottom right), the large areas are 4 x 1nF integrated capacitors. Decoupling capacitors are also integrated underneath the synthesizer IC (bottom left).

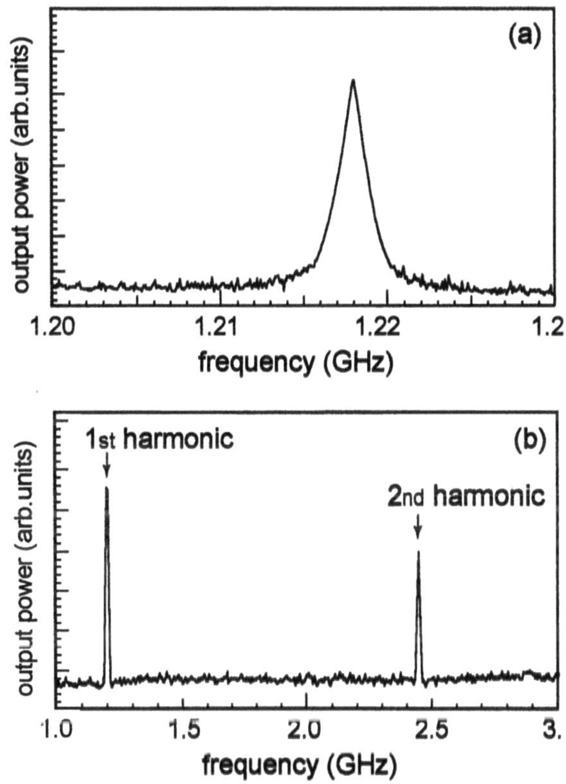

Figure 5. Output power spectrum of the free running VCO showing (a) the 1st harmonic and (b) a wider spectrum including the 2nd harmonic.

The passive integration layout for the VCO was simulated using *Fasterix* [11]. Spice models were added for the transistors and varicap, which enabled time domain simulations of the VCO performance. The measured output spectrum of the free running VCO is shown in figure 5. The principal oscillation frequency is at 1.22 GHz with the second harmonic also clearly visible. The principal oscillation is close to the simulated operating frequency (1.3 GHz) indicating the accuracy of the simulation tool with limited integrated component value spread. The measured phase noise for the VCO was –130 dBc/Hz at 3 MHz offset, which is 13 dB higher than the GSM specification. The higher phase noise factor and thus lower loaded Q-factor has been traced to an error in the electrical design of the module from one of the integrated biasing resistors.

It is interesting to evaluate the phase locked loop module as a potential product. Assuming the error in the electrical design can be solved and the required performance level can be reached, different parts of the module score differently in terms of size, weight, cost and component count. An example of a function, which scores poorly, is the low pass filter on the left of the module. This occupies ~20mm^2 of the passive integration substrate for a filter with 5 passive components (2xR and 3xC). The equivalent circuit in 0402 passive components would be nearly an order of magnitude smaller and over an order of magnitude cheaper than the integrated filter. The integration is inefficient because the high capacitance density required by the low pass filter is not achieved by the integration technology. An example of a function which integration is more effective is the VCO. The passive component density is higher (19 components in ~20mm^2) and their RF performance requirements (high Q inductors and low ESR capacitors) are well suited to the integration technology.

SUMMARY

The integration in the RF front end of a cellular telephone is reviewed and two principal trends to integration in commercial products are identified. Work on the thin film integration of passive components on Si is described and the integration example of a 1.3 GHz phase locked loop module is presented.

ACKNOWLEDGEMENTS

The author wishes to acknowledge Bob Milsom, Marc de Samber, Marc van Delden, Louis Tegelaers and Ton Wagemans for their involvement in the project.

REFERENCES

[1] see for example, G.Heftman, “Are wild times ahead for wireless in 1999?”, Microwaves and RF, Dec 1998, pp 35.

[2] see for example, Philips Semiconductor Handbook

[3] S.Heguri and H.Nakata, “Mobile equipment depends on smaller, better SAW filters”, AEI, Oct 1997, pp 68.

[4] H.Tsumori, “Multilayer ceramics make mobile equipment more mobile”, AEI, May 1998, pp 301.

[5] for overview, visit http://www.dupont.com/mcm/

[6] D.C.Benson, Y.Xuan, J.He, C-M.Lin, C.R.Hodges, E.A.Logan, T.M.Schaefer, and B.K.Gilbert, "Integrated passive components for RF applications", Proc. 1997 Int. Wireless Comm. Conf. pp 175 (Denver, 1997).

[7] R.Frye, K.Tai, M.Lau and A.Lin, "Si-on-Si MCM's with integrated passive components", Proc. IEEE Multichip Module Conf., pp 155 (Santa Cruz, 1992).

[8] P.Pieters, S.Brebels and E.Beyne, "Integration of passive components for microwave filters in MCM-D", 6th Int. Conf. and Exhibition on Multichip Modules, (Denver, 1997)

[9] M.A. de Samber, N.J.Pulsford, M. van Delden and R.F.Milsom, "Low complexity MCM-D technology for high frequency applications", Proc 1998 Int. Conf. Multi-Chip Modules and High Density Packaging (Denver 1998).

[10] J.N.Burghartz "Progress in RF inductors on Si – understanding substrate losses", 1998 Int. Electr. Dev. Meeting (San Francisco 1998).

[11] R.F.Milsom "Electromagnetic simulation of passive IC's", Philips J. Research, Vol. 51, No. 3, pp 449 (1998).

INTEGRATED GHz VOLTAGE CONTROLLED OSCILLATORS

Peter Kinget
Bell Labs - Lucent Technologies
Murray Hill, NJ (USA)

Abstract

The voltage controlled oscillator (VCO) is a critical sub-block in communications transceivers. The role of the VCO in a transceiver and the VCO requirements are first reviewed. The necessity of GHz VCOs and the driving factors towards the monolithic integration of the VCO are examined. VCO design techniques are outlined and design trade-offs are explored. The performance of VCOs in different implementation styles is compared to evaluate when and if VCO integration is desirable.

1 Introduction

The last decade of this century has seen an explosive growth in the communications industry. People want to be connected all the time using wireless communication devices. In addition, the demand for high bandwidth communication channels has exploded with the advent of the internet. Thanks to the high density available on integrated circuits, sophisticated digital modulation schemes can be employed to maximize the capacity of these channels. This has changed the design of wireless and wireline transceivers. We focus on the design of a critical sub-block: the voltage controlled oscillator (VCO). We review the requirements for VCOs and evaluate the advantages and disadvantages of VCO integration.

Voltage controlled oscillators appear in many analog and RF signal processing systems. In this paper we focus on silicon implementations of VCOs for communication applications. First we focus on the three parts of the title: we investigate what the role of a VCO is in a transceiver and discuss the requirements for a VCO by reviewing a typical VCO data-sheet; next, we focus on the necessity of GHz VCOs; and finally, we look into what drives the demand for integrated VCOs.

The trade-offs and techniques for VCO designs are then reviewed and relations between noise performance and power consumption are investigated. The performance of VCOs realized in different implementation styles is compared and finally we try to evaluate if or when integrated GHz VCOs are desirable.

2 Why an Integrated GHz VCO ?

2.1 Why a Controllable Oscillator ?

Most electronic signal processing systems require frequency or time reference signals. To use the full capacity of communication channels, e.g. wireless, wired and optical channels, transmitters modulate the baseband message signal into different parts of the spectrum to exploit better propagation characteristics or to frequency multiplex several messages, and the receivers downconvert them for demodulation. These operations require accurate frequency reference signals. Digital circuits and mixed mode circuits (A/D and D/A converters e.g.) pace and synchronize their operations using a clock signal as a time reference signal.

For the lower end of the spectrum one can use the stable properties of quartz crystals as a resonator to build very accurate fixed frequency or time reference signals. For higher frequencies (> few hunderd MHz) the quality of the crystal resonators degrades due to physical limitations and material properties. Many communications applications require programmable carrier frequencies and the cost and board space of a multitude of crystals would be prohibitive. Indirect frequency synthesis techniques based on a phase-locked-loop (PLL) [1] are preferred to generate programmable carriers and RF frequencies. A less accurate RF oscillator whose frequency can be controlled with a control signal is embedded in a feedback loop and its output frequency is locked to an accurate low frequency reference. These loops are typically implemented as a phase-locked-loop as shown in figure 1.

Two basic types of controlled oscillators exist: voltage controlled oscillators (VCO) with a voltage control signal and current controlled oscillators (ICO) with a current control signal. Whereas we will mainly refer to VCOs in the remainder of the text most concepts are equally applicable to ICOs.

In some instances like data communications, the data rate is very accurately standardized. Still a local clock signal is derived from the incoming data signal with a clock recovery circuit to track small variations in the senders clock rate and to align the phase of the local clock for optimal data recovery [2]. This again requires an oscillator whose frequency is controllable.

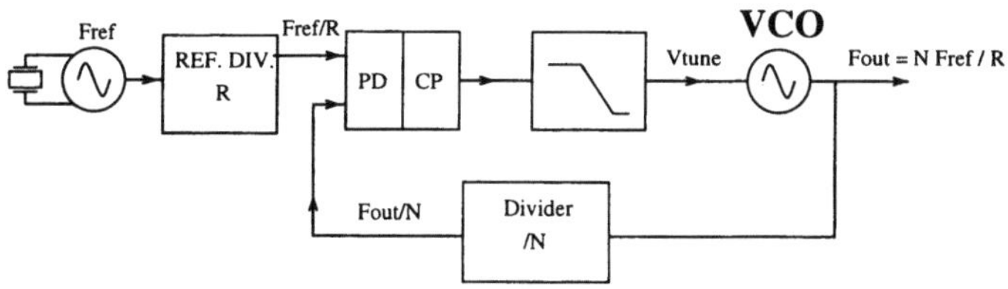

Figure 1: A phase locked loop consists of a voltage controlled oscillator (VCO), frequency divider, phase detector (PD), charge pump (CP) and lead-lag loop filter; the VCO's output frequency F_{out} is set to a multiple of the reference oscillators frequency F_{ref} depending on the divider ratios (N & R).

Another important application of VCOs is for the modulation or demodulation of frequency or angle modulated carriers. Open loop modulation and demodulation as well as closed loop schemes are very popular for portable wireless handsets [3, 4]

2.2 VCO Spec-sheet

Apart from a controllable frequency we now review the other requirements for VCOs; the specification sheet of a VCO typically has the following entries:

Center Frequency: is the output frequency f_0 of the VCO with its control voltage at its center value and is expressed in [Hz]. In this paper we use f_0 or its angular frequency equivalent $\omega_0 = 2\pi f_0$ in [rad/sec] interchangeably.

Tuning Range: is the range of output frequencies that the VCO oscillates at over the full range of the control voltage.

Tuning Sensitivity: is the change in output frequency per unit change in the control voltage, typically expressed in [Hz/V]. VCOs intended for frequency synthesis applications can have a nonlinear relationship between control voltage and oscillation frequency so that several values are quoted or min/max boundaries are given. VCOs for (de)modulation will quote the linearity of the tuning input and the bandwidth of the tuning input.

Spectral Purity: can be specified depending on the application, in the time domain in terms of jitter or in the frequency domain in terms of phase noise or carrier/noise ratio.

Load Pulling: quantifies the sensitivity of the output frequency to changes in its output load. In some applications the output load of the VCO is switched while the VCO must remain at the same frequency to avoid frequency errors

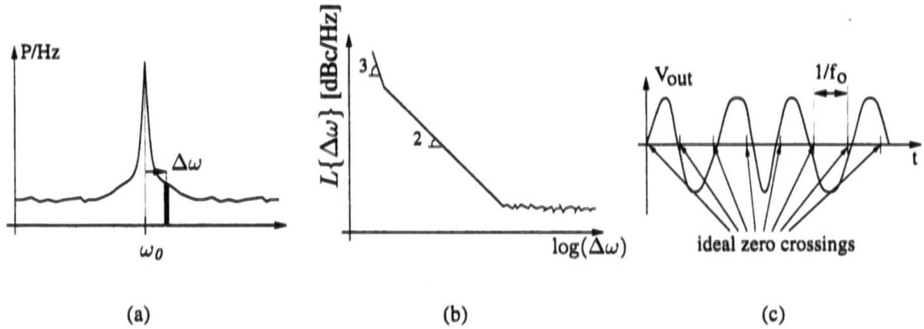

Figure 2: (a) Due to noise sources in the oscillator the output spectrum is not an ideal tone but has noise side bands and a wide-band noise floor; (b) typical plot of phase noise side band as a function of frequency offset from the carrier; (c) in the time domain the zero crossings are not equally spaced because of the effect of phase noise.

when in open loop or to avoid transients in the PLL. This spec depends strongly on the isolation provided by the output stage in the oscillator.

Supply Pulling: quantifies the sensitivity of the output frequency to changes in the power supply voltage and is expressed in [Hz/V]. The power up or down of other circuits can create significant transients in the power supply voltage and it is again desirable that the VCO frequency remains undisturbed.

Power Consumption: specifies the DC power drain by the oscillator and its output buffer circuits.

Output Power: is the power the oscillator can deliver to a specified load. The variation of the output power over the tuning range is also specified.

Harmonic suppression: specifies how much smaller the harmonics of the output signal are compared to the fundamental component and is typically expressed in [dBc].

Spectral Purity

The meaning and relevance of most of the specs is clear from their definition. The spectral purity however requires further discussion especially since it is the key performance measure of a VCO together with its power consumption. Noise sources - thermal, 1/f, supply or substrate interference - cause changes in the amplitude and frequency of oscillation so that the output spectrum of the oscillator is not a pure tone but has noise sidebands (see fig.2(a) and 2(a)); in the time domain this means that there is an amplitude variation and that the zero-crossings of the output waveform are not perfectly equally spaced in time but they exhibit random variations around a nominal value (see fig.2(c)) which are referred to as jitter [5, 6, 7].

The power in the noise sidebands is important for wireless receiver and transmitter applications. The close-in sidebands result in spurious responses of the receiver to nearby interfering channels or blockers [3]; they also contribute to the degradation of the modulation accuracy of the transmitter. The far-out sidebands must be low enough to reduce the spurious emissions by the transmitter to relax its output filter requirements.

The jitter in the zero-crossings of the output waveform of a phase-locked-loop is partly due to the noise or spurious sidebands of the VCO but a large contribution comes from the noise of the other components in the loop. Jitter is a limiting factor in data communications applications since it closes the eye in the eye diagram and so it makes data detection more error prone. In digital circuitry timing jitter reduces the timing margin. For mixed mode applications jitter must be small enough not to affect the accuracy of A/D or D/A conversion e.g.

There are several measures to quantify timing jitter and phase noise which are all mathematically related to each other [5, 7, 6]. In this paper we will use the characterization of the phase noise from the RF spectrum; the RF output power spectrum of an oscillator at ω_0 is symmetrical and the noise in one sideband in a 1Hz bandwidth is used to define $L_{\text{total}}\{\Delta\omega\}$:

$$L_{\text{total}}\{\Delta\omega\} = \frac{\text{noise power in 1Hz BW at } \omega_0 + \Delta\omega}{\text{total carrier power}} \tag{1}$$

In this form L_{total} depends on the effect of phase and amplitude variations. In many circuits the effect of amplitude variations can be eliminated by passing the signal through an explicit or implicit limiting stage but the effect of phase variations cannot be reduced. For most applications only the phase portion L_{phase} of L_{total} is important and is denoted as $L\{\Delta\omega\}$. Figure 2(b) shows the typical phase noise sidebands in an oscillator as a function of the frequency offset from the carrier. White noise sources give rise to a $\Delta\omega^{-2}$ dependence of the phase noise power on frequency offset whereas the $\Delta\omega^{-3}$ dependence is due to the effect of 1/f noise sources in the oscillator [8, 9, 10, 11].

2.3 Why Gigahertz Operation ?

Several evolutions push for the realization of VCOs with center frequencies in the GHz to several GHz range. In the wireless arena, the better propagation characteristics and the larger available bandwidth in the 1 to 2 GHz range have allowed the standardization and exploitation of digital cellular phone systems worldwide. For the fabrication of the wireless phone terminals a large demand for high performance GHz VCOs has emerged. At higher frequencies around 2.5 GHz and 5 GHz new wireless data applications have spurred a lot of interest and large markets are emerging: e.g. short range automation applications in the home, cable replacement wireless links etc.

With the constant shrinking of feature sizes in IC technologies and the increase in clock speeds we are very close to the widespread use of digital systems with clock speeds in the 1 GHz range. The distribution and synchronization of those GHz clocks is very challenging and will rely on on-chip PLL clock multipliers. Already today at clock frequencies of several hunderd MHz these techniques are being used [2]. These applications will also drive the requirement for GHz VCOs.

The same trend exists in data communications were widespread deployment of Gbit/sec data channels is fueled by the Internet growth and the convergence of data and voice communications. These systems rely on clock recovery architectures and also increase the demand for GHz VCOs. It is beyond doubt that GHz VCOs are required in large volumes.

2.4 Why Integrate the Oscillator ?

High volume markets are governed by the *P words*: price - package - performance - power. IC integration reduces production cost since it allows for mass volume production. Integration of the RF components reduces the number of RF pins and thus allows for cheaper packaging solutions. However, integration increases the complexity of the part and thus testing cost/complexity can become a limiting factor or the number of I/O pins can become large which are counterproductive for packaging cost reduction. For low cost and large volume production post-fabrication trimming is to be avoided. By (partly) integrating the VCO on the IC complex automatic calibration techniques become feasible since their is an abundance of cheap active devices along with sophisticated computing power.

Performance is a key factor. In applications governed by standards, meeting the performance specifications is a 'conditio sine qua non'; without standard compliance price, package or power consumption are irrelevant. The benefit of integration for performance is less obvious when the RF functions are combined with the other digital signal processing blocks. This is because the choice of IC technology is driven by the requirements of the majority of the circuits which are digital making the cost of special technology options which enable better analog/RF performance unjustifiable. Consequently integration typically results in a somewhat lower performance of the VCO. In the next section we evaluate the crucial circuit components in a VCO design and how they are affected by integration. Also, we make a comparison of the performance of VCOs implemented at different levels of integration and try to point out if integration makes sense. Power consumption is one of the key performances measures that is discussed.

3 Integrated GHz VCO Design

In this section the design of integrated GHz VCOs is discussed. We look at the different ways to realize a VCO and the design trade-offs between phase noise and power consumption. The design of tuned oscillators is reviewed in detail.

3.1 Classes of VCOs

Oscillators are autonomous circuits that produce a stable periodically time varying waveform. They have at least two states and they cycle through those states at a constant pace. There are three different topologies for controlled oscillators on silicon ICs: ring oscillators, relaxation oscillators and tuned oscillators [3]. Ring oscillators consist of an odd number of single-ended inverters or an even/odd number of differential inverters with the appropriate connections. Relaxation oscillators alternately charge and discharge a capacitor with a constant current between two threshold levels. Tuned oscillators contain a passive resonator - LC tank, transmission line resonator, crystal, SAW - that serves as the frequency setting element.

The first two realizations are very easy to integrate on a monolithic IC and are very compact. Their frequency is controlled by a current or voltage and linear tuning characteristics can be obtained. Moreover, frequency tuning can be done over several orders of magnitude [12]. Tuned oscillators are harder to integrate primarily because of the lack of high quality passive inductors in standard IC technologies and because of their large size. However, tuned oscillators have a much higher frequency stability and spectral purity since it is set by the passive resonator. Relaxation and ring oscillators are typically very sensitive to noise in the switching thresholds and charging currents.

3.2 VCO design trade-offs

In this section we investigate the relation between spectral purity and power consumption of a VCO and how a large tuning range affects the design. The analysis and prediction of phase noise or timing jitter in oscillators is a very difficult task since the oscillator is an autonomous non-linear circuit and the non-linearity is essential for its operation and noise performance evaluation. Recently several techniques have been developed for the accurate simulation of phase noise in oscillators [8, 9, 10] but they are mathematically too involved to intuitively understand the trade-off between the various performance parameters.

Leeson's model [11] has long been the primary analysis and modeling tool for phase noise of oscillators. Recently, [13, 14] have presented other linear noise analyses of oscillators. Although a linear noise analysis of an oscillator has some fundamental problems and inconsistencies - like infinite noise power at the center frequency f_0 - it is analytically treatable and has given a good insight in the trade-offs between parameters.

Using approximated harmonic balance equations a more rigorous analytical derivation of the phase noise of oscillators can be performed [15]; this analysis takes into account the switching behavior of the oscillator. The results allow the evaluation of the contribution of the different noise sources and noise folding effects. Also in [16] frequency domain techniques are presented to study the effects of noise folding in oscillators.

3.2.1 Linear Noise Analysis for Parallel LC Oscillator

It will be beneficial to briefly derive the linear noise analysis of a parallel LC oscillator. First we compute the equivalent parallel resistance of the tank; then, we determine the tank impedance for small offsets from the self resonant frequency. From these results we can determine the necessary negative conductance and the carrier to noise ratio.

Figure 3(a) shows a parallel resonant LC tank with inductor, capacitor and parallel losses. We now compute the admittance of this LC configuration:

$$Y(\omega) = \frac{1}{R_p} + \frac{1}{j\omega L(1+\frac{1}{Q_L^2})} + \frac{j\omega C}{1+\frac{1}{Q_C^2}} + \frac{1}{Q_L \omega L(1+\frac{1}{Q_L^2})} + \frac{\omega C}{Q_C(1+\frac{1}{Q_C^2})} \tag{2}$$

with $Q_L = (\omega L)/R_{sl}$ and $Q_C = 1/(\omega C R_{sc})$ the inductor and capacitor quality factors respectively. Equation (2) represents an equivalent admittance with

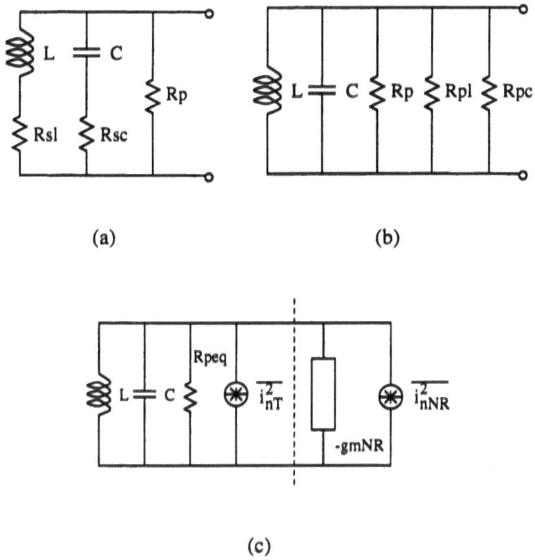

Figure 3: (a) LC tank including series losses for all components and a parallel loss; (b) equivalent LC tank with all losses represented by parallel losses; (c) parallel LC oscillator with ideal negative conductance.

3 real terms and two imaginary terms; for sufficiently large Q_L and Q_C, the tank can be represented by the equivalent circuit of figure 3(b) with $R_{pl} \approx Q_L \omega L$ and $R_{pc} \approx \frac{Q_C}{\omega C}$. It is useful to define the characteristic impedance Z_0 and the quality factor of complete tank circuit Q_T as follows:

$$Z_0 = \left(\frac{1}{\omega C}\right) = \omega L = \sqrt{\frac{L}{C}} \tag{3}$$

$$R_{peq} = R_p // R_{pc} // R_{pl} \tag{4}$$

$$Q_T = \frac{R_{peq}}{Z_0} \tag{5}$$

$$\left(\frac{1}{Q_T}\right) = \frac{Z_0}{R_p} + \left(\frac{1}{Q_L}\right) + \left(\frac{1}{Q_C}\right) \tag{6}$$

Interesting to note is that the total tank quality Q_T is determined by the lowest quality factor component.

To build an oscillator that operates the tank at its parallel resonance ($\omega_0 = 1/\sqrt{LC}$), a negative conductance is added in parallel to compensate the losses and to sustain the oscillation (see figure 3(c)). The necessary minimal negative conductance g_{mc} is given by:

$$g_{mc} = \text{Re}\left\{Y(\omega_0)\right\} = \frac{1}{R_{peq}} = \frac{1}{Q_T Z_0} \tag{7}$$

To guarantee start-up under all conditions the negative conductance is over-designed by a factor $\alpha = 1.5 - 3$ so that $g_{mNR} = -\alpha g_{mc}$.

To evaluate the noise to carrier ratio of the oscillator we compute the tank impedance at an offset frequency $\Delta\omega$ from the oscillation frequency ω_0 from (2):

$$Y(\omega_0 + \Delta\omega) \approx 2j\left(\frac{\Delta\omega}{\omega_0}\right)\frac{1}{Z_0} + \frac{1}{R_{peq}} \tag{8}$$

In an oscillator the effect of real part in the tank admittance - the second term in equation (8) - is cancelled with the negative conductance so that only the first term remains.

We can now calculate the noise to carrier ratio; we assume a given oscillation signal V_{RMS}; the tank losses generate a noise current $\overline{i_{nT}^2} = 4kTg_{mc}$ that is transformed to a noise voltage by the tank impedance; similarly there is a contribution from the noise current of the negative conductance $\overline{i_{nNR}^2} = 4kT\Gamma\alpha g_{mc}$, where Γ is the noise excess factor of the negative conductance implementation (see figure 3(c)). At an offset $\Delta\omega$ from resonance we obtain the following $L_{\text{total}}\{\Delta\omega\}$:

$$L_{\text{total}}\{\Delta\omega\} = \frac{\overline{v_n{}^2}}{V_{RMS}{}^2} = \frac{\overline{i_{nT}^2} + \overline{i_{nNR}^2}}{|Y(\omega_0 + \Delta\omega)|^2}\frac{1}{V_{RMS}{}^2} \tag{9}$$

$$L_{\text{total}}\{\Delta\omega\} = kT\left(\frac{Z_0}{Q_T}\right)\left(\frac{\omega_0}{\Delta\omega}\right)^2\frac{1+\alpha\Gamma}{V_{RMS}{}^2} \tag{10}$$

The classical distribution of $L_{\text{total}}\{\Delta\omega\}$ into phase and amplitude fractions assumes an equal division of the noise so that $L_{\text{phase}}\{\Delta\omega\} = L\{\Delta\omega\} = L_{\text{total}}\{\Delta\omega\}/2$ [6].

3.2.2 Design trade-offs

Figure 4(a) depicts an implementation of the parallel resonant oscillator of figure 3(c) using an nMOS differential pair negative conductance.

We can now derive the oscillation signal V_{RMS} (see also [17, 16, 18]). In figure 4(b) the voltage across the tank ($V_p - V_m$) and the currents in the negative conductance are depicted for transistors with a zero threshold voltage. In

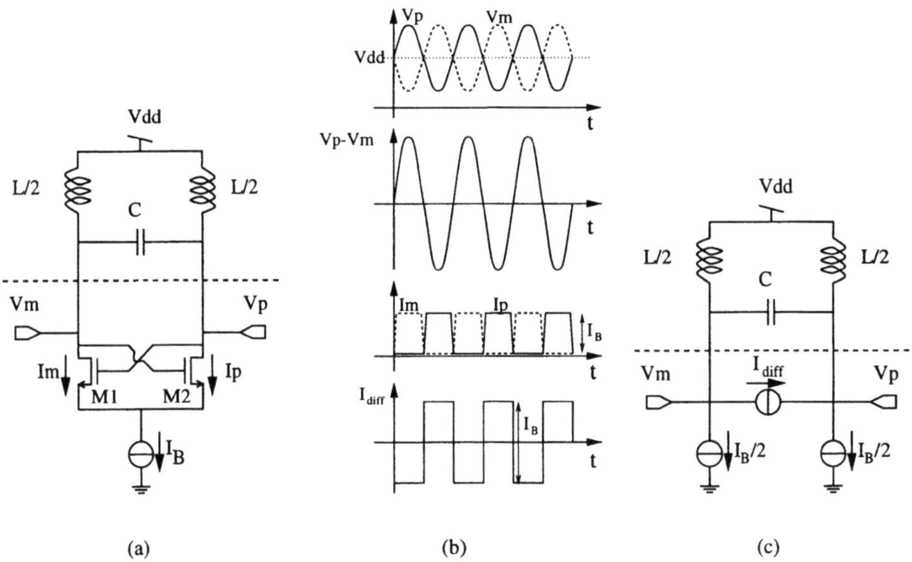

Figure 4: (a) nMOS differential pair implementation of negative conductance; (b) voltage waveforms and current waveforms for zero threshold voltage transistors; (c) equivalent representation of (a) using current sources to model the effect of the switching transistors.

figure 4(c) the effect of the negative conductance circuit is modeled with equivalent common mode and differential mode current sources. The differential output current (I_{diff}) of the negative conductance is a square wave with amplitude $I_B/2$; the tank filters the higher order harmonics of this current signal and only its Fourier component at ω_0 $I_{\omega 0} = 2I_B/\pi$ is converted into a differential voltage by the equivalent impedance at resonance $R_{peq} = Q_T Z_0$ so that the RMS value of the differential voltage across the tank is given by:

$$V_{RMS} = \frac{1}{2\sqrt{2}} R_{peq} I_{\omega 0} = \frac{2}{\sqrt{2}\pi} I_B Q_T Z_0 \tag{11}$$

With real transistors and a real current source similar behavior occurs. Because the small signal negative conductance g_{mNR} is made ($\alpha \geq 1$) times larger than the minimal required g_{mc} to ensure start-up, the oscillator waveforms grow until the negative resistor circuit goes into a switching operation and the amplitude has a similar dependence on I_B, Q_T and Z_0 as in (11). However, the single ended peak value of V_p and V_m must be smaller than V_{DD} to keep the current source working as a current source; for increasing values of the bias current I_B the oscillator will go from a current limited operation – assumed above in (11) – to a voltage limited operation (see e.g. [18]).

The negative conductance g_{mNR} is equal to half the transconductance g_m

of transistors M_1 and M_2 in figure 4(a), which can be expressed as: $g_m = (g_m/I)I_B/2$ with $(g_m/I) = 2/(V_{GS} - V_T)$ [19]. We now rewrite α as:

$$\alpha = \frac{g_{mNR}}{g_{mc}} = \frac{(g_m/I)I_B Q_T Z_0}{4} \tag{12}$$

so that for a fixed start-up gain α we can compute the required bias current as:

$$I_B = \frac{4\alpha}{Q_T Z_0 (g_m/I)} \tag{13}$$

The voltage swing then becomes:

$$V_{RMS} = \frac{8\alpha}{\sqrt{2}\pi(g_m/I)} \tag{14}$$

After substitution of (14) into (10) we obtain the following equivalent expressions:

$$L_{\text{total}}\{\Delta\omega\} = kT\frac{\pi^2}{32}\left(\frac{\omega_0}{\Delta\omega}\right)^2 \frac{1+\alpha\Gamma}{\alpha^2}\frac{(g_m/I)^2 Z_0}{Q_T} \tag{15}$$

$$L_{\text{total}}\{\Delta\omega\} = kT\frac{\pi^2}{8}\left(\frac{\omega_0}{\Delta\omega}\right)^2 \left(\frac{1+\alpha\Gamma}{\alpha}\right)\frac{(g_m/I)}{Q_T{}^2 I_B} \tag{16}$$

Equations (15) and (16) show that the use of a large start-up gain α improves the noise to carrier ratio since increasing the start-up gain increases the noise power contribution of the negative conductance but at the same time increases the carrier power quadratically. For increasing α the $\alpha\Gamma$ in the numerator will exceed 1 and the noise contribution of the negative conductance dominates. Equation (16) can then be rewritten as:

$$L_{\text{total}}\{\Delta\omega\} \propto kT\left(\frac{\omega_0}{\Delta\omega}\right)^2 \frac{(g_m/I)\Gamma}{I_B Q_T^2} \tag{17}$$

In order to arrive at equation (17) we have mixed linear analysis and large signal concepts. It is however noteworthy that more extended non-linear analyses result in a similar relation for $L_{\text{total}}\{\Delta\omega\}$ [15, 16] and the time variant noise effects and noise folding can be taken into account with a proper value for Γ in (17) [15, 16].

Equation (17) provides an important insight in the trade-off between the power consumption and phase noise of an LC oscillator: the power consumption

Specifications	
Center Frequency:	ω_0
Phase Noise:	$L\{\Delta\omega\}$
Technology Constraints	
Resonator quality	Q_T
Design Variables	
Inductance:	L
Capacitance:	C
Tank Characteristic Impedance:	Z_0
Bias Current:	I_B
Transistor :	(g_m/I)
Design Relations	
$\omega_0 = 1/\sqrt{LC}$	
$Z_0 = \sqrt{L/C}$	
$\alpha = \frac{g_{mNR}}{g_{mc}} = (g_m/I)I_B Q_T Z_0/4 \geq 1$ (12)	
$L\{\Delta\omega\} \propto kT\left(\frac{\omega_0}{\Delta\omega}\right)^2 \frac{(g_m/I)\Gamma}{I_B Q_T^2}$ (17)	

Table 1:

is $P = I_B V_{DD}$ so the phase-noise power product ($L\{\Delta\omega\}P$) of the oscillator becomes[1]:

$$L\{\Delta\omega\}P \propto \left(\frac{\omega_0}{\Delta\omega}\right)^2 \frac{(g_m/I)\Gamma V_{DD}}{Q_T^2} \tag{18}$$

It is desirable to minimize both sides of relation (18). Equation (18) also illustrates the importance of the quality factor of the tank circuit Q_T: a high tank quality factor results in low noise and power efficient oscillators. The quality factor is however dependent on the technology available for the fabrication of the tank and is mostly beyond the control of the circuit designer.

3.2.3 LC Oscillator Design Procedure

In table 1 the specifications, constraints, design variables and design relations are summarized for the oscillator in figure 4(a). The designer's task is to choose the appropriate values for L, C, I_B and transistor sizes (or (g_m/I)) so that the center frequency (ω_0), start-up and phase noise requirements $L\{\Delta\omega\}$ are met, with the lowest possible power consumption P.

[1]as discussed in the previous section the phase noise $L\{\Delta\omega\}$ is a fraction of the total noise $L_{\text{total}}\{\Delta\omega\}$.

The center frequency only fixes the product of L and C but their ratio and thus the characteristic impedance Z_0 of the tank is to be chosen. A large Z_0 – large L but small C – for a given choice of bias current I_B yields a large value for V_{RMS} (11) and α (12) which is desirable. For a given start-up gain α the choice of Z_0 does *not* influence the phase noise performance $L\{\Delta\omega\}$ – see equation (16) or equation (17) for large startup gains. $L\{\Delta\omega\}$ is in first order mainly dependent on the choice of I_B (17) and on the available tank quality Q_T. To lower the phase noise the designer has to increase the bias current I_B. This can result in too large voltage swings V_{RMS} (11) over the active devices so that Z_0 must be reduced. Smaller inductors have higher self resonant frequencies and typically somewhat higher quality factors which suggests the choice of low Z_0, however at the expense of smaller voltage swings V_{RMS} and start-up gains α. The final choice of parameters is typically decided with the help of a phase noise simulator.

3.2.4 Tuneable LC Oscillator

For most applications the required tuning range for GHz VCOs is only a few percent since the application bandwidth is much smaller than the center frequency. If tuning is also used to combat the process variations typically much large tuneability is required; a fully integrated VCO e.g. will typically exhibit +/-10% center frequency variations due to on chip capacitance variations. In order to change the center frequency f_0 to build a VCO we have to change the L or the C of the tank electronically. Continuous tuneable inductors are not readily available but varactors – voltage dependent capacitors – can be built using pn junctions or MOS capacitors [20, 21]. To tune between f_{max} and f_{min} at least a fraction ΔC of the total tank capacitance C must be variable:

$$\Delta C \geq 2C\frac{f_{max} - f_{min}}{0.5(f_{max} + f_{min})} \tag{19}$$

In figure 5 the implementation of a parallel resonant LC CMOS VCO with pn varactors or MOS varactors is illustrated.

Tuneability entails three main design issues. Varactors always have a percentage of fixed capacitance associated with the variable capacitance so that the fixed tank capacitance increases. To keep the oscillation frequency high Z_0 has to be lowered which reduces the start-up gain α and the voltage swing. Secondly, varactors have a lower quality factor than fixed capacitors so that the tank quality Q_T reduces if a large tuneability is required. Thirdly, pn junction varactors have to remain reverse biased under all conditions to avoid conduc-

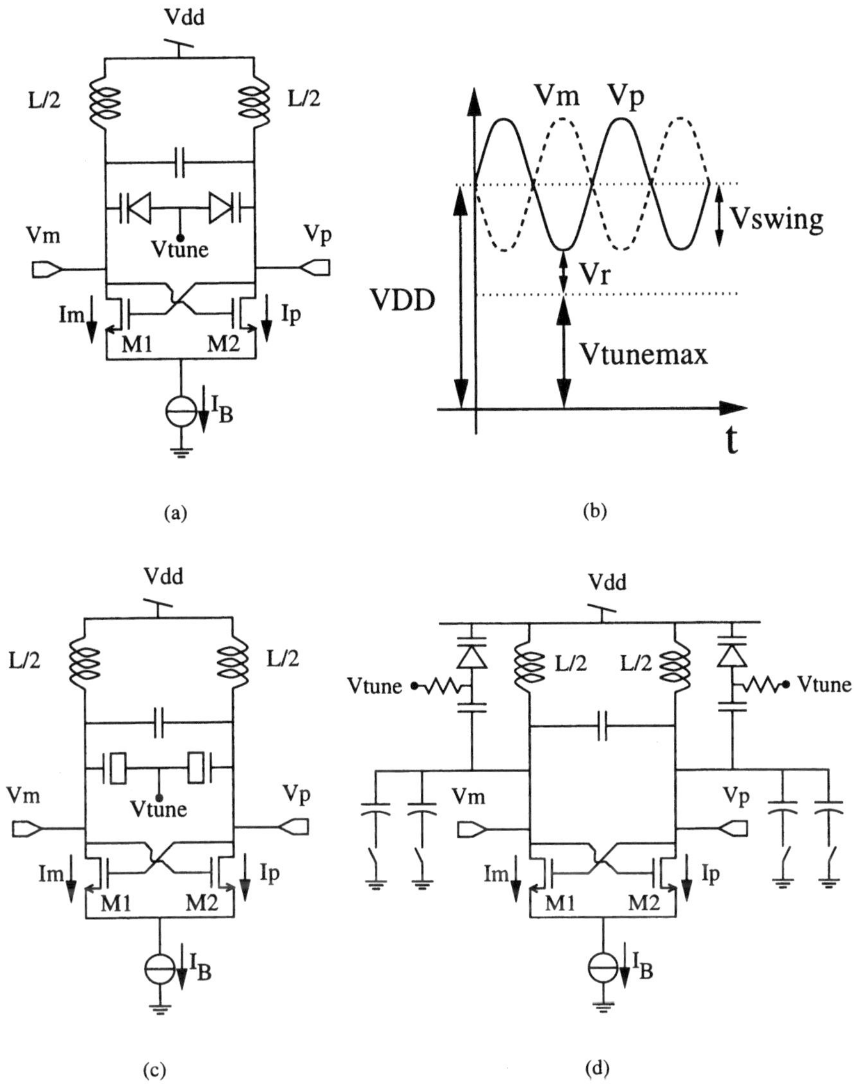

Figure 5: Modification of fixed frequency oscillator in figure 4(a) to a voltage controlled oscillator using (a) pn junction varactors, (c) MOS varactors, or (d) capacitively tapped varactors and discrete tuning capacitors; (b) shows the relation between the maximal tune voltage, diode reverse voltage, voltage swing across the tank and supply voltages.

tion losses and thus further Q_T degradation. As illustrated in figure 5(b), the maximal allowable voltage swing across the varactors for a given tuning voltage range $V_{tunemax}$ is limited by the available supply voltage V_{DD} and the minimal reverse bias across the diode V_R. MOS varactors do not have conduction losses but the voltage swing has be small enough to prevent gate breakdown; also, at larger swings across the MOS varactor the capacitance/voltage relation becomes strongly non-linear and more harmonics are generated.

The last two problems can be alleviated by capacitively tapping the varactors with fixed high Q capacitors (see figure 5(d)). Capacitive tapping results in an impedance transformation [4, 3] and the effective Q of the structure scales up with the tapping ratio but the variability of the varactor ΔC scales down with the tapping ratio[2]. If a fixed capacitor with low parasitic bottom plate and high Q is available, one can thus improve the quality of the varactor at the price of area. At the same time there is no DC connection between the tank and varactor anymore, and the voltage swings across the varactors are scaled down with the tapping ratio so the swing limitations can be remedied.

As process technologies shrink the speed f_0 of the oscillators goes up but the supply voltages scale down. The tuning gain K_V necessary to compensate process variations is ($K_V \propto 20\% \cdot f_0/V_{DD}$) since the tuning voltage must be within the supply voltage for an integrated oscillator. As the oscillator's frequency goes up and the supply voltages go down the tuning gain K_V increases. The noise voltage present at the tuning port of the oscillator results in a frequency modulation of the carrier and thus in phase noise. If we assume a resistor R at the tune port its voltage noise results in phase noise level given by:

$$L\{\Delta\omega\} = 2kTR\left(\frac{K_V}{\Delta\omega}\right)^2 \tag{20}$$

where K_V is in [rad/V]. For increasing K_V the allowable impedance level at the tuning port reduces. In the PLL design this implies a lower impedance level for the loop filter and thus larger capacitors which is counterproductive for chip integration since it increases the area of the PLL. However, the correction for the process variations does not have to be continuous but can be done with discrete tuning [22] (see figure 5(d)) and still be automated so no trimming is required [23]. This approach has the benefit that the tuning sensitivity of the tune port of the VCO is reduced for a fixed tuning voltage range so that less noise from the other PLL components appears at the output [6].

[2]For the series connection of a variable capacitor C_V with a capacitance variation ΔC_V and a quality Q_V, and a high Q capacitor C_{tap}, one can easily obtain that the quality of the series connection is $Q_V(C_{tap} + C_V)/C_{tap}$ and the capacitance variation $\Delta C = \Delta C_V C_{tap}/(C_{tap} + C_V)$.

In summary we note that the introduction of tuneability complicates the oscillator design and consequently has negative implications on the achievable phase noise - power trade-off.

3.2.5 Transistor Sizing

For a given choice of bias current I_B an MOS designer still has to decide on the (g_m/I) of the devices. The (g_m/I) is increased by reducing the $(V_{GS} - V_T)$ at the price of a lower transit frequency f_T ($f_T = g_m/C_{gs}$) of the device [19]. Lower device f_T implies larger capacitive loading by the negative resistance of the tank. These capacitors become part of the tank circuit and the Q of the total tank circuit is then typically denoted as the 'loaded Q' of the oscillator. For high Q tanks is important to keep the transistor parasitics small to avoid degradation of the loaded Q due to the lower quality parasitics. For low Q fully integrated tanks the quality factor of the transistor parasitics can be high enough so that the parasitics can be used as the main part of the tank capacitance and high frequency operation can be achieved [24].

Equation (17) suggests lowering (g_m/I) values to improve the noise performance of the oscillator since a low (g_m/I) results in less noise current for the same bias current I_B. But low (g_m/I) also results in lower start-up gains α (12) and linearizes the negative conductance which results in a reduction of the oscillation amplitude for the same bias current I_B. This causes an increase in the noise to carrier ratio of the oscillator. These effects can be balanced by investigating the best choice for (g_m/I) with a phase noise simulator.

Bipolar devices have a fixed (g_m/I) [19]. The device size is primarily determined by the maximal allowable current density or the base spreading resistance. To reduce the base resistance larger emitter areas and finger layouts are used but this results in the operation of the device below its 'peak f_T'. The parasitic capacitive loading of the tank can become significant and the loaded Q_T can reduce if the tank characteristic impedance was chosen too high.

3.2.6 Other configurations

There a several more oscillator configurations that are well suited for integrated circuit implementations see e.g. [3, 4]. Colpitts configurations are very well established and are very popular for single-ended and bipolar designs. The design trade-offs for those configurations are similar or identical to the ones outlined above.

4 Comparison of VCO implementation styles

There are four major implementation styles for tuned VCOs. Traditionally they are implemented with hybrid modules which are easy to use but are bulky and expensive. A realization with a discrete RF device on the transceiver PCB is more cost effective but requires large board area. A third alternative is to integrate the active parts of the VCO on the transceiver chip and use external resonators. Fully integrated realizations have recently received a lot of (research) attention.

In table 2 the performance of different oscillators implemented with different technologies is listed. Their center frequencies f_0 are different and the phase noise $L\{\Delta f\}$ is specified at different offset frequencies Δf. Equation (17) or more generalized considerations [3] suggest the definition of a figure of merit FM for an oscillator that is independent of its specific oscillation frequency:

$$FM = 10\log\left(\left(\frac{f_0}{\Delta f}\right)^2 \frac{1}{L\{\Delta f\}P}\right) \tag{21}$$

In table 2 FM is given based on P in [mW]. The power consumption P is for the oscillator core only and does not include the output buffer; when only total power consumption numbers were available the core is assumed to account for 1/3 of the total power consumption. For integrated oscillators the number of available metal layers in the process or the availability of a thick metal layer (T) is listed along with the type of substrate material: HR indicates a high resistivity substrate and LR a low resistivity substrate. The types of inductors are inte(grated), discr(ete), bondw(ire) or MCM (on multi-chip-module). Where available the inductor and varactor quality factors QL and QV are given.

In figure 6 the figure of merit is plotted against the oscillation frequency for the different group of implementation styles. A number of interesting observations can be made:

- The *ring and relaxation oscillators* consistently have a very low FM ("Ring-/Relax" in figure 6). This is due to their high noise sensitivity because of the lack of a frequency selective network. Moreover in these non-tuned oscillators, large currents flow through the active devices at the zero-crossings of the oscillation waveform and thus large noise injection occurs around the zero-crossings. Non-linear phase noise simulations [9, 10] show that the oscillator's phase is most sensitive to noise injection close to the zero-crossings.

  The ring and relaxation oscillators are fully integrated oscillators and oc-

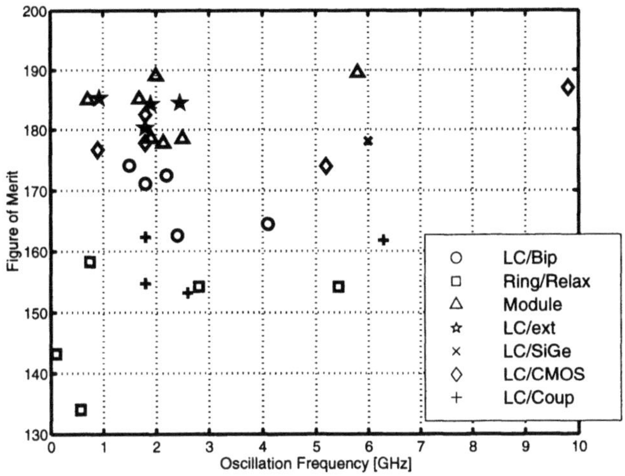

Figure 6: Comparison of the 'Figure of Merit' for several oscillators using different implementation styles

cupy much smaller chip area than integrated tuned oscillators. Ring oscillators also easily provide multi-phase output signals. Because of their compactness ring oscillators are preferred in PLLs for digital clock generation and for clock and data recovery circuits for some data communication applications. Moreover large PLL loop bandwidths can be used in these fixed frequency applications so that the effect of the VCO phase noise on the output jitter is largely reduced [7].

Interestingly the jitter requirements in data communications and digital applications are specified in unit-interval units which are relative units [5]. The jitter requirements in unit-intervals are typically kept almost unchanged when the bit rates are scaled up to conserve timing margins. The relation between the RMS jitter Δt_{RMS} and phasenoise of the PLL $L\{\Delta\omega\}$ is given by [7]:

$$\left(\frac{\Delta t_{RMS}}{T}\right)^2 = \int_{\omega_L}^{\omega_H} 2 \cdot L\{\Delta\omega\} \cdot \mathrm{d}\Delta\omega \tag{22}$$

where T is the period, and $\omega_H - \omega_L$ is the jitter bandwidth. Equation (22) shows that the phase noise requirements $L\{\Delta f\}$ are related to the jitter requirements in unit-intervals $(\Delta t_{RMS}/T)$ so that for constant jitter requirments in unit-intervals the phase noise requirement $L\{\Delta f\}$ remains the same even when the center frequency f_0 has scaled up since the jitter bandwidths do typically not scale as much as the center frequencies. Higher

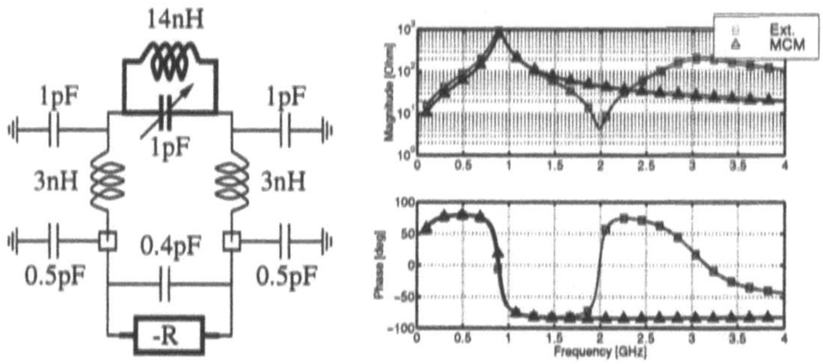

(a) (b)

Figure 7: Typical example of possible spurious oscillations for a 900MHz oscillator with external tank. (a) The connection to the external tank (bold: 14nH and 1pF) from the negative resistor (-R) with its parasitic parallel capacitance (0.4pF) goes through on-chip pad capacitance (0.5pF), bondwires (3nH), and board capacitance (1pF). (b) (square) Besides the wanted parallel resonance at 900MHz, a spurious series resonance at 2 GHz and a spurious parallel resonance at 3.2 GHz exist and can be exited; (b) (triangle) if an MCM or on-chip inductor is used the spurious resonances are eliminated.

bit rate applications thus require PLLs using oscillators with higher FM, or more power needs to be dissipated in the oscillator as indicated by (21). Consequently, higher Q oscillators are (or soon will be) required for the Gbit/sec data communications clocks and digital clocks.

- The *hybrid modules* have farmost the best performance ("Module" in figure 6). They are built with the mix of the best technologies for the tank circuit, varactor and active device and typically use trimming to adjust the center frequency. Whereas they are easy to use, they are large and occupy a lot of board space and are expensive.

- The VCOs built with integrated active components and *external inductors* (discrete, MCM, bondwire) ("LC/ext" in figure 6) also have very good specifications and perform close to the hybrid modules. External inductors have significantly better quality factors than integrated inductors which is the primary reason for the higher FM.

  In order to access the external resonator pins have to be reserved which complicates packaging. Also, coupling between pins or bonding wires as well as coupling on the external PCB can introduce extra interference into the oscillator or can cause unwanted leakage of the large oscillator signal to sensitive nodes in the system.

  For oscillation frequencies in GHz range the tank inductance is at most a few nanohenries to tens of nanohenries. The package lead and bondwire parasitic series inductance are however of the same order of magnitude. In some cases, the high inductance with high Q of bondwires can also be used to the designers advantage by using a bondwire as the tank inductor [13]. With external resonators, the parasitic series inductances and parasitic parallel pin and pad capacitances introduce several spurious resonances in the tank circuit which can result in unwanted modes of oscillation [25] as is illustrated in figure 4. This problem is largely reduced when more advanced packaging techniques are employed such as flip-chip mounting of the silicon die on an MCM - multi-chip-module - substrate. The lead inductance and pad capacitance is greatly reduced and the bondwire inductance is eliminated. Additionally, high quality inductors realized on the MCM substrate can be used for the tank resonator [25, 26]. The use of more advanced packaging increases the packaging cost but enables higher performant RF circuitry and allows the combination of dies in different IC technologies for different parts of the system for optimal performance. The high isolation between the separate dies is an additional benefit.

- The performance of *fully integrated VCOs* strongly depends on the available inductor quality ("LC/CMOS" and "LC/Bipolar" in figure 6). This is not surprising since equation (18) indicates a quadratic dependence on the tank quality factor. Integrated tuned oscillators use planar spirals in the available metal layers to build an inductor[4] (see figure 8). Most standard digital CMOS processes use epi-wafers which consists of low resistive p++ material with a thin lightly doped p-epi layer for the circuitry. The magnetic field of the spirals extends into the substrate and large eddy currents flow which result in a severe degradation of the inductor quality to only 3–5. This low inductor Q limits the overall tank quality to very low values [24]. Some digital CMOS processes and most BiCMOS processes have a high resistivity wafer material so that better inductor quality is achieved. The quality factor is then mainly degraded by the resistive losses in the metalization so that some RF IC technologies have the option of a special thick metalization level to build high Q inductors (QL 10–20). In table 2 the performance of fully integrated MOS LC oscillators is indeed typically better for the high resistivity processes. It is also very promising that the MOS LC oscillators perform well compared to oscillators with external tanks.

  Fully integrated tuned VCOs occupy large chip areas due to the large size of the spiral inductors especially for the low GHz range (see figure 8). The inductor size scales down for higher oscillation frequencies but are still

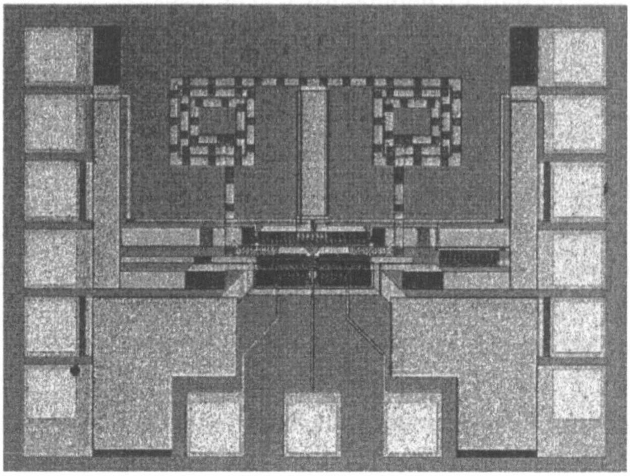

Figure 8: Microphotograph of a fully integrated 0.35um CMOS VCO operating at 5 GHz from [24].

substantial. On the system level, this extra cost in silicon area can be traded for simpler packaging, less external components, smaller board size, and ease of use.

With the down-scaling of the technology feature sizes more interconnection levels must be provided to enable higher digital circuit densities. Also, in deep sub-micron technologies the speed of digital circuits is interconnect-delay limited rather than transistor-delay limited which results in the introduction of low resistivity interconnects (e.g. Cu-based). These trends are favorable for the development of high quality on-chip integrated inductors [27].

- It is remarkable that the fully integrated oscillators with *bipolar active devices* ("LC/Bip" in figure 6) in a BiCMOS technology perform less good than the oscillators with MOS devices ("LC/CMOS" in figure 6). This is in contrast to the general belief that bipolar devices are preferable for RF circuits. A possible explanation lies in one extra complication a bipolar oscillator design has to deal with: possible conduction losses due to base-collector forward bias, whereas in a MOS device the gate terminal is always isolated. Therefore bipolar oscillators either include an extra buffer device for level shifting (see figure 9(a)) or use capacitive tapping (see figure 9(b)) to reduce the voltage swings across base-collector. The extra device introduces more noise sources in the oscillator core and capacitive tapping is unfavorable for the phase noise - power trade-off.

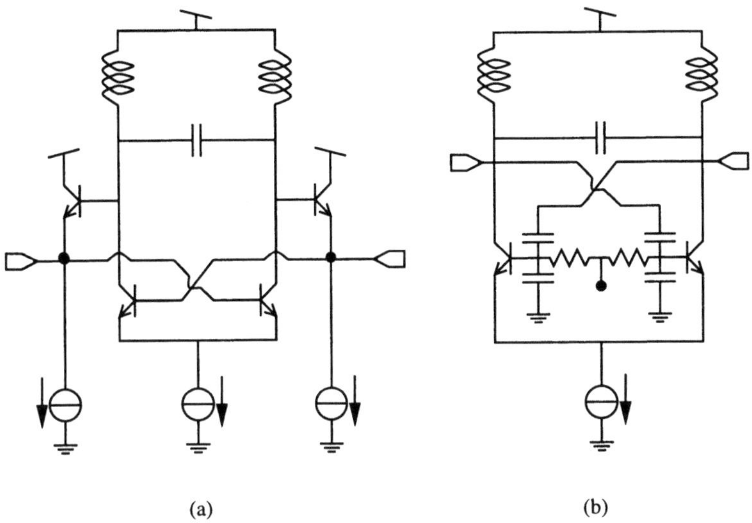

Figure 9: Bipolar implementations of a negative conductance. To avoid lossed due to base-collector forward bias either an emitter follower is used as level shifter (b), or capacitive tapping is used to reduce the signal swing at the base (c).

We have to note however that most of the listed bipolar VCO designs provide high output power and have good output buffering. In some topologies this buffering comes at very little extra power consumption. Most of the listed MOS VCOs do not include strong output buffering and are designed to drive on chip loads only.

Two oscillators using a *SiGe bipolar device* are listed ("LC/SiGe" in figure 6). Interestingly both perform better than the other bipolar VCOs even though they operate at higher frequencies. This could indicate that the BiCMOS VCOs also suffer from the typically fairly high base spreading resistance.

There is one important benefit of bipolar VCOs though that is not apparent from table 2: 1/f noise. Bipolar devices exhibit much lower 1/f noise than MOS devices so that the close-in noise sidebands for bipolar VCOs are lower than for MOS VCOs. The quadratic scaling with Δf assumes only white noise and does not account for the role of 1/f noise.

- Table 2 and figure 6 also include a group of *coupled oscillators* ("LC/Coup"). They consist of two coupled tuned oscillators where the coupling arrangement is typically such that the waveforms of the two oscillators are in quadrature (90 degrees out of phase). Accurate quadrature signals are essential for the realization of image reject mixers for highly integrated transceivers [3, 4] and coupled oscillators can deliver very accurate quadrature signals. Coupled oscillators can also be used to build controllable oscillators without the need for varactors. However, their FM is significantly lower than that of other integrated oscillators even though in table 2 only the power consumption of one oscillator stage of the coupled oscillators is taken into account. This is due to the addition of extra devices to the oscillator core and thus extra noise sources.

5 Conclusions

With the advent of higher communication data rates and digital clock rates and the proliferation of wireless terminals the demand for integrated GHz oscillators is growing. Whereas for digital and data applications fully integrated ring oscillators are being widely used, the use of fully integrated tuned oscillators is only emerging in wireless products. Performance concerns as well as large area still inhibit the widespread acceptance of integrated tuned oscillators. The reduction of the number of RF interfaces in the package, the ease of use of fully integrated parts, compact board size and the implementation of automatic trimming techniques will however outweigh the extra die cost for large volume wireless terminals. The good performance of MOS oscillators and the introduction of better interconnect technologies in deep submicron technologies holds interesting prospects for highly integrated transceivers combining RF and analog front-ends with digital signal processing back-ends. The constant move to higher bit-rates will require a shift from non-tuned ring oscillators to fully integrated tuned oscillators for data and digital applications.

6 Acknowledgments

The author would like to thank the following people for stimulating technical discussions or assistance with oscillator design and testing: T. Aytur, M. Banu, R. Bauder, V. Boccuzzi, N. Belk, P. Davis, A. Demir, A. Dunlop, W. Fischer, M. Frei, J. Glas, P. Feldmann, V. Gopinathan, A. Hajimiri, Q. Huang, S. Kapur, N. Krishnapura, J. Lin, T.P. Liu, S. Logan, D. Long, R. Melville, A. Mehrotra, D. Nelson, J. Roychowdhury, C. Samori, F. Svelto, L. Toth, H. Wang, W. Wilson.

	Type	f_0 [GHz]	$L\{\Delta f\}$ [dBc/Hz]	Δf [kHz]	Tune [%]	Vdd [V]	Idd [mA]	P [mW]	FM	Technology	#M	Sub	L	QL	QV
Vari-L VCO690-5800	module	5.800	-105.0	100	5.5	5.0	2.4	11.9	189.5						
Z-Comm v585me06	module	2.000	-100.0	10	66.7	10.0	5.0	50.0	189.0						
Wang ISSCC 99	LC/CMOS	9.800	-118.0	1000		2.2	5.5	12.0	187.0	CMOS035	4	LR	Inte L	5	
Huang CICC 98	LC/ext	0.926	-112.7	100		3.0	1.6	4.7	185.3	CMOS04			Discr L		
Murata MQH	module	1.686	-103.0	50	2.2	3.0	2.3	7.0	185.1						
Murata MQE	module	0.706	-115.0	100	0.6	3.2	1.6	5.0	185.0						
Kinget ESSCIRC 98	LC/ext	2.450	-124.0	1000	5.0	2.7	2.0	5.4	184.5	CMOS035	4	MCM	MCM L	70	
Dec ISSCC 99	LC/ext	1.900	-126.0	600	9.0	2.5	6.0	15.0	184.3	CMOS			Bondw	80	
Craninckx CICC 97	LC/CMOS	1.800	-113.0	200	20.0	3.0	3.0	9.0	182.5	CMOS04		HR	Inte L	8.6	
Craninckx ISSCC'94	LC/ext	1.800	-115.0	200	5.0			24.0	180.3	CMOS07	2	LR	Bondw	80	
Mini-C JTOS-1910	module	1.910	-132.0	1000	16.1	12.0	6.6	79.2	178.6						
CTI VMS	module	2.500	-90.0	25	23.2	5.0	2.8	14.2	178.5						
Plouchart ESSCIRC 98	LC/SiGe	6.000	-116.0	1000	14.9	3.1	7.1	22.0	178.1	SiGe BiCMOS	3/T	HR	Inte L	11	30
Murata MQG	module	2.138	-90.0	25	3.8	3.0	4.0	12.0	177.8						
Craninckx VLSI 96	LC/CMOS	1.800	-116.0	600	12.0	1.5	4.0	6.0	177.8	CMOS07	2	LR	Inte L	5.7	
Craninckx CICC 97	LC/CMOS	0.900	-108.0	100	20.0	3.0	3.7	11.0	176.7	CMOS04	2	HR	Inte L	5.6	
Plouchart ESSCIRC 98	LC/SiGe	17.380	-104.8	1000	3.6	3.1	7.1	22.0	176.2	SiGe BiCMOS	3/T	HR	Inte L	12	
Dauphinee ISSCC 97	LC/Bip	1.500	-105.0	100	10.0	3.6	7.8	28.0	174.1	BiCMOS08/11GHz			Inte L	5	
Kinget ISSCC 98	LC/CMOS	5.200	-90.0	100	4.0	2.7	4.0	10.8	174.0	CMOS035	4	LR	Inte L	3	
Jansen ISSCC 97	LC/Bip	2.200	-99.0	100	12.0	2.7	8.0	21.6	172.5	Bipolar 15GHz			Inte L	10	30
Waegemans ISSCC 98	LC/Bip	1.800	-112.0	2000	0.7			1.0	171.1	Bipolar		glass	Inte L		
Soyeur ISSCC 96	LC/Bip	4.100	-106.0	1000	9.0	3.0	8.0	24.0	164.5	BiCMOS	5/T		Inte L	7	5
Rofourgaran ISSCC 96	LC/etch	0.820	-100.0	100	14.0			25.0	164.3	CMOS1	2	etch	Inte L	5	
Basedau Esscirc 94	LC/etch	1.000	-95.0	100	0.0			16.0	163.0	CMOS10		etch	Inte L		
Soyeur VLSI 94	LC/Bip	2.400	-92.0	100	0.0	0.0	0.0	50.0	162.6	BiCMOS	4/T		Inte L		
Razavi ISSCC 97	LC/Coup	1.800	-100.0	500	7.0	3.3	2.3	7.6	162.3	CMOS06	3		Inte L		
Liu ISSCC 99	LC/Coup	6.290	-98.4	1000	16.9	1.5	12.0	18.0	161.8	CMOS035	4	LR	Inte L		
Kwasniewski CICC 95	Ring	0.740	-89.0	100	6.0			6.5	158.3	CMOS120					
Nguyen JSSC 90	LC/Coup	1.800	-88.0	100	10.0			70.0	154.7	Bipolar 10GHz			Inte L		

Table 2: Performance of VCOs in different implementation styles

	Type	f_0 [GHz]	$L\{\Delta f\}$ [dBc/Hz]	Δf [kHz]	Tune [%]	Vdd [V]	Idd [mA]	P [mW]	FM	Technology	#M	Sub	L	QL	QV
Hajimiri CICC 98	Ring	2.810	-95.2	1000				10.0	154.2	CMOS025					
Hajimiri CICC 98	Ring	5.430	-98.5	1000	25.0			80.0	154.2	CMOS025					
Lam ISSCC 99	LC/Coup	2.600	-110.0	5000	12.3	2.5	5.2	13.0	153.2	CMOS035	4		Inte L		
Sneep JSSC90	Relax	0.100	-118.0	1000	100.0			30.0	143.2	Bipolar 3GHz					
Banu JSSC 88	Relax	0.560	-90.0	500	100.0			50.0	134.0	CMOS075					

Table 2: Performance of VCOs in different implementation styles

References

[1] W. Egan, *Frequency Synthesis by Phase Lock.* New York, NY (USA): Wiley & Sons, 1981.

[2] in *Monolithic phase-locked loops and clock recovery circuits. Theory and design* (B. Razavi, ed.), IEEE Press, 1996.

[3] B. Razavi, *RF Microelectronics.* Upper Saddle River, NJ (USA): Prentice Hall, 1998.

[4] T. Lee, *The design of CMOS radio-frequency integrated circuits.* Cambridge University Press, 1998.

[5] D. Scherer, "Phase noise instruments," in *Electronic Instrument Handbook* (J. C.F. Coombs, ed.), ch. 22, McGraw-Hill, II ed., 1994.

[6] W. Robins, *Phase Noise in Signal Sources (Theory and applications).* Telecommunications Series, U.K.: Peter Peregrinus Ltd - IEE, 1982.

[7] J. McNeill, *Jitter in Ring Oscillators.* PhD thesis, Boston University, Boston,MA USA, 1994.

[8] F. Kartner, "Analysis of white and $f^{-\alpha}$ noise in oscillators," *International Journal of Circuit Theory and Applications*, vol. 18, no. 5, pp. 485–519, 1990.

[9] A. Hajimiri and T. Lee, "A general theory of phase noise in electrical oscillators," *IEEE Journal of Solid-State Circuits*, vol. 33, Feb 1998.

[10] A. Demir, A. Mehrotra, and J. Roychowdhury, "Phase noise and timing jitter in oscillators," in *Proceedings of the IEEE Custom Integrated Circuits Conference (CICC)*, 1998.

[11] D. Leeson, "A simple model of feedback oscillator noise spectrum," *Proceedings of the IEEE*, vol. 54, pp. 329–330, February 1966.

[12] M. Banu, "MOS oscillators with multi-decade tuning range and gigahertz maximum speed," *IEEE Journal of Solid-State Circuits*, vol. 23, pp. 1386–1393, Dec 1988.

[13] J. Craninckx and M. Steyaert, "Low noise voltage-controlled oscillators using enhanced LC tanks," *IEEE Transactions on Circuits and Systems – II: Analog and Digital Signal Processing*, vol. 42, pp. 794–804, December 1995.

[14] B. Razavi, "A study of phase noise in CMOS oscillators," *IEEE Journal of Solid-State Circuits*, vol. 31, pp. 331–343, March 1996.

[15] Q. Huang, "On the exact design of RF oscillators," in *Proceedings of the IEEE Custom Integrated Circuits Conference (CICC)*, pp. 41–44, May 1998.

[16] C. Samori, A. Lacaita, F. Villa, and F. Zappa, "Spectrum folding and phase noise in LC tuned oscillators," *IEEE Transactions on Circuits and Systems – II: Analog and Digital Signal Processing*, vol. 45, pp. 781–790, July 1998.

[17] D. Pederson and K. Mayaram, *Analog integrated circuits for communications*. Kluwer Academic Publishers, 1991.

[18] A. Hajimiri and T. Lee, "Design issues in CMOS differential LC oscillators," *IEEE Journal of Solid-State Circuits*, vol. 34, pp. 717–724, May 1999.

[19] K. Laker and W. Sansen, *Design of Analog Integrated Circuits and Systems*. McGraw-Hill, 1994.

[20] T. Soorapanth, C. Yue, D. Schaeffer, T. Lee, and S. Wong, "Analysis and optimization of accumulation-mode varactor for RF ICs," in *Digest of Technical Papers Symposium on VLSI circuits*, pp. 32–33, June 1998.

[21] R. Castello, P. Erratico, S. Manzini, and F. Svelto, "A +/-30% tuning range varactor compatible with future scaled technologies," in *Digest of Technical Papers Symposium on VLSI circuits*, pp. 34–35, June 1998.

[22] R. Duncan, K. Martin, and A. Sedra, "A 1-GHz quadrature sinusoidal oscillator," in *Proceedings of the IEEE Custom Integrated Circuits Conference (CICC)*, pp. 91–94, May 1995.

[23] T. Cho, E. Dukatz, M. Mack, D. MacNally, M. Marringa, S. Mehta, C. Nilson, L. Plouvier, and S. Rabii, "A single-chip CMOS direct conversion transceiver for 900MHz spread-spectrum digital cordless phones," in *Digest of Technical Papers IEEE International Solid-State Circuits Conference (ISSCC)*, pp. 228–229, Feb. 1999.

[24] P. Kinget, "A fully integrated 2.7V 0.35um CMOS VCO for 5GHz wireless applications," in *Digest of Technical Papers IEEE International Solid-State Circuits Conference (ISSCC)*, Feb. 1998.

[25] P. Davis, P. Smith, E. Campbell, J. Lin, K. Gross, G. Bath, Y. Low, M. Lau, Y. Degani, J. Gregus, R. Frye, and K. Tai, "Silicon-on-silicon integration

of a GSM transceiver with VCO resonator," in *Digest of Technical Papers IEEE International Solid-State Circuits Conference (ISSCC)*, pp. 248–249, Feb. 1998.

[26] P. Kinget and R. Frye, "A 2.4GHz CMOS VCO with MCM inductor," in *Proceedings European Solid-State Circuits Conference*, Sept. 1998.

[27] J. Burghartz, D. Edelstein, M. Soytter, H. Ainspan, and K. Jenkins, "RF circuit design aspects of spiral inductors on silicon," in *Digest of Technical Papers IEEE International Solid-State Circuits Conference (ISSCC)*, pp. 246–247, Feb. 1998.

MIX
Papier aus verantwortungsvollen Quellen
Paper from responsible sources
FSC® C105338

If you have any concerns about our products,
you can contact us on
ProductSafety@springernature.com

In case Publisher is established outside the EU,
the EU authorized representative is:
Springer Nature Customer Service Center GmbH
Europaplatz 3, 69115 Heidelberg, Germany

Printed by Libri Plureos GmbH
in Hamburg, Germany